Lecture Notes in Computer Science

Lecture Notes in Artificial Intelligence   14502

Founding Editor

Jörg Siekmann

Series Editors

Randy Goebel, *University of Alberta, Edmonton, Canada*
Wolfgang Wahlster, *DFKI, Berlin, Germany*
Zhi-Hua Zhou, *Nanjing University, Nanjing, China*

The series Lecture Notes in Artificial Intelligence (LNAI) was established in 1988 as a topical subseries of LNCS devoted to artificial intelligence.

The series publishes state-of-the-art research results at a high level. As with the LNCS mother series, the mission of the series is to serve the international R & D community by providing an invaluable service, mainly focused on the publication of conference and workshop proceedings and postproceedings.

Hiram Calvo · Lourdes Martínez-Villaseñor ·
Hiram Ponce · Ramón Zatarain Cabada ·
Martín Montes Rivera · Efrén Mezura-Montes
Editors

Advances in Computational Intelligence

MICAI 2023 International Workshops

WILE 2023, HIS 2023, and CIAPP 2023
Yucatán, Mexico, November 13–18, 2023
Proceedings

Editors
Hiram Calvo (iD)
Instituto Politecnico Nacional
Center for Computing Research
Ciudad de México, Mexico

Hiram Ponce (iD)
Universidad Panamericana
Ciudad de México, Mexico

Martín Montes Rivera (iD)
Universidad Politécnica de Aguascalient
Aguascalientes, Mexico

Lourdes Martínez-Villaseñor (iD)
Panamerican University
Ciudad de México, Mexico

Ramón Zatarain Cabada (iD)
Instituto Tecnologico de Culiacan
Culiacán, Sinaloa, Mexico

Efrén Mezura-Montes (iD)
Universidad Veracruzana
Veracruz, Mexico

ISSN 0302-9743   ISSN 1611-3349 (electronic)
Lecture Notes in Artificial Intelligence
ISBN 978-3-031-51939-0   ISBN 978-3-031-51940-6 (eBook)
https://doi.org/10.1007/978-3-031-51940-6

LNCS Sublibrary: SL7 – Artificial Intelligence

This Springer imprint is published by the registered company Springer Nature Switzerland AG
The registered company address is: Gewerbestrasse 11, 6330 Cham, Switzerland

Paper in this product is recyclable.

Preface

The Mexican International Conference on Artificial Intelligence (MICAI) is a yearly international conference series that has been organized by the Mexican Society for Artificial Intelligence (SMIA) since 2000. MICAI is a major international artificial intelligence (AI) forum and the main event in the academic life of the country's growing AI community.

This year, MICAI 2023 was graciously hosted by the Instituto de Investigaciones en Matemáticas Aplicadas y en Sistemas (IIMAS) and the Universidad Autónoma del Estado de Yucatán (UAEY). The conference presented a cornucopia of scientific endeavors. From incisive keynote lectures and detailed paper presentations to hands-on tutorials, thought-provoking panels, and niche workshops, the spectrum of activities aimed to cater to a wide audience. Moreover, we continued the legacy of announcing the José Negrete Award, the SMIA Best Thesis in Artificial Intelligence Contest's results. This year, the historic and culturally rich city of Mérida, Yucatán was our chosen rendezvous.

MICAI conferences publish high-quality papers in all areas of AI and its applications. The proceedings of the previous MICAI events have been published by Springer in its Lecture Notes in Artificial Intelligence (LNAI) series (volumes: 1793, 2313, 2972, 3789, 4293, 4827, 5317, 5845, 6437, 6438, 7094, 7095, 7629, 7630, 8265, 8266, 8856, 8857, 9413, 9414, 10061, 10062, 10632, 10633, 11288, 11289, 11835, 12468, 12469, 13067, 13068, 13612, and 13613). This year, the proceedings of the MICAI 2023 main conference were published in volumes 14391 and 14392. Since its foundation in 2000, the conference has grown in popularity and improved in quality.

Three workshops were held jointly with the conference. The proceedings of the MICAI 2023 workshops are published in this volume, and it contains 34 papers structured into three sections:

- WILE 2023: 16th Workshop on Intelligent Learning Environments
- HIS 2023: 16th Workshop of Hybrid Intelligent Systems
- CIAPP 2023: 5th Workshop on New Trends in Computational Intelligence and Applications

This volume will be of interest for researchers in all fields of artificial intelligence, students specializing in related topics, and the general public interested in recent developments in AI.

The MICAI workshops received for evaluation 54 submissions from 2 countries: Iran and Mexico. From these submissions, 34 papers were selected for publication in this volume after a double-blind peer-reviewing process and three reviews per submission. It was carried out by the international Program Committee of workshops, from 3 countries: Mexico, Spain, and USA. The acceptance rate was 63%.

WILE 2023

Artificial intelligence has had a constant influence on a wide range of activities and areas of our life. New technologies have been developed as a result, and numerous industries have seen improvements in business procedures and overall efficiency. However, AI has the power to completely change education by providing more individualized instruction and better learning outcomes. Learning environments and tutoring systems that make learning easier and more effective for students have been around for a while. Intelligent tutoring systems (ITSs) and, more recently, intelligent learning environments (ILEs) are two examples of these developing systems or environments.

Intelligent learning environments are areas that can be designed to enhance cooperative learning activities in the classroom by promoting student engagement and involvement. In order to do this, intelligent learning environments must employ both established AI technologies, like artificial neural networks, Bayesian networks, and fuzzy logic, as well as newly developed ones, like conversational agents (Avatars) in smart environments, affective computing (Sentiment Analysis, for example), generative AI (ChatGPT), extended reality with immersion detection, etc., all of which aim to raise students' academic performance.

The aim of this workshop is to bring together active researchers and students in the field of Intelligent Learning Environments so that they can present and discuss innovative theoretical work and original applications, exchange ideas, establish collaboration links, discuss important recent achievements, and talk over the significance of results in the field to AI in general. Our goal for WILE 2023 was to give researchers a platform to showcase their work while also investigating new approaches to integrate AI techniques in the creation of educational systems.

We invited authors to submit papers presenting original and unpublished research in all areas related to Intelligent Learning Environments. Suggested topics included but were not limited to:

- Web-based intelligent tutoring systems
- Intelligent learning management systems
- Affective tutoring systems
- Modeling, enactment, and intelligent use of emotion and affect
- Natural language and dialogue approaches to ILE design and construction
- Authoring tools in intelligent tutoring systems
- Learning companions
- Applications of cognitive science
- Semantic Web technologies
- Student modeling
- Sentiment analysis in educational applications
- Gamification and game-based learning
- Educational data mining
- Learning analytics

The entire submission, reviewing, and selection process, as well as preparation of the proceedings, was supported by Microsoft's Conference Management Toolkit.

We want to thank Ramón Zatarain Cabada, our keynote speaker, for his valuable contribution to this workshop. Many people took part in WILE 2023, and we are thankful for the cooperation of the RedICA (Conacyt Thematic Network in Applied Computational Intelligence) members who served on the Technical Committee, as well as members of the Mexican Society for Artificial Intelligence (SMIA Sociedad Mexicana de Inteligencia Artificial). As every year, SMIA 2023 was the appropriate host for this event.

HIS 2023

Hybrid Intelligent Systems (HIS) offer a compelling solution to complex challenges in various application domains, including biology, medicine, logistics, management, engineering, technology, social, and humanities. These intricate systems require a multidisciplinary approach, utilizing various artificial intelligence techniques.

This workshop features 16 selected papers (15 by Mexican authors and one by an Iranian author) from a total of 30 received papers in Microsoft's Conference Management Toolkit, covering topics such as Machine Learning, Fuzzy Systems, Reasoning, Intelligent Control, Computer Vision, Optimization, and the Internet of Things.

The workshop HIS 2023 was an event showcasing cutting-edge research in hybrid intelligent systems. HIS fosters discussions on the latest advancements in the field with two track sessions: the first features articles with finished research, while the second consists of posters that showcase proposals and prototypes for future research. The workshop's primary objective is to present expert model systems applied to specific research topics.

We want to convey our deepest gratitude to the reviewers and members of the program committee and the four members of the organizing committee, all of them Mexican researchers who have played a pivotal role in ensuring the quality of the 16th HIS. We are immensely grateful for their hard work and dedication in making this event a success.

We sincerely thank Oscar Castillo from the Tecnológico Nacional de México - Instituto Tecnológico de Tijuana, our keynote speaker, for his valuable contribution to our event. His magisterial talk on Type-2 and Type-3 Fuzzy Systems: Theory and Applications was enriched by his extensive experience and remarkable achievements in the field. We greatly appreciate his presence and the insight he shared with our audience.

We would also like to express our appreciation to SMIA and the MICAI organizers for their continued collaboration in developing this prestigious event. It was our pleasure to contribute to MICAI with research in Hybrid Intelligent Systems.

CIAPP 2023

Computational Intelligence (CI) paradigms have become a critical factor in the resurgence of Artificial Intelligence which is now part of daily life. Therefore, basic and applied CI research have substantially grown, and more spaces for discussion on these topics are required.

This workshop aims to put together researchers, practitioners, students, and those interested in presenting novel findings and applications related to computational intelligence techniques. The workshop also looks to be a means to establish possible collaborations among the attendees. CIAPP 2023 had oral presentations of the accepted papers.

Interested authors were invited to submit original papers, including novel CI research and findings. The topics included but were not limited to the following:

- Machine Learning
- Data Mining
- Statistical Learning
- Automatic Image Processing
- Intelligent Agents/Multi Agent Systems
- Evolutionary Computing
- Swarm Intelligence
- Combinatorial and Numerical Optimization
- Parallel and Distributed Computing in Computational Intelligence

We want to thank all the people involved in the organization of the workshops: the authors of the papers published in this volume –it is their research work that gives value to the proceedings– and the organizers for their work. We thank the reviewers for their great effort spent on reviewing the submissions and the Program and Organizing Committee members of the workshops.

A special acknowledgment goes to the local committee, led by Antonio Neme, whose meticulous coordination was instrumental in realizing MICAI 2023 in Mérida, Yucatán, Mexico. Our thanks extend to IIMAS's director, Ramsés Mena, and its academic secretary, Katya Rodríguez. We are also indebted to Anabel Martín from the Faculty of Mathematics at the UADY for her invaluable assistance in securing the university facilities.

The entire submission, reviewing, and selection process, as well as preparation of the proceedings, was supported by Microsoft's Conference Management Toolkit (https://cmt3.research.microsoft.com/). Last but not least, we are grateful to Springer for their patience and help in the preparation of this volume.

In conclusion, MICAI 2023 is more than just a conference. It is a confluence of minds, a testament to the indefatigable spirit of the AI community, and a beacon for the future of Artificial Intelligence. As you navigate through these proceedings, may you find inspiration, knowledge, and connections that propel you forward in your journey.

The MICAI series website is www.MICAI.org. The website of the Mexican Society for Artificial Intelligence, SMIA, is www.SMIA.mx. Contact options and additional information can be found on these websites.

November 2023

Hiram Calvo
Lourdes Martínez-Villaseñor
Hiram Ponce
Ramón Zatarain-Cabada
Martín Montes Rivera
Efrén Mezura-Montes

Organization

Conference Committee

General Chair

Hiram Calvo — Instituto Politécnico Nacional, Mexico

Program Chairs

Hiram Calvo — Instituto Politécnico Nacional, Mexico
Lourdes Martínez-Villaseñor — Universidad Panamericana, Mexico
Hiram Ponce — Universidad Panamericana, Mexico

Workshop Chair

Hiram Ponce — Universidad Panamericana, Mexico

Tutorials Chair

Roberto Antonio Vázquez Espinoza de los Monteros — Universidad La Salle, Mexico

Doctoral Consortium Chairs

Miguel González Mendoza — Tecnológico de Monterrey, Mexico
Juan Martínez Miranda — Centro de Investigación Científica y de Educación Superior de Ensenada, Mexico

Keynote Talks Chairs

Gilberto Ochoa Ruiz — Tecnológico de Monterrey, Mexico
Iris Méndez — Universidad Autónoma de Ciudad Juárez, Mexico

Publication Chair

Hiram Ponce — Universidad Panamericana, Mexico

Financial Chairs

Hiram Calvo   Instituto Politécnico Nacional, Mexico
Lourdes Martínez-Villaseñor   Universidad Panamericana, Mexico

Grant Chair

Leobardo Morales   IBM, Mexico

Local Organizing Committee

Abigail Uribe Martínez   Universidad Autónoma del Estado de Yucatán,
                                   Mexico
Candy Sansores   Universidad Autónoma del Estado de Yucatán,
                                   Mexico
Abraham Mandariaga Mazón   Universidad Autónoma del Estado de Yucatán,
                                   Mexico
Carlos Bermejo Sabbagh   Universidad Autónoma del Estado de Yucatán,
                                   Mexico
Ali Bassam   Universidad Autónoma del Estado de Yucatán,
                                   Mexico
Eric Ávila Vales   Universidad Autónoma del Estado de Yucatán,
                                   Mexico
Anabel Martin   Universidad Autónoma del Estado de Yucatán,
                                   Mexico
Erik Molino Minero Re   Universidad Autónoma del Estado de Yucatán,
                                   Mexico
Antonio Aguileta   Universidad Autónoma del Estado de Yucatán,
                                   Mexico
Fernando Arámbula Cosío   Universidad Autónoma del Estado de Yucatán,
                                   Mexico
Antonio Neme   Universidad Autónoma del Estado de Yucatán,
                                   Mexico
Helena Gomez Adorno   Universidad Autónoma del Estado de Yucatán,
                                   Mexico
Blanca Vázquez   Universidad Autónoma del Estado de Yucatán,
                                   Mexico
Israel Sánchez Domínguez   Universidad Autónoma del Estado de Yucatán,
                                   Mexico
Joel Antonio Trejo Sánchez   Universidad Autónoma del Estado de Yucatán,
                                   Mexico

Norberto Sánchez	Universidad Autónoma del Estado de Yucatán, Mexico
Jorge Perez-Gonzalez	Universidad Autónoma del Estado de Yucatán, Mexico
Paul Erick Méndez Monroy	Universidad Autónoma del Estado de Yucatán, Mexico
Julián Bravo Castillero	Universidad Autónoma del Estado de Yucatán, Mexico
Ramón Aranda	Universidad Autónoma del Estado de Yucatán, Mexico
Karina Martínez	Universidad Autónoma del Estado de Yucatán, Mexico
Vicente Carrión	Universidad Autónoma del Estado de Yucatán, Mexico
Mauricio Orozco del Castillo	Universidad Autónoma del Estado de Yucatán, Mexico
Victor Manuel Lomas Barrie	Universidad Autónoma del Estado de Yucatán, Mexico
Nidiyare Hevia Montiel	Universidad Autónoma del Estado de Yucatán, Mexico
Víctor Sandoval Curmina	Universidad Autónoma del Estado de Yucatán, Mexico
Nora Cuevas Cuevas	Universidad Autónoma del Estado de Yucatán, Mexico
Victor Uc Cetina	Universidad Autónoma del Estado de Yucatán, Mexico
Nora Pérez Quezadas	Universidad Autónoma del Estado de Yucatán, Mexico
Yuriria Cortés Poza	Universidad Autónoma del Estado de Yucatán, Mexico

Program Committee

Alberto Ochoa-Zezzatti	Universidad Autónoma de Ciudad Juárez, Mexico
Aldo Marquez-Grajales	Instituto Tecnológico Superior de Xalapa, Mexico
Alexander Bozhenyuk	Southern Federal University, Russia
Andrés Espinal	Universidad de Guanajuato, Mexico
Angel Sánchez García	Universidad Veracruzana, Mexico
Anilu Franco	Universidad Autónoma del Estado de Hidalgo, Mexico
Antonieta Martinez	Universidad Panamericana, Mexico
Antonio Neme	UNAM, Mexico

Ari Barrera Animas	Universidad Panamericana, Mexico
Asdrúbal López Chau	Universidad Autónoma del Estado de México, Mexico
Belém Priego Sánchez	Universidad Autónoma Metropolitana Azcapotzalco, Mexico
Bella Martinez Seis	Instituto Politécnico Nacional, Mexico
Betania Hernandez-Ocaña	Universidad Juárez Autónoma de Tabasco, Mexico
Carlos A. Reyes García	INAOE, Mexico
Claudia Gómez	Instituto Tecnológico de Ciudad Madero, Mexico
Daniela Alejandra Moctezuma Ochoa	CentroGEO-CONACyT, Mexico
Dante Mújica-Vargas	CENIDET, Mexico
Diego Uribe	Tecnológico Nacional de México - ITL, Mexico
Eddy Sánchez-DelaCruz	Tecnológico Nacional de México - Campus Misantla, Mexico
Eduardo Valdez	Instituto Politécnico Nacional, Mexico
Efrén Mezura-Montes	Universidad Veracruzana, Mexico
Eloísa García-Canseco	Universidad Autónoma de Baja California, Mexico
Elva Lilia Reynoso Jardon	Universidad Autónoma de Ciudad Juárez, Mexico
Eric Tellez	CICESE-INFOTEC-CONACyT, Mexico
Ernesto Moya-Albor	Universidad Panamericana, Mexico
Félix Castro Espinoza	Universidad Autónoma del Estado de Hidalgo, Mexico
Fernando Gudino	UNAM, Mexico
Garibaldi Pineda Garcia	Applied AGI, UK
Genoveva Vargas-Solar	Grenoble Alpes University, CNRS, France
Gilberto Ochoa-Ruiz	Tecnológico de Monterrey, Mexico
Giner Alor-Hernandez	Tecnológico Nacional de México - ITO, Mexico
Guillermo Santamaría-Bonfil	BBVA México, Mexico
Gustavo Arroyo	Instituto Nacional de Electricidad y Energías Limpias, Mexico
Helena Gómez Adorno	IIMAS-UNAM, Mexico
Hiram Ponce	Universidad Panamericana, Mexico
Hiram Calvo	Center for Computing Research - Instituto Politécnico Nacional, Mexico
Hugo Jair Escalante	INAOE, Mexico
Humberto Sossa	Instituto Politécnico Nacional, Mexico
Iris Iddaly Méndez-Gurrola	Universidad Autónoma de Ciudad Juárez, Mexico
Iskander Akhmetov	Institute of Information and Computational Technologies, Kazakhstan
Ismael Osuna-Galán	Universidad de Quintana Roo, Mexico

Israel Tabarez	Universidad Autónoma del Estado de México, Mexico
Jaime Cerda	Universidad Michoacana de San Nicolás de Hidalgo, Mexico
Jerusa Marchi	Federal University of Santa Catarina, Brazil
Joanna Alvarado Uribe	Tecnológico de Monterrey, Mexico
Jorge Perez Gonzalez	UNAM, Mexico
José Alanis	Universidad Tecnológica de Puebla, Mexico
José Martínez-Carranza	INAOE, Mexico
Jose Alberto Hernandez-Aguilar	Universidad Autonoma del Estado de Morelos, Mexico
José Carlos Ortiz-Bayliss	Tecnológico de Monterrey, Mexico
Juan Villegas-Cortez	UAM - Azcapotzalco, Mexico
Juan Carlos Olivares Rojas	Tecnológico Nacional de México - ITM, Mexico
Karina Perez-Daniel	Universidad Panamericana, Mexico
Karina Figueroa Mora	Universidad Michoacana de San Nicolás de Hidalgo, Mexico
Leticia Flores Pulido	Universidad Autónoma de Tlaxcala, Mexico
Lourdes Martinez-Villaseñor	Universidad Panamericana, Mexico
Luis Torres-Treviño	Universidad Autónoma de Nuevo León, Mexico
Luis Luevano	Institut National de Recherche en Informatique et en Automatique, France
Mansoor Ali Teevno	Tecnológico de Monterrey, Mexico
María Lucía Barrón Estrada	TecNM-Instituto Tecnológico de Culiacán, Mexico
Masaki Murata	Tottori University, Japan
Miguel Gonzalez-Mendoza	Tecnológico de Monterrey, Mexico
Miguel Mora-Gonzalez	Universidad de Guadalajara, Mexico
Mukesh Prasad	University of Technology Sydney, Australia
Omar López-Ortega	Universidad Autónoma del Estado de Hidalgo, Mexico
Rafael Guzman-Cabrera	Universidad de Guanajuato, Mexico
Rafael Batres	Tecnológico de Monterrey, Mexico
Ramon Brena	Instituto Tecnológico de Sonora, Mexico
Ramón Zatarain Cabada	Tec Culiacán, Mexico
Ramón Iván Barraza-Castillo	Universidad Autónoma de Ciudad Juárez, Mexico
Roberto Antonio Vasquez	Universidad La Salle, Mexico
Rocio Ochoa-Montiel	Universidad Autónoma de Tlaxcala, Mexico
Ruben Carino-Escobar	Instituto Nacional de Rehabilitación - Luis Guillermo Ibarra Ibarra, Mexico
Sabino Miranda	INFOTEC-CONACyT, Mexico
Saturnino Job Morales	Universidad Autónoma del Estado de México, Mexico

Workshops Organization

WILE 2023 Organizing Committee

General Chairs

Ramón Zatarain	Tecnológico Nacional de México-Instituto Tecnológico de Culiacán, Mexico
Lucía Barrón	Tecnológico Nacional de México-Instituto Tecnológico de Culiacán, Mexico
Yasmín Hernández	Tecnológico Nacional de México-Cenidet, Mexico
Carlos Alberto Reyes García	INAOE, Mexico
Karina Figueroa	Universidad Michoacana de San Nicolás de Hidalgo, Mexico

Program Committee

Ramón Zatarain	Tecnológico Nacional de México-Instituto Tecnológico de Culiacán, Mexico
Lucía Barrón	Tecnológico Nacional de México-Instituto Tecnológico de Culiacán, Mexico
Yasmín Hernández	Tecnológico Nacional de México-Cenidet, Mexico
Karina Mariela Figueroa Mora	UMSNH, Mexico
Carlos A. Reyes García	Instituto Nacional de Astrofísica, Óptica y Electrónica, Mexico
Giner Alor Hernández	Tecnológico Nacional de México-Instituto Tecnológico de Orizaba, Mexico
Miguel Pérez Ramírez	Instituto Nacional de Electricidad y Energías Limpias, Mexico
Jaime Muñoz Arteaga	Universidad Autónoma de Aguascalientes, Mexico
Rafael Morales Gamboa	Universidad de Guadalajara, Mexico
Guillermo Santamaría Bonfil	BBVA, Mexico
Carlos Alberto Lara Álvarez	CIMAT Zacatecas, Mexico
Hugo Arnoldo Mitre Hernández	CIMAT Zacatecas, Mexico
María Elena Chávez Echeagaray	Arizona State University, USA

María Blanca Ibáñez Espiga	Universidad Carlos III de Madrid, Spain
Alicia Martínez Rebollar	Tecnológico Nacional de México-Cenidet, Mexico
María Lucila Morales Rodríguez	Tecnológico Nacional de México-Instituto Tecnológico de Ciudad Madero, Mexico
Héctor Rodríguez Rangel	Tecnológico Nacional de México-Instituto Tecnológico de Culiacán, Mexico
Julieta Noguez Monroy	Tecnológico de Monterrey, Mexico
Samuel González López	Tecnológico Nacional de México-Instituto Tecnológico de Nogales, Mexico
Raúl Oramas Bustillos	Universidad Autónoma de Occidente, Mexico
José Mario Ríos Félix	Tecnológico Nacional de México-Instituto Tecnológico de Culiacán, Mexico
Luis Alberto Morales Rosales	Universidad Michoacana de San Nicolás de Hidalgo, Mexico
Maritza Bustos López	Tecnológico Nacional de México-Instituto Tecnológico de Orizaba, Mexico

HIS 2023 Organizing Committee

General Chairs

Martín Montes Rivera	Universidad Politécnica de Aguascalientes, Mexico
Carlos Alberto Ochoa Zezzatti	Universidad Autónoma de Ciudad Juárez, Mexico
Julio Cesar Ponce Gallegos	Universidad Autónoma de Aguascalientes, Mexico
José Alberto Hernández Aguilar	Universidad Autónoma del Estado de Morelos, Mexico

Program Committee

Edgar Gonzalo Cossio Franco	Instituto de Información Estadística y Geográfica de Jalisco, Mexico
Humberto Velasco Arellano	Universidad Politécnica de Aguascalientes, Mexico
Daniela Paola López Betancourt	Universidad Politécnica de Aguascalientes, Mexico
Carlos Alejandro Guerrero Méndez	Universidad Autónoma de Zacatecas, Mexico

Humberto Muñoz Bautista	Universidad Tecnológica Metropolitana de Aguascalientes, Mexico
Miguel Ángel Ortiz Esparza	Universidad Autónoma de Aguascalientes, Mexico
Himer Avila George	Universidad de Guadalajara, Mexico
Alejandro Padilla Díaz	Universidad Autónoma de Aguascalientes, Mexico
Carlos Alberto Lara Alvarez	CIMAT Zacatecas, Mexico
Roberto Antonio Contreras Masse	Instituto Tecnológico Ciudad Juárez, Mexico
Irma Yazmín Hernández Báez	Universidad Politécnica el Estado de Morelos, Mexico

CIAPP 2023 Organizing Committee

General Chairs

Héctor-Gabriel Acosta-Mesa	Universidad Veracruzana, Mexico
Rocío-Erandi Barrientos-Martínez	Universidad Veracruzana, Mexico
Efrén Mezura-Montes	Universidad Veracruzana, Mexico
Marcela Quiroz-Castellanos	Universidad Veracruzana, Mexico

Program Committee

Martha Lorena Avendaño	Universidad Veracruzana, Mexico
Aldo Márquez Grajales	Universidad Veracruzana, Mexico
Nancy Pérez Castro	Universidad de Papaloapan, Mexico
Rafael Rivera-López	Tecnológico Nacional de México-Instituto Tecnológico de Veracruz, Mexico
Guillermo Hoyos-Rivera	Universidad Veracruzana, Mexico
Adriana L. López Lobato	Universidad Veracruzana, Mexico
Octavio Ramos Figueroa	Universidad de Xalapa, Mexico
Jesús Adolfo Mejía de Dios	Universidad Autónoma de Coahuila, Mexico
Mario Graff Guerrero	INFOTEC Aguascalientes, Mexico
José Luis Morales Reyes	Universidad de Xalapa, Mexico

External Reviewers

Adriana Laura López-Lobato	Universidad Veracruzana, Mexico
Carlos-Alberto Lopez-Herrera	Universidad Veracruzana, Mexico

David Herrera-Sánchez   Universidad Veracruzana, Mexico
Jesús-Arnulfo Barradas-Palmeros Universidad Veracruzana, Mexico
José Fuentes-Tomás   Universidad Veracruzana, Mexico
Jose-Luis Llaguno-Roque   Universidad Veracruzana, Mexico
Saul Dominguez-Isidro   Universidad Veracruzana, Mexico

Contents

WILE 2023

DinoApp, an Augmented Reality Application for Learning About
Dinosaurs 3
  Aldo Uriarte-Portillo, Luis-Miguel Sánchez-Zavala,
  Ramon Zatarain-Cabada, María-Lucia Barrón-Estrada,
  and Victor-Manuel Batiz-Beltran

School Dropout Prediction with Class Balancing and Hyperparameter
Configuration 12
  P. Alejandra Cuevas-Chávez, Samuel Narciso,
  Eduardo Sánchez-Jiménez, Itzel Celerino Pérez, Yasmín Hernández,
  and Javier Ortiz-Hernandez

Method to Identify Emotions in Immersive Virtual Learning Environments
Using Head and Hands Spatial Behavioral Information 21
  Jorge Enrique Velázquez-Cano, Juan Gabriel Gonzáles-Serna,
  Leonor Rivera-Rivera, Nimrod Gonzáles-Franco,
  José Alejandro Reyes-Ortiz, Máximo López-Sánchez,
  and Blanca Dina Valenzuela-Robles

A New Approach for Counting and Identification of Students Sentiments
in Online Virtual Environments Using Convolutional Neural Networks 29
  José Alberto Hernández-Aguilar, Yasmín Hernández,
  Lizmary Rivera Cruz, and Juan Carlos Bonilla Robles

Automated Facial Expression Analysis for Cognitive State Prediction
During an Interaction with a Digital Interface 41
  Maricarmen Toribio-Candela, Gabriel González-Serna,
  Andrea Magadan-Salazar, Nimrod González-Franco,
  and Máximo López-Sánchez

Can We Take Out CARLA from the Uncanny Valley? Analyzing Avatar
Design of an Educational Conversational Agent 50
  Pablo Isaac Macias-Huerta, Carlos Natanael Lecona-Valdespino,
  Guillermo Santamaría-Bonfil, and Fernando Marmolejo-Ramos

HIS 2023

A GPT-Based Approach for Sentiment Analysis and Bakery Rating
Prediction 61
   Diego Magdaleno, Martin Montes, Blanca Estrada,
   and Alberto Ochoa-Zezzatti

Brake Maintenance Diagnostic with Fuzzy-Bayesian Expert System 77
   Misael Perez Hernández, Martín Montes Rivera,
   Ricardo Perez Hernández, and Roberto Macias Escobar

Use of IoT-Based Telemetry via Voice Commands to Improve
the Gaudiability Rate of a Generation Z Pet Habitation Experience 102
   Alberto Ochoa-Zezzatti, Jose De los Santos, Maylin Hernandez,
   Ángel Ortiz, Joshuar Reyes, Saúl González, and Luis Vidal

Analysis of Convolutional Neural Network Models for Classifying
the Quality of Dried Chili Peppers (Capsicum Annuum L) 116
   David Navarro-Solís, Carlos Guerrero-Méndez,
   Tonatiuh Saucedo-Anaya, Daniela Lopez-Betancur, Luis Silva,
   Antonio Robles-Guerrero, and Salvador Gómez-Jiménez

Fuzzy-Bayesian Expert System for Assistance in Bike Mechanical Issues 132
   Roberto Macías Escobar, Martín Montes Rivera,
   and Daniel Macias Escobar

Application of the Few-Shot Algorithm for the Estimation of Bird
Population Size in Chihuahua and Its Ornithological Implications 152
   Jose Luis Acosta Roman, Carlos Alberto Ochoa-Zezzatti,
   Martin Montes Rivera, and Delfino Cornejo Monroy

Searcher for Clothes on the Web Using Convolutional Neural Networks
and Dissimilarity Rules for Color Classification Using Euclidean Distance
to Color Centers in the HSL Color Space 159
   Luciano Martinez, Martín Montes, Alberto Ochoa Zezzatti,
   Julio Ponce, and Eder Guzmán

Identifying DC Motor Transfer Function with Few-Shots Learning
and a Genetic Algorithm Using Proposed Signal-Signature 170
   Martín Montes Rivera, Marving Aguilar-Justo,
   and Misael Perez Hernández

Real-Time Emotion Recognition Using Convolutional Neural Network:
A Raspberry Pi Architecture Approach 191
   Antonio Romero and Ángel Armenta

Optimization of CO_2 Capture Efficiency Through Analysis of Temperature
Variables in a Packed Absorption Column 201
    Rafael Terrero Mariano, José Ismael Ojeda Campaña,
    Carlos Alberto Ochoa Ortiz, Miriam Navarrete Procopio,
    and Víctor Manuel Zezatti Flores

Fuzzy-Bayesian Expert System for Suggesting Personalized Training
Plans with Exercises and Routines 218
    Rosa Lizeth Estrada Ortega

Proposal of a Storage Methodology for Asset Management, Using
Artificial Intelligence Techniques, to Make Unit Load Work Processes
More Efficient 242
    Ismael Cardona and Edgar Gonzalo Cossio Franco

Smart Geo-Reference System for the Prediction of the Mercalli Scale
for Hurricanes in Los Cabos Area an Approximation of a Risk Atlas 259
    Luis Fernando Bernal Sánchez, José Ismael Ojeda Campaña,
    and Alberto Ochoa Zezzatti

Intelligent Applications for the Inclusion of People with Hearing
Disabilities in the Communication 274
    Marco Antonio Martínez, Julio Cesar Ponce Gallegos,
    Alejandro Padilla Diaz, and Francisco Javier Álvarez Rodríguez

TICCAD: A Resource for Higher Education During and After Sars-CoV2 285
    María Esmeralda Arreola Marín, Mariela Chávez Marcial,
    José Iraic Alcantar Alcantar, and Edgar Gonzalo Cossio Franco

Implementation of Time Series to Determine Purchase and Use of Electric
Cars in a Smart City Considering Generation Z as Target Population 298
    Shaban Mousavi Ghasemlou, Alberto Ochoa-Zezzatti, Vianey Torres,
    Erwin Martinez, and Victor Lopez

CIAPP 2023

Blood Cell Image Segmentation Using Convolutional Decision Trees
and Differential Evolution 315
    Adriana-Laura López-Lobato, Héctor-Gabriel Acosta-Mesa,
    and Efrén Mezura-Montes

Multi-objective Evolutionary Algorithm Based on Decomposition
to Solve the Bi-objective Internet Shopping Optimization Problem
(MOEA/D-BIShOP) 326
  Miguel A. García-Morales, José A. Brambila-Hernández,
  Héctor J. Fraire-Huacuja, Juan Frausto-Solis, Laura Cruz-Reyes,
  Claudia Guadalupe Gómez-Santillan, Juan Martín Carpio Valadez,
  and Marco Antonio Aguirre-Lam

A Surrogate-Assisted Differential Evolution Approach for the Optimization
of Ben's Spiker Algorithm Parameters 337
  Carlos-Alberto López-Herrera, Héctor-Gabriel Acosta-Mesa,
  and Efrén Mezura-Montes

Auto Machine Learning Based on Genetic Programming for Medical
Image Classification 349
  David Herrera-Sánchez, Héctor-Gabriel Acosta-Mesa,
  and Efrén Mezura-Montes

Use of a Surrogate Model for Symbolic Discretization of Temporal Data
Sets Through eMODiTS and a Training Set with Varying-Sized Instances 360
  Aldo Márquez-Grajales, Efrén Mezura-Montes,
  Héctor-Gabriel Acosta-Mesa, and Fernando Salas-Martínez

Estimation of Anthocyanins in Homogeneous Bean Landraces Using
Neuroevolution 373
  José-Luis Morales-Reyes, Elia-Nora Aquino-Bolaños,
  Héctor-Gabriel Acosta-Mesa, and Aldo Márquez-Grajales

Representation of Expert Knowledge on Product Design Problems Using
Fuzzy Cognitive Maps 385
  Hector-Heriberto Rodriguez-Martinez, Jesus-Adolfo Mejia-de Dios,
  and Irma-Delia García-Calvillo

Neural Architecture Search for Placenta Segmentation in 2D Ultrasound
Images 397
  José Antonio Fuentes-Tomás, Héctor Gabriel Acosta-Mesa,
  Efrén Mezura-Montes, and Rodolfo Hernandez Jiménez

Experimental Study of the Instance Sampling Effect on Feature Subset
Selection Using Permutational-Based Differential Evolution 409
  Jesús-Arnulfo Barradas-Palmeros, Rafael Rivera-López,
  Efrén Mezura-Montes, and Héctor-Gabriel Acosta-Mesa

Computational Learning in Behavioral Neuropharmacology 422
  Isidro Vargas-Moreno, Héctor Gabriel Acosta-Mesa,
  Juan Francisco Rodríguez-Landa, Martha Lorena Avendaño-Garido,
  and Socorro Herrera-Meza

Analysis of Proteins in Microscopic Skin Images Using Machine Vision
Techniques as a Tool for Detecting Alzheimer's Disease 432
  Sonia Lilia Mestizo-Gutiérrez, Héctor Gabriel Acosta-Mesa,
  Francisco García-Ortega, and María Esther Jiménez-Cataño

Comparative Study of the Starting Stage of Adaptive Differential
Evolution on the Induction of Oblique Decision Trees 439
  Miguel Ángel Morales-Hernández, Rafael Rivera-López,
  Efrén Mezura-Montes, Juana Canul-Reich,
  and Marco Antonio Cruz-Chávez

Correction to: Use of IoT-Based Telemetry via Voice Commands
to Improve the Gaudiability Rate of a Generation Z Pet Habitation
Experience C1
  Alberto Ochoa-Zezzatti, Jose De los Santos, Maylin Hernandez,
  Ángel Ortiz, Joshuar Reyes, Saúl González, and Luis Vidal

Author Index 453

WILE 2023

DinoApp, an Augmented Reality Application for Learning About Dinosaurs

Aldo Uriarte-Portillo[1,2]([✉]) [iD], Luis-Miguel Sánchez-Zavala[2],
Ramon Zatarain-Cabada[1] [iD], María-Lucia Barrón-Estrada[1] [iD],
and Victor-Manuel Batiz-Beltran[1] [iD]

[1] Tecnológico Nacional de México-Instituto Tecnológico de Culiacán, 80210 Culiacán, Sinaloa, México
{aldo.up,ramon.zc,lucia.be,victor.bb}@culiacan.tecnm.mx
[2] Coordinación General Para el Fomento a la Investigación Científica e Innovación del Estado de Sinaloa (CONFIE), 80050 Culiacán, Sinaloa, México
{aldouriarte,luissanchez}@ccs.edu.mx

Abstract. Museums have exhibition halls where visitors can perform activities to learn about various areas of knowledge. Taking advantage of mobile devices' popularity, museums implement applications based on augmented reality technology so that visitors can live a unique experience. This work presents a mobile application called DinoApp based on Google AR Foundation ARCore technology that uses augmented reality with markers and without markers using ground detection. The visitor can explore the different 3D models, visualize a descriptive card of each extinct creature, learn about the most principal features of dinosaurs, listen to their sounds, design their scenario, and capture a memento, thus living a unique and enriching experience.

Keywords: Augmented Reality · dinosaurs · interactive learning environments

1 Introduction

Augmented reality (AR) as an emerging technology has gained popularity, thanks to the fact that it offers the user a unique experience by combining virtual elements with the real world [1]. AR emerged decades ago, however, it has had a positive impact on entertainment, video games, marketing, advertising, education, tourism, architecture, medicine, languages, and culture [2, 3]. In education, it has been shown to positive effect on the motivational state and learning gains of students fostering learners' interest in the activities, increased retention, greater understanding of learning topics [4], and a favorable motivational state in students due to the competitive advantages it offers by complementing real space with 3D elements, either using a head-mounted display, a Lens, or a mobile device by geolocation or marker-based technique [5].

On the other hand, museums are considered learning sites that offer a wide range of educational experiences, especially in science, culture, anthropology, history, arts, and technology [6, 7].

© The Author(s), under exclusive license to Springer Nature Switzerland AG 2024
H. Calvo et al. (Eds.): MICAI 2023 Workshops, LNAI 14502, pp. 3–11, 2024.
https://doi.org/10.1007/978-3-031-51940-6_1

The main purpose of a museum is to offer an exceptional and innovative experience for visitors, which can be difficult for museums that do not have a variety of cultural and entertainment exhibits. To improve communication and strengthen the dissemination of culture, science, and technology, museums have been integrating their content with the use of emerging technologies [7], such as AR and virtual reality (VR) and extended reality (XR) or other technologies to attract the public's attention: In Italy, 16% of museums have started to implement augmented reality, 12% implemented virtual reality technology and 10% prefer offer video games [8].

The main contribution of this work is the creation of a mobile application designed for Android devices that uses augmented reality (AR) technology to enrich the experience of museum visitors. The application offers users the opportunity to explore and learn about diverse types of extinct dinosaurs, such as Velociraptors, Triceratops, Pterodactyles, and T-Rex's, among others. Users can see a 3D model and see a short animation of the dinosaur. The application uses two main approaches to AR-marker-based and ground detection using ARCore technology. This allows users to interact with virtual content by pointing their devices to specific markers or by placing them on flat surfaces in the real environment. The main attraction is that the user can design his scenario and capture the picture of the memory.

2 Related Work

This section discusses works that implement AR with topics related to museums around the world. In the first instance, in [9] an analysis of the applications developed with emphasis on the cultural heritage field that use AR in the period from 2012 to 2021 is carried out, compiled from the Scopus and Clarivate Web of Science databases. Within this analysis, Google's Tango project stands out, which focuses on the 3D representation of the Prejmer Church, a UNESCO monument. Its main attraction is the 3D reconstruction of cultural artifacts, digital heritage, virtual museums, and tourism, among others.

The Guilin Museum [10] has an AR application developed in Unity. This application is based on the detection and tracking of the user's face, as well as the detection and tracking of images. In addition, a marketing application has been created to promote the museum, using Schmitt's model, which considers the users' emotions. The results of this study demonstrate that marketing combined with AR experience manages to integrate the rationality and sensitivity of users. Users participate voluntarily through the application's features, which arouses their curiosity through sensory stimulation and emotional projection. This in turn triggers a change in their mindset to better understand the products offered. In addition, these experiences generate perceptual impulses in users, adding value to the overall experience.

The Cleveland Museum of Art offers a unique experience to its visitors using the AR-based ArtLens 2.0 app, available for iOS and Android, extending its reach to a wide audience. ArtLens 2.0 uses sophisticated image recognition software that allows users to explore and discover fascinating details about the artworks on display at the museum. In addition, the application provides detailed maps to facilitate a more enriching visiting experience at the institution. ArtLens 2.0 has achieved significant success, with more than

73,000 downloads on iOS and more than 9,000 downloads on Android. This achievement is due in large part to the developers' focus on listening to visitor feedback and making continual improvements to the app, adapting it to the needs and expectations of the public [11].

The History Museum of the University of Pavia houses historical pieces related to the fields of physics and medicine. They developed an AR application that offers historical and scientific material: stories, 3D animations, images, and video. The app was designed to be as unobtrusive and unobtrusive as possible, preserve the historical ambiance of the museum, combine social and educational aspects, record user behavior, and make the museum experience more vibrant and active, and therefore captivating. The topics that most capture the attention of viewers are the explanation of the voltaic pile, X-rays, the lantern microscope, Volta's pistol, and zero-plastics [12].

In Wei [13] the authors focused on dinosaurs' topic in science courses for middle-high school students. They developed an AR marker-based application to show animations, video, images, and audio to the students. They utilized instruments to measure learning experiences and interests' experiences in two times: prior to and after the experiment. They categorized users into two groups: the experimental group that utilized augmented reality application and the control group without AR. The findings demonstrate that the learners in the experimental group obtained higher scores in both learning experiences and interests compared to the control group users.

3 Methodology

This section presents the stages that make up the development of the proposal:

- In this stage of Requirements gathering, the information pertinent information was gathered from Centro de Ciencias de Sinaloa (CCS) museum staff to determine the areas of interest, which were prioritized according to the needs of the institution based on their functionality and quality.
- Proposal design: involves creating the application's interfaces, i.e., how the platform will look and feel to users, and the use cases were proposed.
- Development and implementation: with the proposal design approved, the development of the mobile application begins. This is where the designs are converted into a functional and navigable platform.
- Evaluation by experts: once the first version of the platform has been developed, it is evaluated by experts familiar with the museum industry, and user experience to provide feedback to improve the quality and functionality of the platform.
- End-user testing: after implementing the experts' suggestions and making the necessary adjustments, DinoApp is ready to be assessed by the end-users, i.e., the people who will visit and use the platform. The general method of the project development process is presented in Fig. 1.

Fig. 1. Methodology of the project development process

4 DinoApp Learning Tool

DinoApp was developed using Unity 2021, and as AR it used AR Foundation's ARCore technology. First, the learning tool has three main scenes: the first one, "Create your stage", "Information" and "Quiz". Users can visit the privacy notice at any time, as well as being able to download bookmarks if they wish. The main advantage of this technology is that the camera of the mobile device only needs to detect a plane. Once detected, it can place on the scene one or more 3D elements referring to an extinct dinosaur or creature. Subsequently, the styles of each scene were created, and their visual elements (buttons, colors, links) were added.

4.1 DinoApp Architecture

The software architecture was designed using the layers model (Presentation, Business Logic, Data, and Services) to ease communication between its components, which is presented in Fig. 2. Once each view was elaborated, the respective functionalities were added to each element using C# scripting.

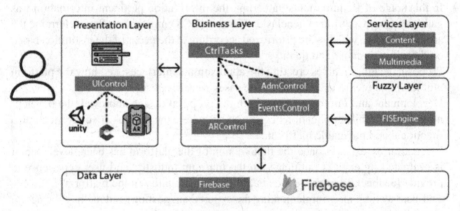

Fig. 2. DinoApp architecture

The Presentation layer through its **UIControl** component displays the information to the user through the graphical user interface. We sought to minimize the use of hardware resources. C# scripting was used to detect the resolution of the devices since users can access them from cell phones, computers, and tablets.

The Business Logic layer has **CtrlTasks** as its main component, which is responsible for processing the information and coordinating the controllers that will allow it to provide functionality to each component. This layer allows the communication between the learning tool through the EventsControl component, to provide the resources to display the augmented content through the ARControl component and to send and receive the information from the database through the AdmControl component.

The Data layer refers to the place where the information is stored, for data persistence Google's Firebase technology was implemented.

The Services Layer oversees managing the multimedia content that will be loaded according to the requests received from the business logic layer, which subsequently sends the request to update the graphical user interface of the Presentation layer.

The Fuzzy Layer determines the next level of complexity for the following quiz question. According to the performance of the user's answers, this module will adapt the next question based on the information produced by the student's interaction. The FISEngine subcomponent oversees fuzzy inference to adapt the content (it controls the fuzzy rules). FISEngine contains the linguistic variables, the fuzzy sets, and several labels. The input variables are the number of correct answers, the number of errors made, and the difficulty level of the previous question. To present the next question complexity, we added a fuzzy logic module. Forty-eight rules were defined to conduct the fuzzy inference, in which success, errors, and degree of complexity of the previous question were adopted. We implemented the AFORGE.NET library to perform the fuzzy inference. An example of a fuzzy rule is:

IF success is LOW and errors is HIGH and previous is LOW Then nextLevel is LOW.

4.2 System Interfaces

This section describes the most important interfaces used in the development of the "DinoApp" learning tool. The main menu contains three main buttons and a secondary button (see Fig. 3), the first button links the user to the "create your scenario" scene, where the user can select 3D elements through ground detection provided by ARCore to superimpose a dinosaur onto the screen. The "información dinosaurios" button provides the user with a technical sheet with a description of each 3D element and its major features. In this scene, the user uses the markers to select the element he wants to visualize. The "Quiz" button links the user to a short quiz where the user must answer ten questions from the same information that the application has provided. At the bottom of the menu there is an icon, where when pressed, the user will see a panel with the privacy notice and a QR code which, when selected, will redirect the user to the URL to download the available markers.

DinoApp learning tool scenes are described as follows:

– Create your scenario: In this section, users must focus their mobile device on the ground to detect a plane. Once the plane is detected, eight 3D models are available in a Box Slider, the user can select the 3D models, place them on the screen, rotate them, turn them, and zoom in or out of the screen, each 3D model a built-in audio that refers to the creature. If the user wishes, he can save a photo of the souvenir. Figure 4 shows the GUI of this scene.

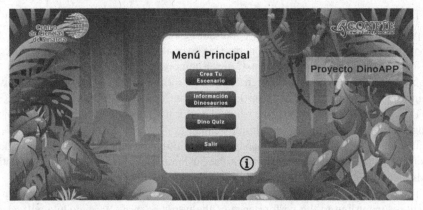

Fig. 3. DinoApp main menu

Fig. 4. Creating a scenario using ground detection.

- Dino Information: In this scene, the user can scan a marker to over-pose a 3D element in the mobile screen. The user can rotate the 3D element and capture a screenshot that is saved directly on the mobile device. Figure 5 shows an example of this scene.

- Dino Quiz. In this scene, the user can respond to ten questions about the content of the learning tool. The questions that will appear to the user have three levels of difficulty, which will be adapted according to the answers provided by the user. The more correct the answers, the more difficult the questions will appear, thanks to the fuzzy logic technique integrated by the learning tool. Figure 6 shows an example of the Quiz.

Once the software development was completed, it was evaluated by staff from the CCS Museum to validate the reliability and quality of the information displayed to the public. Regarding the functionality of the learning tool, there was an exhaustive review

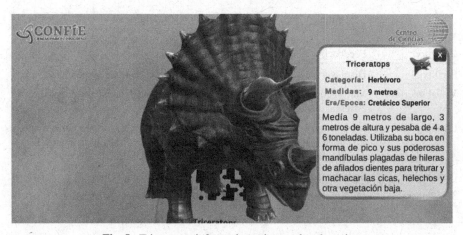

Fig. 5. Triceratops information using marker detection.

Fig. 6. Question from the Quiz test.

conducted on computers, cell phones, tablets, and iPads. Once evaluated, they were published on the Vercel platform.

5 Evaluation and Results

Tests were performed each time a scene of the learning tool was completed, to check its operation and detect any errors from the beginning. The learning tool was evaluated by six users, who provided feedback to improve the user experience, the design of the visual interfaces, and the operability of the learning tool. Table 1 shows the feedback on the platform from the different users who assessed it.

The DinoApp learning tool can be utilized during museum visits. Additionally, certain guides offer instructions on downloading and operation. DinoApp may be directly downloaded from the Google Play Store.

Table 1. User evaluation of the learning tool.

Category	Feedback
Facility	Display a message with a brief description of each scene visited by the user
Visual interface	Multimedia content (images and videos) must be dynamically adapted to different devices
Usability	Keep the size of individual scenes as small as possible for best performance It is great that you can combine markers and ground detection to attract people's attention
Persistence	Save the user performances of responding to the dino Quiz
3D Elements	Modify the size and textures of 3D models for better visualization

6 Conclusions and Future Work

AR is a technology that has shown favorable results in educational [15], cultural, tourism, and museum [16] fields; especially since they have been developed for mobile devices. An AR application was developed for use in the "Centro de Ciencias de Sinaloa" Museum as an alternative to augmented reality mobile applications that cover the same content to complement user satisfaction and offer a complementary alternative to visitors. This not only complements user satisfaction but also provides an additional option for visitors who can explore the augmented reality content even if they are not physically in the museum. This accessibility adds significant value to the museum and its visitor experience.

This work could serve as a model to explore and expand the use of this technology in different fields and sectors, offering a richer and more engaging experience for users and visitors in various industries.

As future work, this project has the potential to be applied in various domains such as gastronomy, tourism, entertainment, architecture, and personnel training. In the near future, our primary objective is to incorporate sentiment analysis, utilizing artificial intelligence techniques to identify emotional states based on user-provided sentences, for improved comprehension of user feedback while utilizing the learning tool.

References

1. Azuma, R.: A survey of augmented reality. Presence Teleoperators Virtual Environ. **6**, 355–385 (1997). https://doi.org/10.1162/pres.1997.6.4.355
2. Yovcheva, Z., Buhalis, D., Gatzidis, C.: Smartphone augmented reality applications for tourism. E-Rev. Tour. Res. **10**, 63–66 (2012)
3. Ibañez, M.B., Delgado-Kloos, C.: Augmented reality for STEM learning: a systematic review. Comput. Educ. **123**, 109–123 (2018). https://doi.org/10.1016/j.compedu.2018.05.002
4. Fidan, M., Tuncel, M.: Integrating augmented reality into problem based learning: the effects on learning achievement and attitude in physics education. Comput. Educ. **142**, 103635 (2019)
5. Jasche, F., Hoffmann, S., Ludwig, T., Wulf, V.: Comparison of different types of augmented reality visualizations for instructions. In: Proceedings of the 2021 CHI Conference on Human Factors in Computing Systems, pp. 1–13 (2021)

6. Bachiller, C., Monzo, J.M., Rey, B.: Augmented and virtual reality to enhance the didactical experience of technological heritage museums. Appl. Sci. **13**, 3539 (2023)
7. Pallud, J.: Impact of interactive technologies on stimulating learning experiences in a museum. Inf. Manag. **54**, 465–478 (2017)
8. Riva, J.: The assessment of the level of digital innovation in Italian museums (2020)
9. Boboc, R.G., Băutu, E., Gîrbacia, F., Popovici, N., Popovici, D.M.: Augmented reality in cultural heritage: an overview of the last decade of applications. Appl. Sci. **12**, 9859 (2022). https://doi.org/10.3390/APP12199859
10. Zhu, Y., Wang, C.: Study on virtual experience marketing model based on augmented reality: museum marketing (example). Comput. Intell. Neurosci. **2022** (2022). https://doi.org/10.1155/2022/2485460
11. Ding, M.: Augmented Reality in Museums (2017)
12. Falomo Bernarduzzi, L., Bernardi, E.M., Ferrari, A., Garbarino, M.C., Vai, A.: Augmented reality application for handheld devices: how to make it happen at the Pavia university history museum. Sci. Educ. **30**, 755–773 (2021). https://doi.org/10.1007/S11191-021-00197-Z/FIGURES/7
13. Guo, W., Xue, Y., Sun, H., Chen, W., Long, S.: Utilizing augmented reality to support students' learning in popular science courses. In: 2017 International Conference of Educational Innovation through Technology (EITT), pp. 311–315 (2017)
14. Schönhofer, A., Hubner, S., Rashed, P., Aigner, W., Judmaier, P., Seidl, M.: Viennar: user-centered-design of a bring your own device mobile application with augmented reality. In: De Paolis, L.T., Bourdot, P. (eds.) AVR 2018. LNCS, vol. 10851, pp. 275–291. Springer, Cham (2018). https://doi.org/10.1007/978-3-319-95282-6_21
15. Koparan, T., Dinar, H., Koparan, E.T., Haldan, Z.S.: Integrating augmented reality into mathematics teaching and learning and examining its effectiveness. Think. Ski. Creat. **47**, 101245 (2023)
16. Zhou, Y., Chen, J., Wang, M.: A meta-analytic review on incorporating virtual and augmented reality in museum learning. Educ. Res. Rev. **36**, 100454 (2022). https://doi.org/10.1016/j.edurev.2022.100454

School Dropout Prediction with Class Balancing and Hyperparameter Configuration

P. Alejandra Cuevas-Chávez[1]([✉])[iD], Samuel Narciso[1],
Eduardo Sánchez-Jiménez[1][iD], Itzel Celerino Pérez[2], Yasmín Hernández[1][iD],
and Javier Ortiz-Hernandez[1][iD]

[1] Computer Science Department, Tecnológico Nacional de México/Cenidet,
62490 Cuernavaca, Mexico
{d18ce074,m22ce051,m22ce005,yasmin.hp,javier.oh}@cenidet.tecnm.mx
[2] Tecnológico Nacional de México/campus Cerro azul, 92519 Veracruz, Mexico
LA20500611@cerroazul.tecnm.mx

Abstract. School dropout and academic underachievement have significant effects on economic growth and employment in society. This phenomenon impacts not only the intellectual development of students but also their access to desirable job opportunities, which can improve their quality of life. This paper focuses on school dropout in the Predict Students' Dropout and Academic Success dataset. We use three resampling techniques for the imbalanced problem: Adaptive Synthetic, Support Vector Machine-Synthetic Minority Oversampling Technique, and Synthetic Minority Oversampling Technique+Edited Nearest Neighbor. We also compare the performance of the Random Forest, Support Vector Machine, and XGBoost classifiers between the default hyperparameter configuration and Bayesian configuration. The Synthetic Minority Oversampling Technique+Edited Nearest Neighbor technique obtained the best average performance using the Support Vector Machine classifier, achieving an accuracy of 93.55% and a precision of 94.11%.

Keywords: academic performance · bayesian optimization · class imbalanced · educational data mining · hyperparameter configuration · machine learning · resampling techniques · school dropout

1 Introduction

Education is a crucial factor in personal and social development. It offers knowledge and opportunities that impact individual lives and the wider community. The intellectual growth of individuals has an affirmative effect on the society they inhabit, resulting in a more prosperous and developed population. However, school dropout presents a major issue during academic education that affects all academic grades. This phenomenon has a significant impact on both

H. Calvo et al. (Eds.): MICAI 2023 Workshops, LNAI 14502, pp. 12–20, 2024.
https://doi.org/10.1007/978-3-031-51940-6_2

students' intellectual development and their access to favorable job opportunities and higher quality of life [1]. Education yields positive outcomes for individuals and society, whereas dropping out of school has negative consequences. The absence of education can result in a less educated, less productive population that is more susceptible to social difficulties. Therefore, it is crucial to identify the determining factors and potential outcomes associated with this phenomenon and the necessary preventive measures. The use of machine learning models to analyze data enables an efficient and precise comprehension of this phenomenon. Machine learning is a promising method for predicting student dropout based on their performance data [2]. The goal is to predict and prevent dropout in a successful way.

The main contribution of this paper is to identify the best conditions in machine learning to predict school dropout in the Dropout and Academic Success dataset, focusing mainly on the prediction of dropout. In this sense, machine learning can facilitate timely intervention when a student is at risk of dropping out.

The remainder of this paper is structured as follows. Section 2 presents related work. Section 3 introduces the research methodology applied to the Dropout and Academic Success dataset. This includes descriptions of the dataset's characteristics, data preprocessing techniques, resampling methods, hyperparameter configuration (HC), and the performance of the machine learning classifiers. In Sect. 4, we briefly discussed the results of the classifier's performance according to the resampling techniques. Finally, a conclusion is offered in Sect. 5.

2 Related Work

This section reviews three research using a school dropout dataset we based on for this research.

Wan Yaacob et al. [3] compare different data mining methods, such as K-Nearest Neighbors (KNN), Decision Tree, Random Forest, Logistic Regression, and Neural Networks, to determine the most effective classification model for anticipating student dropout in higher education. The performance of each model was evaluated and compared using various metrics, including accuracy, area under the curve, recall, precision, and the ROC curve. The results indicate that all five models had an accuracy rate of over 80%. However, Logistic Regression achieved the best performance, with an accuracy rate of 90%.

Realinho et al. [4] introduce the Predicting Student Dropout and Academic Success dataset. The dataset exhibits significant imbalance, as indicated by the authors. Thus, the authors propose addressing this issue by implementing data or algorithm-level solutions. In the data-level approach, they recommend using sampling techniques like Synthetic Minority Oversampling (SMOTE) or Adaptive Synthetic sampling approach (ADASYN). On the algorithmic level, it is advisable to use methods like Balanced Random Forest, Easy Ensemble, or classifiers that have additional balancing strategies, such as Exactly Balanced Bagging, Roughly Balanced Bagging, Over-Bagging, or SMOTE-Bagging.

Niyogisubizo et al. [5] proposed a two-level hybrid algorithm ensemble approach. The first level comprises the Random Forest, Extreme Gradient Boosting,
and Gradient Boosting algorithms. While the second level used a Feed-Forward
Neural Network. They optimized hyperparameters for the assembled algorithms,
aiming for the best performance of the proposed hybrid algorithm ensemble at
each level. The proposed algorithm's performance underwent separate comparisons with at least four other algorithms using different classification metrics,
including accuracy, precision, recall, AUC, and F1-score. All algorithms achieved
metrics above 75%. However, the ensemble outperformed the individual algorithms, achieving 92% accuracy and excelling in the other evaluated metrics.

Although Niyogisubizo et al. [5] uses hyperparameter optimization in an
ensemble based on hybrid Random Forest, XGBoost, Gradient Boosting, and
Feed-forward Neural Networks to obtain an accuracy of 92%, it is possible to
improve this result by finding an optimal scaler instead of a standard scaler.
Also, it can be useful to balance the dataset with a proper technique for a specific data set. For the case of Wan Yaacob et al. [3], the accuracy rate (90%)
of their Logistic Regression model could improve with hyperparameter tuning
as same as Realinho et al. [4], which focused on balancing the dataset with
ADASYN technique.

3   Material and Methods

This section is organized into five subsections: the dataset, the preprocessing,
the resampling techniques, the hyperparameter optimization, and the machine
learning classifiers.

3.1   Dataset

The Predict students' Dropout and Academic Success dataset is publicly available on the website Kaggle [6]. This dataset contains demographic, macroeco-
nomic, and socioeconomic data, along with student enrollment information, for
the first and second semesters of 17 undergraduate degrees across various fields
of study. The dataset consists of 4424 instances with 34 attributes and one multiclass target. The dataset has the following attributes: 20 discrete attributes, eight
binary attributes, five continuous attributes, and one ordinal attribute. The target has three classes labeled as dropout (0), enrolled(1), and graduate(2).

The data is imbalanced concerning the enrolled students. For the class labeled
dropout, there are 1421 instances (32.1%). While for the class labeled enrolled,
there are 794 instances (17.9%). Finally, the class labeled graduate has 2209
instances (49.9%).

3.2   Preprocessing

Once the data is collected, the next step is to prepare the it before building
the machine learning model. In the Predict Students' Dropout and Academic

Success dataset, there were no missing values, and no value was imputed. We proceed to scale the dataset using the RobustScaler technique. Then, we split the data into training and testing sets and resampled them with three different techniques. For outliers, we decided not to treat them. This decision is based on critical information that will be lost if we decide to treat them.

3.3   Resampling Techniques

In the state-of-the-art, resampling techniques have a variety of approaches. We applied ADASYN, SVM-SMOTE, and the hybrid technique SMOTE+ENN.

- ADASYN: Adaptive Synthetic uses a weighted distribution for different minority class examples based on how difficult they are to learn. It generates more synthetic data for minority class examples that are harder to learn compared to minority class examples that are easier to learn [7].
- SVM-SMOTE: This method focuses only on the minority class instances around the borderline, based on the Synthetic Minority Oversampling Technique with SVM as the base classifier [8].
- SMOTE+ENN: The Synthetic Minority Oversampling Technique in combination with the Edit Nearest Neighbor Technique is a hybrid technique that combines over-and under-sampling methods. Samples with a class that is different from the class of at least two of their three nearest neighbors are removed from the sample space [9].

3.4   Hyperparameter Configuration and Machine Learning Classifiers

Hyperparameter configuration not only influences the performance of the classifiers but also the efficiency of a model, as they can control aspects of the learning process (overfitting and underfitting). There are diverse hyperparameter optimization techniques, such as Random Search, Grid Search, Particle Swarm Optimization, and Genetic Algorithm, as the best known. However, the criterion for selecting the right set of hyperparameters for this research was based on Bayesian Optimization for three different classifiers: Random Forest, Support Vector Machine (SVM), and XGBoost.

Bayesian optimization is a model-based approach to sequentially solving problems. The framework of Bayesian optimization comprises two crucial components. The first is a probabilistic surrogate model that captures the preconceptions about the unknown objective function using a prior distribution. The second component is an observation model that describes how data is generated and involves a loss function that specifies the optimal sequences of query. Typically, such loss functions are expressed through regret, whether it be simple or cumulative. The goal is to minimize the expected loss to identify the most optimal sequence of queries. Following each query output, the prior is updated to generate a more informative posterior distribution across the objective function space [10]. In addition, we used 10-cross validation for the classifiers performance. The hyperparameters were fixed by classifier and are shown in Table 1.

Table 1. Hyperparameter configuration.

Classifier	Hyperparameter	Adasyn	SVM-SMOTE	SMOTE+ENN
Random Forest	max_depth	None	25	35
	min_samples_split	2	2	3
	n_estimators	200	200	200
	criterion	gini	gini	entropy
SVM	C	5	12.5	2.0
	kernel	rbf	rbf	rbf
	gamma	0.03	0.02	0.03
XGBoost	n_estimators	200	50	90
	colsample_bytree	0.8	0.8	0.8
	learning_rate	0.2	0.25	0.6
	max_depth	7	15	7
	subsample	0.8	0.8	0.8

4   Results and Discussion

In this section, we have detailed the results of the experiments using three resampling techniques, one hyperparameter configuration technique, and three classifiers. For the experiments, we used the Python language implemented on Jupyter Notebook 6.4.8 on a MacBook Pro (2017) with 2.9 GHz Quad-Core Intel® Core™ i7, 16 GB RAM 2133 MHz LPDDR3, Intel® HD Graphics 630 1536 MB, and macOS Ventura as the operating system.

We split the dataset into 80% training and 20% testing. Also, we trained the model without hyperparameter configuration and with Bayesian Optimization using the evaluation metrics of accuracy, precision, recall, and F1-score. The averaged test results for each classifier with and without hyperparameter configuration are shown in Table 2.

The original dataset has 1421 instances for class 0, 794 for class 1, and 2209 for class 2. Using the ADASYN technique, the resampled dataset was updated to 2257 instances for class 0, 2416 for class 1, and 2209 for class 2. Comparing the two performances, it is noticeable that the performance for Bayesian Optimization obtained better results in all the metrics, highlighting the precision value by XGBoost (from 84.51% to 86.43%). The results of ADASYN classifiers without hyperparameter configuration improved from 69.66%-70.61% to 84.78%-85.66% with Bayesian Optimization.

SVM-SMOTE technique resampled the dataset to 2209 instances for class 0, class 1, and class 2. Similarly to the ADASYN results, the performance for Bayesian Optimization got better results in all the metrics, specifically on the precision by XGBoost (from 84.72% to 85.97%). The results of SVM-SMOTE classifiers without hyperparameter configuration improved from 73.99%-84.83% to 82.39%-85.97% with Bayesian Optimization.

SMOTE+ENN resampled the dataset to 1272 instances for class 0, 1574 for class 1, and 951 for class 2. In the contrary to ADASYN and SVM-SMOTE, SMOTE+ENN achieved the best results, because in all the metrics (accuracy, precision, recall, and F1-score) the results varies between 92.30% to 94.11% with Bayesian Optimization. The results without hyperparameter configuration varies between 87.23% to 92.62%.

From the three resampling techniques we used to balance the dataset (ADASYN, SVM-SMOTE, and SMOTE+ENN), the best one for the characteristics of the Predict Students' Dropout and Academic Success dataset was the hybrid one: SMOTE+ENN. Comparing the evaluation metrics results, we found that removing instances, as SMOTE+ENN does, can positively improve the performance of the classifiers. But, it could also remove critical information or lead to a misclassification of whether the student will drop out, enroll in a university, or graduate in the best of cases.

On the other hand, combining hyperparameter configuration (Bayesian Optimization) with resampling techniques obtained better results as SMOTE+ENN achieved using SVM as a classifier. This technique produced remarkably high levels of accuracy at 93.55%, as well as precision at 94.11%. The XGBoost classifier obtained a recall of 93.11% and an F1-score of 93.39%. However, as our main goal was to identify student dropouts effectively, we found that the SVM model was particularly adept at identifying the class label dropout with an 87% recall, 95% accuracy, and an F1-score of 92%.

The improvement and results from the SMOTE+ENN model can be influenced by the scaling technique and the treatment of the outliers. Since we decided not to treat them, the classifier's performance could be affected by this decision, although we use the RobustScaler technique, which uses statistics that are robust to outliers. We notice an improvement in all the metrics (accuracy, precision, recall, and F1-score) by comparing the results from the models without hyperparameter tuning to the models using Bayesian Optimization. In a particular case, for example, there is an improvement of .59% for the Random Forest classifier in combination with the SMOTE+ENN resampling technique.

We firmly believe that the performance could be improved using other approaches and techniques for the whole process of building a machine learning model. However, the model SMOTE+ENN, in combination with the SVM classifier and Bayesian Optimization for hyperparameter tuning, has an optimal performance for the type of imbalance data and the non-treatment for outliers. Also, this model could be improved by focusing on other evaluation metrics for dealing with imbalanced data, such as F2-score and area under the precision-recall curve (PR-AUC).

Table 2. Testing classifiers performance.

Resampling	Classifier	HC	Accuracy	Precision	Recall	F1-score
ADASYN	Random Forest	without	83.80%	84.05%	83.75%	83.82%
		with bayesian	85.11%	85.32%	85.05%	85.11%
	SVM	without	69.71%	70.61%	69.66%	69.71%
		with bayesian	85.11%	85.66%	85.05%	84.78%
	XGBoost	without	84.31%	84.51%	84.51%	84.28%
		with bayesian	86.27%	86.43%	86.22%	86.26%
SVM-SMOTE	Random Forest	without	84.61%	84.83%	84.57%	84.63%
		with bayesian	85.52%	85.75%	85.46%	85.53%
	SVM	without	74.20%	74.64%	74.01%	73.99%
		with bayesian	82.42%	82.47%	82.50%	82.39%
	XGBoost	without	84.61%	84.72%	84.55%	84.59%
		with bayesian	85.82%	85.97%	85.76%	85.80%
SMOTE+ENN	Random Forest	without	91.84%	92.54%	91.72%	92.08%
		with bayesian	92.50%	93.16%	92.30%	92.67%
	SVM	without	87.23%	88.00%	87.36%	87.57%
		with bayesian	93.55%	94.11%	93.06%	93.38%
	XGBoost	without	92.10%	92.62%	92.21%	92.40%
		with bayesian	93.15%	93.73%	93.11%	93.39%

In future work, we will focus on other hyperparameter optimization techniques, such as Random Search, Grid Search, or Particle Swarm Optimization. For HPO, we will add other hyperparameters, such as max_features and min_samples_leaf for Random Forest, degree and coef0 for SVM, and gamma for XGBoost. We will implement the research by Hernández et al. [11] based on personality recognition relying on the dominance, influence, steadiness, and compliance (DISC) personality model, which may contribute to our future approach based on student personality to predict dropout rates. With this improvement, the model could identify the student's personality more affected by school dropout. This could be an opportunity for the school to focus on the potential students with the risk of dropping out of school for planning strategies for an early intervention to avoid students doing it. This type of machine learning model can provide insights into why the students are dropping out of school, identify the students at risk, and apply the targeted interventions depending on the student's needs. It could also help with the school's resource planning, prioritizing the students who need the most support. Finally, if this kind of machine learning model is implemented in frameworks, applications, or systems, it could be relevant to a decision-making support system. It will lead the academic staff or educational institutions to timely interventions and evaluate the effectiveness of dropout prevention programs.

5   Conclusions

The classifiers used in this paper to predict college student dropout based on their academic performance were Random Forest (RF), Support Vector Machine (SVM), and Extreme Gradient Boosting (XGBoost). The Predict Students' Dropout and a Academic Success dataset analyzed consisted of 4422 records, 34 attributes, and a target class with three possible labels: graduated, enrolled, and dropout. Since there was an imbalance in the number of instances across the three labels, we applied three resampling techniques (ADASYN, SMOTE-SVM, SMOTE+ENN) to ensure the balance in the three possible labels of the target.

The three classifiers (RF, SVM, XGBoost) were applied to the dataset resampled by ADASYN, SVM-SMOTE, and SMOTE+ENN using default hyperparameters and Bayesian optimization to determine which classifier benefited the most from optimization. We trained the machine learning models with a 10-cross validation. Although the dataset includes three possible labels as a target, we focused on the dropout label to identify the students at risk of dropping out. The Random Forest algorithm achieved the highest performance in recall (89%) with SMOTE+ENN without hyperparameter configuration. Both Random Forest and XGBoost achieved the same value in recall (82%) with the SVM-SMOTE resampling technique. Random Forest achieved 81% in recall with the ADASYN technique. On the contrary, using the SVM classifier with ADASYN, SVM-SMOTE, and SMOTE+ENN, the values obtained in recall were 92%, 86%, and 89%, respectively.

The importance of configuring hyperparameters for predicting dropout rates is evident compared to non-optimized models. Because the performance of the classifiers suffered an increase in dropout prediction than the machine learning models trained with the default hyperparameter configuration. The SVM model, with its configured hyperparameters, serves as a valuable prediction tool for college student dropout rates based on academic performance. Its use enables the anticipation of student dropout, providing opportunities to implement measures in order to prevent it. By utilizing this predictive methodology, educational institutions can proactively apply tailored academic interventions and student support programs to decrease this phenomenon, improving student retention in the education system.

References

1. Rochin Berumen, F. L.: Deserción escolar en la educación superior en México: revisión de literatura. RIDE. Rev. Iberoam. Investig. Desarro. **11**(22) (2021)
2. Kuz, A., Morales, R.: Ciencia de Datos Educativos y aprendizaje automático: Un caso de estudio sobre la deserción estudiantil universitaria en México. Educ. Knowl. Soc. (EKS) **e30080** (2023)
3. Wan Yaacob, W.F., et al.: Predicting student drop-out in higher institution using data mining techniques. J. Phys: Conf. Ser. **1496**, 012005 (2020)
4. Realinho, V., Machado, J., Baptista, L., Martins, M.V.: Predicting student dropout and academic success. Data **7**(11), 146 (2022)

5. Niyogisubizo, J., Liao, L., Nziyumva, E., Murwanashyaka, E., Nshimyumukiza, P.C.: Predicting student's dropout in university classes using two-layer ensemble machine learning approach: a novel stacked generalization. Comput. Educ.: Artif. Intell. **3**, 100066 (2022)
6. Kaggle   Homepage.   https://www.kaggle.com/datasets/thedevastator/higher-education-predictors-of-student-retention/data. Accessed 14 Oct 2023
7. He, H., Bai, Y., Garcia, E.A., Li, S.: ADASYN: adaptive synthetic sampling approach for imbalanced learning. In: IEEE International Joint Conference on Neural Networks (IEEE world congress on computational intelligence), pp. 1322–1328. IEEE, Hong Kong (2008)
8. Nguyen, H.M., Cooper, E.W., Kamei, K.: Borderline over-sampling for imbalanced data classification. Int. J. Knowl. Eng. Soft Data Paradig. **3**(1), 4–21 (2009)
9. Batista, G.E.A.P.A., Prati, R.C., Monard, M.C.: A study of the behavior of several methods for balancing machine learning training data. J. SIGKDD Explor. Newsl. **6**(1), 20–29 (2004)
10. Shahriari, B., Swersky, K., Wang, Z., Adams, R.P., De Freitas, N.: Taking the human out of the loop: a review of Bayesian optimization. In: Proceedings of the IEEE, pp. 148–175. IEEE (2015)
11. Hernández, Y., Martínez, A., Estrada, H., Ortiz, J., Acevedo, C.: Machine learning approach for personality recognition in Spanish texts. Appl. Sci. **12**(6), 2985 (2022)

Method to Identify Emotions in Immersive Virtual Learning Environments Using Head and Hands Spatial Behavioral Information

Jorge Enrique Velázquez-Cano[1]([✉]), Juan Gabriel Gonzáles-Serna[1],
Leonor Rivera-Rivera[2], Nimrod Gonzáles-Franco[1], José Alejandro Reyes-Ortiz[3],
Máximo López-Sánchez[1], and Blanca Dina Valenzuela-Robles[1]

[1] Computer Science Department TecNM/CENIDET, 62490 Cuernavaca, Morelos, México
{d19ce066,gabriel.gs,nimrod.gf,maximo.ls,
blanca.vr}@cenidet.tecnm.mx
[2] Instituto Nacional de Salud Pública, 62100 Cuernavaca, Morelos, México
lrivera@insp.mx
[3] Universidad Autónoma Metropolitana, Mexico City, México
jaro@azc.uam.mx

Abstract. Emotion detection has been carried out using different channels and sensors, facial recognition being one of the most frequent. Likewise, physiological signals such as EEG, ECG, EMG, and GSR along with various artificial intelligence classification techniques, such as SVM and NN, are recurrent for this task. However, to achieve this, multiple devices are used, most are considered invasive, expensive, not very portable and require large computing capacity. Therefore, and supported by the literature that suggests that different cognitive states are reflected in behavior and body postures, this work proposes an emotional detection approach during the use of an Immersive Virtual Learning Environment (IVLE), using the information provided by the Immersive Virtual Reality (IVR) device in conjunction with data obtained during interactions and adapted self-report instruments. Since only the data extracted from the device itself is used, we consider this as a single-modal approach. This seeks to be a portable-low-cost solution with precision enough so it can be implemented in an affective tutoring system in an IVR IVLE.

Keywords: IVLE · Virtual Reality · Affective tutoring system · Emotion detection · Affective state · Child Sexual Abuse · CSA

1 Introduction

During a training process, one important aspect is the motivation of the trainee. Following this premise, mechanisms that aim to activate and keep the student motivated are established [1]. However, less attention is paid to the learner's emotional state, which given its importance in the learning process [2], could be the cause of the results of a training program, whether favorable or not.

H. Calvo et al. (Eds.): MICAI 2023 Workshops, LNAI 14502, pp. 21–28, 2024.
https://doi.org/10.1007/978-3-031-51940-6_3

There is concern among parents and teachers about the impact of Child Sexual Abuse (CSA) prevention programs on the emotional state of the participants, who are typically between the ages of 6 and 12 years old. The sensitive content of these programs is a cause for worry, and it is important to ensure that children feel safe and comfortable while learning about these important issues. Although some studies suggest that cases of anxiety, fear, and trauma resulting from CSA prevention programs are rare [3, 4], they are still recognized as significant issues [3, 5, 6]. However, CSA survivors have also pointed out that the benefits of CSA prevention programs can outweigh the negative effects, and that these effects may depend more on the environment in which the children develop [7].

Although information and communication technologies can be part of prevention programs [6, 8–11], IVR technologies have been scarcely mentioned. Traditional teaching methods such as: images, videos and theatrical sketches [3, 12] are still commonly used in prevention programs. These methods can be replaced with VR which has the added benefits of being reusable, more interesting and allowing active knowledge acquisition [13].

2 Background

2.1 Emotions and Emotion Detection During the Learning Process

According to cognitive theory, an emotion is a synchronized sequence of changes in the states of our cognitive, neurophysiological, motivational, motor expression, and subjective feeling subsystems, in response to an event's evaluation. The stimulus, whether external or internal, is so relevant that it concerns the organism [14]. Madera-Carillo [15] summarizes this as systematic responses that reflect the central activation of various body systems in preparation for a physical reaction. These responses involve a set of experiences, attitudes, and beliefs about the world that influence how we perceive events and situations [16]. Emotions are an essential part of every human being and can influence our behavior, cognitive abilities, decision-making, resilience, well-being, and the way people communicate with each other [17].

During learning, emotions are categorized into four quadrants based on valence and activation [18]: 1) Positive valence and activation generate motivation in the student. 2) Positive valence and negative activation generate a state of momentary relaxation. 3) Negative valence and positive activation can put pressure on students. 4) Negative valence and activation decrease the student's will to continue with the learning process.

Positive emotions don't always have a positive impact and negative emotions don't always generate adverse situations during learning [19]. Emotions can trigger actions and be expressed through movements, gestures, and body postures. Specific emotions or their intensity are associated with differences in posture and movement patterns, some of which are shared between emotions of the same valence [20].

Detecting emotions during the learning process involves analyzing patterns in behavior and physiological responses of the organism. This approach has been widely covered by research studies [18] and [21]. Facial recognition is commonly used, followed by speech/voice recognition, recollection of physiological signals, and text analysis. Postural and behavioral analysis is not mentioned. However, the selection of signals depends on the context, algorithm, or intended use of the information. In some cases, students may need to wear various sensors, as in [22] where the signals were used to identify their emotional state and predict the perceived difficulty of the learning process. However, it is important to note that using sensors may cause discomfort and a feeling of intrusion [23] which can introduce bias into the results. The study implicitly acknowledges this issue. In [19, 24], and [25] emotions were examined taking advantage of non-verbal behavior such as gestures, position and head movements, clicks, and keystroke interactions during readings and tests of different difficulties. Kinect sensors have also been used due to being a low-cost equipment for postural detection, activity recognition, and manual gestures.

Some of the techniques acquire greater complexity as the age of the study group decreases (e.g. trying to analyze texts from users who barely know how to write).

2.2 Affective Tutoring Systems

An affective tutoring system (ATS) is a type of computer system that employs affective computing, based on Picard's work [26], to mimic a human tutor. The system adapts to the emotional state of the learner and reacts in situations that could negatively impact the learning process [25]. The basic architecture of an ATS is depicted in Fig. 1.

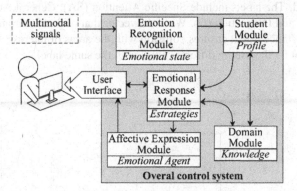

Fig. 1. Basic architecture of an ATS

The emotion recognition module uses various physical and physiological signals to identify the student's emotional state. Based on this, the emotional response module evaluates the tutoring situation and implements emotional regulation strategies. It is possible to incorporate an affective expression module to extend the user interface with a virtual tutor that expresses emotions in response to the student's actions and emotional state.

According to [18] taxonomy of computerized learning, the ATS is an improvement/replacement of the Intelligent Tutoring System (ITS). However, the literature suggests that these are different systems [25] that can complement each other and that sometimes share resources and techniques.

3 Proposed Method

The objective is to establish an ATS in an IVLE oriented to CSA prevention already developed, and help in regulating the user's emotions during exposure to sensitive topics. To do so and as pointed out in the previous section, a method to recognize emotions is needed. The system employs portable IVR devices and intends to not rely on external equipment to sense, gather, or process data. Furthermore, a qualitative tool that enables the user to communicate their emotions after each scene is to be incorporated as an interactive VR scene.

In order to implement the ATS, the extension of the IVLE will take place in three stages: registration, classification, and regulation. This work will cover the first two due that is in these parts that the proposed method takes place. The registration stage involves collecting raw data from which characteristics for further analysis will be processed and extracted. To perform this, the scenes incorporate mechanisms that gather and store the information associated with the position and orientation of the IVR devices, along with a timestamp. Additionally, a record of the user's visual focus and the completion time of the scenes with interactions will also be implemented.

To determine where the user is looking, the model shown in Fig. 2 is followed, which is an adaptation of [27]. In our 3D scenario, areas of interest that consist of an invisible collider and a label indicating the type of attention expected from the user are incorporated. The labels include Specific Attention (SA), General Attention (GA), Avoidance (A), and Distraction (D). We use a vector that originates from the camera c and ends at a distance d that ensures collision with the areas of interest. We keep track of the time spent by the user in different areas with the same label.

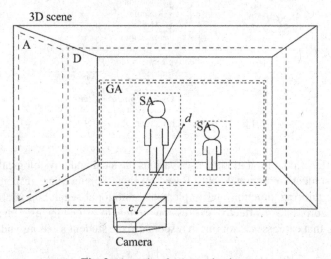

Fig. 2. Attention focus mechanism.

The Self-Assessment Manikin (SAM) [28] will be adapted and integrated into the system as a qualitative tool. The adaptation will only show essential information to the user, as seen in Fig. 3. To help users understand each dimension, a question that promotes introspection will be included. Only Valence and Arousal dimensions are being considered due to previous qualitative results with the IVLE that showed that children of the selected age group have difficulty understanding the Dominance dimension. The SAM will be implemented between scenes.

In the classification stage, we will collect data with a test group that will follow a simple protocol of answering entry and exit questionnaires and use the IVLE for a given duration.

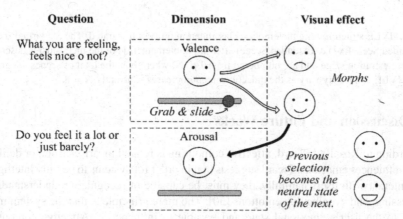

Fig. 3. Adaptation of the Self-Assessment Manikin for use in IVR.

For each interaction, the IVLE will keep a record of the collected data using the mechanisms described in the registration stage. In addition to the basic characteristics of location and rotation others will be calculated, for example, a distance vector between the visor and the haptic controls. It is also intended to estimate and record events. An event is defined by one or several sequential or simultaneous movements that may be associated with a particular emotion [20]. We sought to generate a data set that allows, through an automatic classification process using artificial intelligence, to explore relations between these and the reported emotional states. The aim is to take advantage of the fact that the dimensions of the SAM are similar to the dimensions of Russell's circumplex model of affect [29], which locates emotions on a two-dimensional plane composed of valence and activation. This emotional model is one of the most used in affective computing [21].

Classification is a process that does not require user intervention to be carried out and, therefore is integrated into the SAM scene after the user finishes the interactions and as a background process. The Fig. 4 shows the resulting structure of the IVLE. At this point, it is expected that the application can provide information about the emotional state induced by each of the IVLE scenes. This can be used as a guide to improve the user experience or modify the content of a scene if it is inducing emotional states other than those intended.

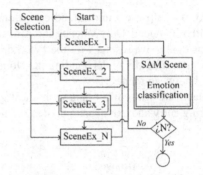

Fig. 4. IVLE structure with the emotion classification process integrated. The 3D virtual scenes are named SceneEx (i.e. Extended Scene) after the implementation of the mechanisms described in the registration stage and are numbered from 1 to N, where N is the total of scenes integrated in the IVLE and it may vary as the didactic content increases. Currently N = 8.

4   Discussion and Future Works

According to Rosalind Picard, affective computing is related to, arises from, or deliberately influences emotions. Picard suggests that in order for systems to be truly intelligent and interact with users naturally, they must be capable of recognizing, understanding, expressing, and even having emotions [30]. The main principle is that the system must adapt to the user's emotional state and be able to influence it. Affective computing involves both emotional classification and emotional induction.

The discussion of how and where emotions originate and how they should be classified has led to the emergence of various models with different numbers of emotions.

In [31] a dataset with 10 emotions was used, while [32] focused only on emotions involved in the learning process, and classified them using Russell's approach. Despite the differences in models, all of them acknowledge the existence of affective states, which can be identified through the analysis of physical and physiological data. However, it is a challenging task to determine the patterns that follow each emotion, as they can vary for each person [30].

Unlike conventional input devices like a mouse or keyboard, a wearable can offer various forms of interaction that take advantage of gestures and movements. Even though IVR devices are not a typical wearable by Picard's definition, the signals provided by these offer similar information and can be used for similar purposes. In fact, according to [33], IVR will play a crucial role in affective computing.

Several studies have linked the movements, posture, face orientation, and facial expressions of students to their level of attention and emotional and cognitive state [24, 25, 32]. In addition to facial expressions, hand movements can also provide valuable information about a person's emotional state [34]. Therefore, it is possible to infer a person's intentions or level of interest by analyzing the posture of these body parts. However, and even when [35] points out that body expressions are as good as facial expressions at conveying emotions, can a specific emotion, like serene or content that are so close within Russell's circumplex model, be identified with this information? or only its valence can be determined?

The proposed method of using the information provided by IVR devices is further supported by the fact that these devices are designed to accurately determine the location and orientation of each element. This information can be directly used instead of inferring it through an intermediate method such as artificial vision. We hypothesize that is within the subtleties where the answers to the previous questions lies.

Although some options, like random forest and support vector machines, have already been considered based on the findings in the literature, the artificial intelligence technique to be used for emotional classification in the IVR device is still being determined. The decision will depend on whether the IVR device has sufficient processing power to efficiently execute the developed algorithms and provide results in real-time, which is crucial for the successful implementation of an ATS in the IVLE of a CSA prevention program.

References

1. Plass, J.L., Homer, B.D., Kinzer, C.K.: Foundations of game-based learning. Educ. Psychol. **50**(4), 258–283 (2015)
2. D'Mello, S., Graesser, A.: Dynamics of affective states during complex learning. Learn. Instr. **22**, 145–157 (2012)
3. Tutty, L.M.: Listen to the children: kids' impressions of who do you tell™. J. Child Sex. Abus. **23**(1), 17–37 (2014)
4. Kenny, M.C., Helpingstine, C., Long, H.: College students' recollections of childhood sexual abuse prevention programs and their potential impact on reduction of sexual victimization. Child Abuse Negl. **104**, 104486 (2020)
5. Allen, K.P., Livingston, J.A., Nickerson, A.B.: Child sexual abuse prevention education: a qualitative study of teachers' experiences implementing the second step child protection unit. Am. J. Sex. Educ. **15**(2), 218–245 (2020)
6. Malamsha, M.P., Sauli, E., Luhanga, E.T.: Development and validation of a mobile game for culturally sensitive child sexual abuse prevention education in Tanzania: mixed methods study. JMIR Serious Games **9**(4), e30350 (2021)
7. Gubbels, J., Assink, M., Prinzie, P., Van der Put, C.E.: What works in school-based programs for child abuse prevention? The perspectives of young child abuse survivors. Soc. Sci. **10**(10), 404 (2021)
8. Jones, C., Scholes, L., Rolfe, B., Stieler-Hunt, C.: A serious-game for child sexual abuse prevention: an evaluation of orbit. Child Abuse Negl. **107**, 104569 (2020)
9. Moon, K.J., Park, K.M., Sung, Y.: Sexual abuse prevention mobile application (SAP_MobAPP) for primary school children in Korea. J. Child Sex. Abus. **26**(5), 573–589 (2017)
10. Kim, S.-J., Kang, K.-A.: Effects of the child sexual abuse prevention education (C-SAPE) program on South Korean fifth-grade students' competence in terms of knowledge and self-protective behaviors. J. Sch. Nurs. **33**(2), 123–132 (2016)
11. Scholes, L., Jones, C., Stieler-Hunt, C., Rolfe, B.: Serious games for learning: games-based child sexual abuse prevention in schools. Int. J. Incl. Educ. **18**(9), 934–956 (2014)
12. Walsh, K., Berthelsen, D., Hand, K., Brandon, L., Nicholson, J.M.: Sexual abuse prevention education in Australian primary schools: a national survey of programs. J. Interpers. Violence **34**(20), 4328–4351 (2019)
13. Schweiger, M., Wimmer, J., Chaudhry, M., Siegle, B.A., Xie, D.: Lernerfolg in der schule durch augmented und Virtual Reality? Eine quantitative synopse von wirkungsstudien

zum einsatz virtueller realitäten in grund- und weiterführenden schulen, MedienPädagogik zeitschrift für theorie und praxis der medienbildung 47(AR/VR - Part 1):1–25 (2022)
14. Scherer, K.R.: What are emotions? And how can they be measured? Soc. Sci. Inf. **44**(4), 695–729 (2015)
15. Madera-carrillo, H., Zarabozo, D., Ruíz-díaz, M., Berriel-saez, P.: El sistema internacional de Imágenes Afectivas (IAPS) en población mexicana. (2015)
16. Institutional Repository of the University of Alicante, Psychology and health department. https://rua.ua.es/dspace/bitstream/10045/3834/33/TEMA%209_PROCESOS%20PSICOL%C3%93GICOS%20BASICOS.pdf. Accessed 10 Oct 2023
17. Morrish, L., Rickard, N., Chin, T.C., Vella-Brodrick, D.A.: Emotion regulation in adolescent well-being and positive education. J. Happiness Stud. **19**(5), 1543–1564 (2018)
18. Hasan, M.A., Noor, N.F.M., Rahman, S.S.B.A., Rahman, M.M.: The transition from intelligent to affective tutoring system: a review and open issues. IEEE Access **8**, 204612–204638 (2020)
19. Taub, M., Azevedo, R., Rajendran, R., Cloude, E.B., Biswas, G., Price, M.J.: How are students' emotions related to the accuracy of cognitive and metacognitive processes during learning with an intelligent tutoring system? Learn. Instr. **72**, 101200 (2021)
20. Wallbott, H.G.: Bodily expression of emotion. Eur. J. Soc. Psychol. **28**(6), 879–896 (1998)
21. Yadegaridehkordi, E., Noor, N.F.B.M., Ayub, M.N.B., Affal, H.B., Hussin, N.B.: Affective computing in education: a systematic review and future research. Comput. Educ. **142**, 103649 (2019)
22. Alqahtani, F., Katsigiannis, S., Ramzan, N.: Using wearable physiological sensors for affect-aware intelligent tutoring systems. IEEE Sensors J. **21**(3), 3366–3378 (2020)
23. Petrovica, S., Anohina-Naumeca, A., Ekenel, H.K.: Emotion recognition in affective tutoring systems: collection of ground-truth data. Procedia Comput. Sci. **104**, 437–444 (2017)
24. Behera, A., Matthew, P., Keidel, A., Vangorp, P., Fang, H., Canning, S.: Associating facial expressions and upper-body gestures with learning tasks for enhancing intelligent tutoring systems. Int. J. Artif. Intell. Edu. **30**(2), 236–270 (2020)
25. Zaletelj, J., Košir, A.: Predicting students' attention in the classroom from Kinect facial and body features. EURASIP J. Image Video Process. **2017**(1), 1–12 (2017)
26. Picard, R.W.: Affective Computing. MIT Press, Cambridge (2000)
27. Lin, Y., Lan, Y., Wang, S.: A method for evaluating the learning concentration in head-mounted virtual reality interaction. Virtual Reality **27**, 863–885 (2022)
28. Bradley, M.M., Lang, P.J.: Measuring emotion: the self-assessment manikin and the semantic differential. J. Behav. Ther. Exp. Psychiatry **25**(1), 49–59 (1994)
29. Russell, J.A.: A circumplex model of affect. J. Pers. Soc. Psychol. **39**(6), 1161–1178 (1980)
30. Picard, R.W., Healey, J.: Affective wearables. Pers. Technol. **1**(4), 231–240 (1997)
31. Happy, S., Patnaik, P., Routray, A., Guha, R.: The Indian spontaneous expression database for emotion recognition. IEEE Trans. Affect. Comput. **8**(1), 131–142 (2017)
32. Ashwin, T.S., Guddeti, R.M.R.: Automatic detection of students' affective states in classroom environment using hybrid convolutional neural networks. Educ. Inf. Technol. **25**(2), 1387–1415 (2020)
33. Marín-Morales, J., et al.: Affective computing in virtual reality: emotion recognition from brain and heartbeat dynamics using wearable sensors. Sci. Rep. **8**(1), 13657 (2018)
34. Corneanu, C., Noroozi, F., Kaminska, D., Sapinski, T., Escalera, S., Anbarjafari, G.: Survey on emotional body gesture recognition. IEEE Trans. Affect. Comput. **12**(2), 505–523 (2018)
35. Kleinsmith, A., Bianchi-Berthouze, N.: Affective body expression perception and recognition: a survey. IEEE Trans. Affect. Comput.Comput. **4**(1), 15–33 (2013)

A New Approach for Counting and Identification of Students Sentiments in Online Virtual Environments Using Convolutional Neural Networks

José Alberto Hernández-Aguilar[1]([×]) (iD), Yasmín Hernández[2] (iD),
Lizmary Rivera Cruz[1], and Juan Carlos Bonilla Robles[1]

[1] Autonomous University of Morelos State, Av. Universidad 1001, Col. Chamilpa, 62209
Cuernavaca, Morelos, México
jose_hernandez@uaem.mx
[2] Centro Nacional de Investigación y Desarrollo Tecnológico, Internado Palmira S/N, Col.
Palmira, 62490 Cuernavaca, Morelos, México
yasmin.hp@cenidet.tecnm.mx

Abstract. In this paper, we discuss the importance of counting students and the identification of their sentiments using convolutional networks, specifically YoloV3 and Deepface. First, we discuss the importance of counting and identifying students' sentiments in online virtual environments; then, we discuss the related work. Later, we present the proposed methodology; we use a repository of fifty images obtained from online virtual classrooms, then we preprocess the images by transforming them to grayscale and normalizing their values to reduce the complexity of processing, and later, we identify and count the number of students by using Convolutional Neural Networks (Yolo-V3), and then by using DeepFace, we recognize the dominant emotion in each student, and in the group, finally, we obtain performances metrics in both stages. The preliminary results show that the identification and counting of students has a precision of 96.7% with a threshold > 50%. The prevalent sentiments in virtual classrooms are neutral 35%, sad 23%, and fear 17%.

Keywords: Counting of Students · sentiment analysis · virtual environments · deep learning · Yolov3 · DeepFace

1 Introduction

Detection of faces using a computer system, in its simple form, consists of identifying faces in images or frames. This process has gained significant importance in security, marketing, social media, and education, among others [8]; this importance grew more due to the COVID-19 pandemic due to our activities moving remote.

According to [9], facial recognition is essential for the automation and customization of learning. In schools and educational environments, facial recognition is a preliminary

H. Calvo et al. (Eds.): MICAI 2023 Workshops, LNAI 14502, pp. 29–40, 2024.
https://doi.org/10.1007/978-3-031-51940-6_4

step required for more complex tasks such as automatic attendance roll management to reduce the burden of carrying the roll call and prevent fake attendance.

Facial recognition is used for e-assessment for identifying intruders during virtual examinations and in school security monitoring for identifying intruders and other threats such as gun-shaped objects [9]. Similarly, the analysis of facial actions, microexpressions, eye tracking, and other facial landmarks are used to detect academic emotions (e.g., contentment, anxiety, hope, etc.), cues of learning, and engagement, which adaptive systems can exploit [10].

According to [9], Facial features have been used to detect emotions like boredom, confusion, delight, engagement, frustration, and surprise when students are involved in deep-level learning [11, 12] use computer vision and machine learning techniques to detect emotions from data collected in a real-world environment of a school computer lab; up to thirty students at a time participated in the class, and results were cross-validated to ensure generalization to new students; the classification was successful (AUC = .816). Facial sentiment analysis is a trending topic in computer vision [15].

Identifying and counting students using artificial intelligence is an innovative and promising application with numerous potential benefits, particularly in educational institutions and other settings where student attendance tracking is crucial. According to [16, 17] main benefits are:

Accuracy and Efficiency: AI-based student counting systems offer higher accuracy and efficiency regarding manual methods. They can identify and count students in real-time, reducing errors caused by human oversight or bias.

Improved Attendance Monitoring: These systems provide a more reliable way to monitor student attendance. They can be used in classrooms, libraries, cafeterias, or even significant events to record who is present accurately.

Cost-Effective: Over time, AI-based student counting can be cost-effective. It reduces the need for manual attendance taking, which can be labor-intensive and prone to errors. This can lead to potential savings in terms of staff time. This time can be used for educational purposes.

Data Analysis: AI systems can count students and collect data about attendance patterns. This data can be analyzed to identify trends, such as frequently absent students or times of peak attendance. This information can be valuable for educators, administrators, and students.

Scalability: These systems can be scaled to accommodate different settings and institutions, from small classrooms to large lecture halls. They are versatile and adaptable to various scenarios.

Integration with Other Systems: AI-based student counting systems can be integrated with other educational software, such as attendance management systems, which can streamline administrative tasks further.

Real-Time Alerts: These systems can provide real-time alerts to administrators or teachers if certain thresholds are met, such as a low attendance rate in a class or the presence of potentially dangerous objects, which allows for prompt action.

Machine Learning and Deep Learning Advancements: With machine learning and Deep Learning, these systems can improve their accuracy over time by learning from previous data, making them more reliable and efficient.

While AI-based student counting systems offer many advantages, they also face technical issues, like initial setup costs and the need for regular maintenance. **Privacy Concerns** need to be considered; implementing AI-based systems should be done with careful consideration of privacy issues. Cameras and other sensors used for counting students should comply with privacy regulations, and the data collected should be handled responsibly.

1.1 Problem Statement

It is beneficial to apply computer vision techniques and convolutional neural networks (CNN) for automatically detecting and counting students to help teachers move through the list, monitor (supervisory functions), and analyze their sentiments during class to adapt educational content, among other activities being carried out online.

1.2 Paper Contribution

In this research, we propose a methodology based on computer vision and deep learning to detect students and analyze their sentiments by using a set of images obtained from online classrooms; for this purpose, we use state-of-the-art computer vision techniques and convolutional neuronal networks like Yolo-V3 (for detecting and counting) and DeepFace (for analyzing dominant sentiment). Our approach differs from others because we transform the frames (images) in color to grayscales, and then we normalize the frames to deal with the problem of different light conditions in online classrooms. Both preprocessing stages reduce processing complexity when counting students and identifying sentiments. With our proposal, it is possible to identify individual and collective sentiments in virtual classrooms, which can help adapt educational content or strategies.

2 Related Work

In [4], is presented a methodology based on machine learning to identify faces. It is divided into five stages. In the first stage, the input image is received. The RGB image is converted to grayscale and normalized in the second stage. The third stage consists of extracting characteristics (features). The process continues with the fourth stage, where the classifying is done using machine learning. Finally, face detection is performed, and its metrics are obtained.

Most research in human-environment perception is limited to single-person environments, failing to address the complexity of spaces occupied by multiple individuals. To tackle the challenge of human behavior perception in multi-human environments, [2] introduces a novel approach for crowd counting, termed "DeepCount." This approach utilizes deep learning techniques within closed environments equipped with WiFi signals, representing a pioneering step in multi-human environment analysis. Rather than directly counting the crowd using WiFi, they leverage Convolutional Neural Networks (CNNs) to automatically establish the relationship between the number of people and WiFi channel data. A Long Short-Term Memory (LSTM) network – a kind of CNN- also manages dependencies between the headcount and Channel State Information (CSI). To mitigate the need for extensive labeled data required by deep learning, they introduce an online learning mechanism to detect human activity and fine-tune the deep learning model, reducing the training data requirements and enabling the model to evolve. DeepCount is evaluated using commercial WiFi devices and exhibits an average prediction accuracy of 86.4% in environments with up to 5 individuals. Moreover, through an activity recognition model, which judges door switches to adjust the crowd variance and correct Deep Learning predictions, the accuracy reaches 90%.

[16] describes the application of CNNs to identify sentiments, used initially in areas like computer vision, recommender systems, and natural language processing, and how, recently, CNNs have been used to increase the efficiency of sentiment analysis tasks. This research mentions a broad domain of applications of sentiment analysis, including business, government, and biomedicine. Sentiment Analysis can be used to infer users' opinions in different contexts. Recommender systems benefit from users' behavior, and in medicine, sentiment analysis can be used to detect diseases (mental or physical).

In [18], sentiment-driven features are discussed to classify the user's sentiment reaction. Instead of using standard computer vision features or CNN features to recognize objects and scenes, authors used the features provided by Deep-SentiBank and others extracted from models that exploit deep networks trained on face expressions. Two datasets of short videos with sentiment annotations were tested: LIRIS-ACCEDE and MEDIAEVAL-2015. Classification accuracy outperforms similar systems but uses a smaller number of features.

According to [15], Facial Expression Recognition (FER) is complex but valuable in healthcare applications, social marketing, emotionally driven robots, and human-computer interaction. A human's six most generic emotions are anger, happiness, sadness, disgust, fear, and surprise. According to these authors, the preferred method for emotion classification is CNN. In the FER process, the first step is facing detection, which provides a bounding box that is put over the face. The FER pipeline includes 1) Dataset, 2) Face detection – dimension reduction – normalization, and 3) Feature extraction and emotion classification. FER accuracy depends on the parameters selected, such as illumination factors and facial obstruction like hand, age, and sunglasses.

2.1 Convolutional Neural Networks

CNNs have shown significant success and innovation in computer vision and image processing. CNN consists of various layers, including input, convolutional, pooling, and fully connected layers. Researchers around the world are using CNNs in the field of sentiment analysis [6]. For instance, [7] proposed the most popular CNN model for sentence-level sentiment classification and experimented with CNN built on top of pre-trained word2vec. The experimental results show that pre-trained vectors can serve as an excellent feature extractor for tasks related to Natural Language Processing (NLP) using deep learning.

2.2 Yolo-V3

CNNs are used not only for analyzing texts or audio but also for frames and videos. A well-known CNN to identify objects is YOLO-V3, based on ResNet and FPN (Feature-Pyramid Network) architectures [1]. The YOLO-V3 feature extractor, called Darknet-53 (it has 52 convolutions), contains skip connections (like ResNet) and three prediction heads (like FPN)—each processing the image at a different spatial compression [6].

According to [1], Yolo-V3 performs well over different input resolutions; when tested with input resolution 608×608 on the COCO-2017 validation set, Yolo-V3 scored 37 mAP (mean Average Precision). This score matches GluonCV's trained version of Faster-RCNN-ResNet50 (a faster-RCNN architecture that uses ResNet-50 as its backbone) but is seventeen times faster.

2.3 DeepFace

According to [6], DeepFace is a Lightweight Face Recognition and Facial Attribute Analysis Framework that allows for identifying Age, Gender, Emotion, and Race. It is a project in Python that is easy to integrate into Python Projects. In the analysis proposed by [6], the experiments show that humans have 97.53% accuracy on facial recognition tasks, whereas those models already reached and passed that accuracy level. DeepFace also comes with a robust facial attribute analysis module including age, gender, facial expression (angry, fear, neutral, sad, disgust, happy, and surprise), and race (including asian, white, middle eastern, indian, latino, and black) predictions. The result will be the size of the faces appearing in the source image.

In this research, we will name indistinct emotions and sentiments.

3 Methodology

The proposed methodology (pipeline) is divided into five stages, as shown in Fig. 1. In the first stage, the input images or frames obtained from online classrooms are compiled. In the second stage, preprocessing, each image is transformed from a Red-Green-Blue RGB format to gray scales, and normalization is applied to deal with the problem of different light conditions. A convolutional neuronal network identifies and counts the students in the third stage; for this purpose, we use Yolo-V3. The fourth stage consists of extracting the dominant sentiments of students; for this purpose, we use another state-of-the-art CNN called DeepFace. Finally, in the fifth stage, the performance metrics of counting students and dominant sentiments are obtained.

Dataset. The dataset was obtained from a virtual classroom during the pandemic on different days of the week and in other sessions. We processed fifty frames with varying numbers of occupation and lighting conditions.

Preprocessing. At this preprocessing stage, the input image in Red-Green-Blue (RGB) format is converted to grayscale to facilitate feature extraction and data storage, which reduces complexity [13]. In this stage, we also apply normalization to correct images that are too dark or excessively light. All images are subjected to preprocessing color-to-grayscale conversion and normalization. OpenCV libraries in the Python language were used for this purpose.

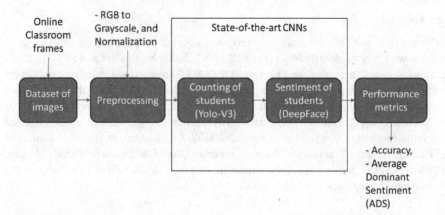

Fig. 1. Methodology for counting students and identification of dominant sentiment in virtual classrooms

Counting of Students. We identify and count the number of students using Yolo-V3. Normalization allows identifying students and event objects like books, cell phones, and laptops. We consider two measures for counting students, one with no threshold and another with a threshold > 0.50, to avoid false positives.

Sentiments identification. By using DeepFace [5], we identify the dominant sentiment (emotion) in each student and the group of students for the fifty images. Identification of emotions is shown in Fig. 2. The first row shows the number of images being processed; the second row shows the path where the normalized and preprocessed image is, and then the identification of the dominant sentiment is detected for each student.

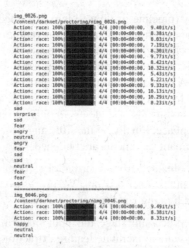

Fig. 2. Identification of dominant emotions using DeepFace.

Metrics

For counting students, we used Precision to address the problem at hand. Precision attempts to answer the following question: What proportion of positive identifications were correct? Eq. (1) shows the formula for Precision [14]:

$$\frac{TP}{(TP + FP)} \tag{1}$$

where: TP = True Positive, and FP = False Positive

For the identification of Dominant Sentiment, we use the Average Dominant Sentiment (ADS) formula:

$$\frac{\sum_{i=1}^{n-emotions} i}{Total\ of\ identified\ emotions} \tag{2}$$

According to [5], the emotions are six: anger, happiness, disgust, fear, neutral, sadness, and surprise, and the total of identified emotions is the sum of the emotions detected by DeepFace.

3.1 Design of the Experiments

We execute the proposed methodology shown in Fig. 1 on fifty images obtained from online classrooms; each image has a different number of students and lighting conditions.

4 Results and Discussion

Table 1 shows the results of the measurement metrics obtained from the experiments performed. The first column describes the image analyzed; the second column has the number of students in the frame determined by a human. The third column shows the True Positives (TP), the fourth column the False Positives (FP), the fifth column TP with a Threshold > 0.50, the sixth column the precision, and the seventh column precision with a Threshold > 0.50 of the students identified by Yolo-V3. The last row of the table contains the average Precision in Bold.

Table 1. Precision in counting students.

imagen	persons in frame	TP	FP	TP Th > 0.5	Precision	Precision Th > 0.50
img_0001.png	16	15	1	14	0.9375	0.9333
img_0002.png	8	7	0	7	1.0000	1.0000
img_0003.png	24	20	0	18	1.0000	1.0000

(continued)

Table 1. (*continued*)

imagen	persons in frame	TP	FP	TP Th > 0.5	Precision	Precision Th > 0.50
img_0004.png	5	5	0	5	1.0000	1.0000
img_0005.png	3	3	0	3	1.0000	1.0000
img_0006.png	3	3	0	3	1.0000	1.0000
img_0007.png	7	8	0	7	1.0000	1.0000
img_0008.png	8	8	0	8	1.0000	1.0000
img_0009.png	6	6	0	6	1.0000	1.0000
img_0010.png	5	5	0	5	1.0000	1.0000
img_0011.png	3	3	0	3	1.0000	1.0000
img_0012.png	6	6	0	6	1.0000	1.0000
img_0013.png	5	5	0	5	1.0000	1.0000
img_0014.png	2	2	0	2	1.0000	1.0000
img_0015.png	2	2	0	2	1.0000	1.0000
img_0016.png	5	4	0	4	1.0000	1.0000
img_0017.png	3	3	0	3	1.0000	1.0000
img_0018.png	2	2	0	2	1.0000	1.0000
img_0019.png	5	5	0	5	1.0000	1.0000
img_0020.png	7	6	1	6	0.8571	0.8571
img_0021.png	3	3	1	3	0.7500	0.7500
img_0022.png	7	6	0	6	1.0000	1.0000
img_0023.png	3	3	0	3	1.0000	1.0000
img_0024.png	3	3	0	3	1.0000	1.0000
img_0025.png	5	5	1	5	0.8333	0.8333
img_0026.png	24	20	0	18	1.0000	1.0000
img_0027.png	7	6	2	6	0.7500	0.7500
img_0028.png	6	5	1	5	0.8333	0.8333
img_0029.png	7	7	0	7	1.0000	1.0000
img_0030.png	10	9	0	7	1.0000	1.0000
img_0031.png	8	7	0	5	1.0000	1.0000
img_0032.png	6	6	0	5	1.0000	1.0000
img_0033.png	7	7	1	7	0.8750	0.8750
img_0034.png	6	3	0	3	1.0000	1.0000

(*continued*)

Table 1. (*continued*)

imagen	persons in frame	TP	FP	TP Th > 0.5	Precision	Precision Th > 0.50
img_0035.png	9	8	1	7	0.8889	0.8750
img_0036.png	2	2	0	2	1.0000	1.0000
img_0037.png	7	7	0	7	1.0000	1.0000
img_0038.png	3	3	0	3	1.0000	1.0000
img_0039.png	1	1	0	1	1.0000	1.0000
img_0040.png	1	1	0	1	1.0000	1.0000
img_0041.png	2	2	0	2	1.0000	1.0000
img_0042.png	12	10	1	10	0.9091	0.9091
img_0043.png	1	1	0	1	1.0000	1.0000
img_0044.png	11	11	0	11	1.0000	1.0000
img_0045.png	9	9	0	9	1.0000	1.0000
img_0046.png	8	8	1	8	0.8889	0.8889
img_0047.png	1	1	0	1	1.0000	1.0000
img_0048.png	3	3	0	3	1.0000	1.0000
img_0049.png	22	22	3	22	0.8800	0.8800
img_0050.png	15	15	0	1	1.0000	1.0000
Average					**0.9681**	**0.9677**

Results of precisions (in bold) shown at the end of Table 1 are like those reported in [3], where the use of Deep Learning, especially YOLO, provided satisfactory results, with F1 scores ranging from 0.94 to 1, and its precision ranges from 97.37% to 100% involving three datasets of higher resolution. We obtained 0.9681 (96.81%) with no threshold and 0.9677 (96.77%) with a threshold > 50. Which means the obtained results at this stage are good.

Continuing with our analysis, Table 2 shows the identified dominant sentiments for each image in the dataset using DeepFace sentiment analysis. For each identified student in the frame, the prevailing emotion of the student is considered. The last row of the table counts each column's dominant sentiments: angry, happiness, disgust, fear, neutral, sadness, and surprise.

Table 2. Identified dominant sentiments.

image	angry	happy	disgust	fear	neutral	sad	surprise
img_0001.png	1	4	0	0	2	2	0
img_0002.png	0	3	0	3	7	5	0
img_0003.png	4	2	0	2	7	5	0
img_0004.png	1	1	0	1	0	1	0
img_0005.png	0	1	0	2	0	0	0
img_0006.png	0	0	0	0	1	1	0
img_0007.png	0	1	0	0	2	2	0
img_0008.png	0	0	1	2	2	2	0
img_0009.png	0	0	0	0	1	3	0
img_0010.png	1	1	0	1	3	0	0
img_0011.png	2	0	0	0	1	0	0
img_0012.png	1	0	0	0	1	3	0
img_0013.png	2	0	0	0	0	1	0
...	...	...	...	...	...	...	...
img_0040.png	0	0	0	0	1	0	0
img_0041.png	0	0	0	0	2	0	0
img_0042.png	0	1	0	2	2	2	0
img_0043.png	0	0	0	1	0	0	0
img_0044.png	0	0	0	1	2	4	1
img_0045.png	2	0	0	2	3	0	1
img_0046.png	1	1	0	0	3	2	0
img_0047.png	0	0	0	0	0	1	0
img_0048.png	0	2	0	0	2	0	0
img_0049.png	2	0	0	4	2	5	1
img_0050.png	1	1	0	1	3	3	1
sum	26	29	1	43	87	59	7

Using Eq. 2, we calculate Average Dominant Sentiments (ADS) for online classrooms under study. The results are shown in Fig. 3.

Figure 3 shows that the dominant average sentiment is neutral (35%), followed by sadness (23%), fear (17%), and happiness (12%). The images used for this research obtained from online classrooms during the COVID-19 pandemic could explain the high percentages of sadness and fear emotions.

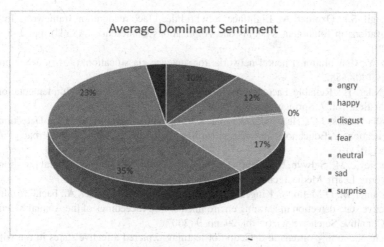

Fig. 3. Average Dominant Sentiments in online classrooms.

5 Conclusions and Future Works

Yolo-V3 is a fast and helpful tool for counting people in online environments with high accuracy (96% or more). Yolo-V3 performs well for counting individual students or a set of students in virtual classrooms.

DeepFace is a promising tool to detect sentiments in online virtual classrooms. However, some tuning of parameters must be done to improve the detection for the whole virtual classroom. In general, the proposed Methodology proves to perform well.

Future works include segmenting images with several students to improve the detection of sentiments for a whole online classroom. Besides, we want to assess the methodology with more images with different numbers of students and light conditions. Another direction is to try the methodology with images generated in real-time. We want to evaluate Yolo-V5 and compare the results obtained with those obtained with Yolo-V3 for the identification and counting of students.

References

1. Almog, U.: Yolov3 Explained (2020). https://towardsdatascience.com/yolo-v3-explained-ff5 b850390f. Accessed 14 Oct 2023
2. Liu, S., Zhao, Y., Xue, F., Chen, B., Chen, X.: DeepCount: Crowd counting with WiFi via deep learning. (2019). arXiv preprint arXiv:1903.05316 . Accessed 15 Oct 2023
3. Hardjono, B., Tjahyadi, H., Rhizma, M.G., Widjaja, A.E., Kondorura, R., Halim, A.M.: Vehicle counting quantitative comparison using background subtraction, viola jones and deep learning methods. In: 2018 IEEE 9th Annual Information Technology, Electronics and Mobile Communication Conference (IEMCON), pp. 556–562. IEEE (2018)
4. Mehta, D., Siddiqui, M.F.H., Javaid, A.Y.: Facial emotion recognition: a survey and real-world user experiences in mixed reality. Sensors 18(2), 416 (2018)
5. DeepFace: Deepface 0.0.79. (2023). https://pypi.org/project/deepface/. Accessed 16 Oct 2023

6. Serengil, S.I., Ozpinar, A.: Lightface: a hybrid deep face recognition framework. In: 2020 Innovations in Intelligent Systems and Applications Conference (ASYU), pp. 1–5. IEEE (2020)

7. Kim, Y.: Convolutional neural networks for sentence classification, (2014). arXiv preprint arXiv:1408.5882

8. Wechsler, H.: Reliable Face Recognition Methods: System Design, Implementation and Evaluation, vol. 7. Springer, Boston (2009)

9. Bonilla-Robles, J.C., Hernández-Aguilar, J.A., Santamaría-Bonfil, G.: Face Detection with Applications in Education. In: Online Learning Analytics, pp. 213–228. Auerbach Publications (2021)

10. Andrejevic, M., Selwyn, N.: Facial recognition technology in schools: critical questions and concerns. Learn. Media Technol. **45**(2), 115–128 (2020)

11. McDaniel, B., D'Mello, S., King, B., Chipman, P., Tapp, K., Graesser, A.: Facial features for affective state detection in learning environments. In: Proceedings of the Annual Meeting of the Cognitive Science Society, vol. 29. no. 9 (2007)

12. Bosch, N., et al.: Automatic detection of learning-centered affective states in the wild. In: Proceedings of the 20th International Conference on Intelligent User Interfaces, pp. 379–388 (2015)

13. Raveendran, S., Edavoor, P.J., YB, N.K. Vasantha, M.: Design and implementation of reversible logic-based RGB to grayscale color space converter. In: TENCON 2018–2018 IEEE Region 10 Conference, IEEE (2018)

14. Google, D.: Classification and recovering. (2023). https://developers.google.com/machine-learning/crash-course/classification/precision-and-recall?hl=es-419. Accessed 31 Oct 2023

15. Patel, K., et al.: Facial sentiment analysis using AI techniques: state-of-the-art, taxonomies, and challenges. IEEE Access **8**, 90495–90519 (2020)

16. Martin, S.M.: Artificial Intelligence, Mixed Reality, and the Redefinition of the Classroom. Rowman & Littlefield, Lanham (2019)

17. Ciolacu, M., Tehrani, A.F., Binder, L., Svasta, P.M.: Education 4.0-artificial intelligence assisted higher education: early recognition system with machine learning to support students' success. In: 2018 IEEE 24th International Symposium for Design and Technology in Electronic Packaging(SIITME), pp. 23–30. IEEE (2018)

18. Baecchi, C., Uricchio, T., Bertini, M., Del Bimbo, A.: Deep sentiment features of context and faces for affective video analysis. In: Proceedings of the 2017 ACM on International Conference on Multimedia Retrieval, pp. 72–77, (2017)

Automated Facial Expression Analysis for Cognitive State Prediction During an Interaction with a Digital Interface

Maricarmen Toribio-Candela$^{(\boxtimes)}$, Gabriel González-Serna, Andrea Magadan-Salazar, Nimrod González-Franco, and Máximo López-Sánchez

Computer Science Department, TecNM/CENIDET, Cuernavaca, Morelos, México
{m22ce054,gabriel.gs,andrea.ms,nimrod.gf,
maximo.ls}@cenidet.tecnm.mx

Abstract. Automated multimodal facial expression analysis is an advanced technology used in user experience research to predict cognitive states during UX evaluation. It involves analyzing user is facial expressions as they interact with digital interfaces using various tools and techniques. The information obtained through this analysis can help evaluate the user experience, thereby leading to the development of more efficient and higher-quality digital products. The multimodal extraction strategy involves detecting 46 points related to head movement, hand position, and facial expressions. Three classification algorithms were analyzed in conjunction with the Cam3D and Pandora data sets. The results indicate that Random Forest achieved an accuracy of 98%, KNN achieved an accuracy of 97%, and SVM achieved an accuracy of 95% for the detection of attention, concentration, and distraction. Incorporating cognitive state detection during UX assessment represents a valuable opportunity to improve the quality and efficiency of digital products.

Keywords: cognitive state · computer vision · facial action code system · facial recognition · hand-in-face

1 Introduction

Humans communicate their mental, emotional, and cognitive states through facial expressions. However, identifying facial expressions that relate to cognitive states like concentration, attention, and distraction is a complex issue in computer vision. This is because it demands sophisticated techniques to analyze and interpret multimodal biometric data.

A systematic review was conducted to analyze works [1–19, 20] related to the recognition and classification of cognitive states using machine learning and deep learning algorithms. In [1], it is explained that detecting hand gestures on the face, head movements, and eye scans is crucial for recognizing cognitive states along with facial expressions. The state-of-the-art is divided into two categories: 1) Multimodal Data Detection

H. Calvo et al. (Eds.): MICAI 2023 Workshops, LNAI 14502, pp. 41–49, 2024.
https://doi.org/10.1007/978-3-031-51940-6_5

and 2) Cognitive State Recognition. The first section of the analysis reviewed articles [1–19, 20] which primarily aimed to recognize cognitive states. Many of these works used preprocessing techniques, Machine Learning, and Deep Learning algorithms. These techniques were instrumental in building the cognitive state classification system.

2   Motivation

To provide a more satisfying user experience, it is essential to understand the cognitive states of users, such as attention, concentration, and distraction. By tailoring and personalizing interactions based on these states, we can create digital products that align optimally with our users' needs and preferences. To improve the accuracy in recognizing cognitive states, computer vision techniques can be utilized to analyze multimodal biometric data (facial expressions, head movement, and hand position) of the user during software interaction. These techniques help to identify cognitive states expressed by users unconsciously and can provide valuable information, resulting in improved accuracy levels compared to conventional methods. In conventional methods, the results depend on the user's ability to evaluate post-interaction and the interpretation of the UX evaluator, as shown in Fig. 1.

- **Difficult to pinpoint** user feedback.
- They **depend** on the **will** of the user.
- **Easy** to **alter** information.
- Relatively **simple** and **lacking** in objective data evidence.
- It cannot be applied during the evaluation.

Fig. 1. Tools used in a conventional UX assessment.

The articles published in [1–19, 20] tend to explore expression differences mainly in the pixel space of the complete facial image. This approach considers facial expression as a global expressive feature. However, some researchers have aimed to improve the accuracy of facial expression identification by using complementary techniques. They choose to select representative facial areas as objects of analysis and propose local feature extraction methods to calculate specific descriptors for these areas. Methods such as Haar waterfalls [9] and the use of the oriented gradient histogram (HOG) [8] focus on local features of the face. These approaches help capture specific details of facial expressions, which in turn, improve recognition accuracy.

In contrast to the first category of methods, the second category involves a geometry-based approach, which focuses on identifying and tracking facial feature points. These points correspond to the movements of the face and body, which are useful in capturing expressive features. This approach is similar to the methods used in previous studies [10] and [11] for analyzing facial expressions and body postures. The geometry-based approach offers greater robustness against external factors like luminance and occlusion, making it a more reliable method.

2.1 Objective

Developing a method to identify three cognitive states attention, concentration, and distraction, through the processing of multimodal behavioral data related to facial expression, head posture, and hand movement in videos of users participating in the process of evaluating the user experience of digital products.

3 Method

Body language is considered the primary visual system for displaying cognitive states and is the most important and complex part of nonverbal communication. Cognitive states are a set of skills and competencies that enable individuals to establish logical relationships, a task that can only be performed by humans [12]. Attention is essential to our interaction with the world outside and how we organize our thoughts and actions accordingly. Concentration is the force with which an individual focuses their attention on a particular activity, object, or task, abstracting it from everything else. Distraction occurs when we expend resources and effort to focus our attention on relevant stimuli and then have difficulty returning to our original task, which can reduce our performance.

3.1 Units of Action

Observable facial movements are referred to as Action Units (AUs). These AUs represent various aspects of one or multiple cognitive states and can be used to express a cognitive state. The most effective way to measure cognitive states in the human face is through a facial action coding system. This system correlates specific regions of the face with AUs, which in turn helps algorithms learn from this coding. The Facial Action Coding System, created by psychologists Paul Ekman and Wallace V. Friesen [8], has become a standard for measuring facial expressions. The system is based on a taxonomy of human facial movements originally developed by Swedish anatomist Carl-Herman Hjortsjö. FACS enables the systematic categorization of physical expressions of emotions and has become a valuable tool for both psychologists and animators.

3.2 Hand-over-Face

In our daily conversations, our hands play an important role in nonverbal communication. We use them for simple actions such as pointing at objects, as well as more complex gestures that express our emotions and thoughts. Moreover, we tend to bring our hands closer to our partially covered face, which helps us infer cognitive states.

Studies indicate that people's ability to understand communication that involves both the face and body is only 35%. Human interpretation of different social interactions in various situations is most accurate when both the face and body are observable.

3.3 Landmarks

During the development of the algorithm, the most critical stage involves describing the face in detail. This includes identifying landmarks on the face, such as the eyes, nose, mouth, and other essential facial structures. We also undertake additional tasks, such as facial alignment, head posture estimation, and identifying facial expressions associated with cognitive states [14].

3.4 Proposed Solution

The algorithm was developed using the solution scheme depicted in Fig. 2. It involves 8 stages, out of which the first five are focused on extracting characteristics through the calculation of geometric distances. The remaining 3 stages are dedicated to classifier training, validation, and testing. Each step is described in detail below.

Fig. 2. Schematic of the classification system of cognitive states.

3.5 Image Repositories

In the first stage, we conducted a state-of-the-art analysis to select the most commonly used public image repositories, which resulted in the selection of Cam3D [14] and Pandora [15]. Table 1 shows the characteristics of the selected datasets.

Table 1. Datasets for training and testing algorithms.

BD	Cognitive States	Images by class
Cam3D	Attention, distraction, concentration	600
Pandora	Attention and distraction	700

The analysis of each dataset has light variation and is considered from torso to head in controlled environments with sufficient luminosity to be able to recognize facial expressions. Examples of each dataset are shown in (Fig. 3 and 4).

Facial expressions are described using Euclidean distances of certain muscle movements, as specified by Ekman [8]. These muscle movements are used to identify cognitive

Fig. 3. Cognitive states from the Cam3D dataset [14].

Fig. 4. Cognitive states from the Pandora dataset [15].

states associated with each expression. To carry out this process, the coordinates of facial features, hand position, and head position are collected by taking the starting and end points of each feature. In order for the detector to make accurate predictions, grayscale images are inputted and the face, hand, and shoulder are located. The predictor points are obtained from the parts of interest of the user who is presenting a cognitive state, as shown in Fig. 5.

Fig. 5. 543 points placed in face, hand and torso, present attention cognitive status.

3.6 Selection of Important Characteristic Points

We selected 47 key points from a total of 543 to calculate essential distances and store them in a CSV file for identifying cognitive states. We also evaluate the angle of rotation of the head using facial and body points, similar to how we assess the position of the hands. These points play a crucial role in our analysis and are shown in Table 2. The selected points can be visualized (Fig. 6).

Based on the facial action coding system for cognitive states proposed by Ekman, the points were selected. This system provides the units of action that occur in each cognitive state. Action units refer to the fact that Ekman assigned an identifier with the acronym AU # to the movements of each of the muscles of the face that correspond to

Table 2. Datasets for training and testing algorithms

No	Regions of the face	Facial points
1	Right eyelid	[258,386,259,295,283]
2	Left eyelid	[29,159,53,28,66]
3	Right eyebrow	[276,342, 384,441,285,8]
4	Left eyebrow	[46,113,157,221,55,8]
5	Mouth	[11,16, 2,164]
6	Nose	[6]
7	Left hand	[12,0]
8	Right Hand	[12,0]
9	Pose	[12,11,13,14,0]
10	Eyes	[145,159,374,386,474,476,362,469, 471,133]

Fig. 6. The 47 facial and body landmarks placed on the face and body, expressing concentration.

each emotion. For example, in the cognitive state of attention, the action units AU1, AU2, AU5, and AU17 are involved. The first corresponds to the lifting of the outer eyebrow, the second corresponds to raising the inner eyebrow, the third corresponds to raising the upper eyelid, and the fourth corresponds to puckering the chin.

3.7 Calculation of Euclidean Distance

To identify the UAs, we first calculate the distance between the previously selected points that help us obtain the UAs proposed by Ekman, we incorporate the use of the Euclidean distance (indicated in Formula 1). In which 31 combinations of points corresponding to each UA were obtained to recognize cognitive states.

$$d_{e\,(p_1,p_2)} = \sqrt{(x_2-x_1)^2+(y_2-y_1)^2} \tag{1}$$

where $d_e\,(p_1, p_2)$ is the distance that is intended to be calculated from a characteristic point of (face, hand or head) to another characteristic point of (face, hand or head); For example, to obtain the UA5 proposed by Ekman, we calculated the distance the left eyelid, using the points [159,29,66,28,53] as shown in (Fig. 7).

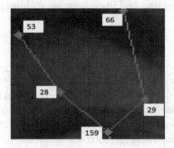

Fig. 7. Points [159,29,66,28,53] left eyelid distance.

4 Results

We started by training the classification algorithm using a pre-labeled dataset named Cam3D, which contained 600 images for each cognitive state. To avoid an imbalance of classes in the dataset, we took 350 images for each class. Next, we established the optimal hyperparameters for each classifier, which enabled us to achieve an accuracy of 97% with Random Forest, 95% with KNN, and 98% with SVM. All three classification algorithms were trained on the same dataset. This section presents the precision obtained from each classifier algorithm, see Table 3.

Table 3. Precision of each algorithm.

Algorithms	Accuracy
Support Vector Machines	98.0%
Random Forest	97.0%
K-Nearest Neighbors	95.0%

4.1 Conclusion

We have developed a method to recognize cognitive states by analyzing facial expressions, and hand and head positions, using specific points on the hand, face, and body as reference. Our system was tested using static images from Cam3d, and the results have been encouraging. However, to validate the system, it is crucial to analyze its performance in different environments using videos instead of static images. This will help us determine the consistency of our results and identify any challenges that may arise in a dynamic context. Moving forward, we plan to test and adapt our model in various environments and conditions to assess its versatility and robustness. Once the system is validated, we intend to implement it on the UXLab platform to identify cognitive states in users participating in the evaluation of the user experience of digital products. This will lead to a better understanding and improvement of human-computer interaction in different contexts.

References

1. Wardana, A.Y., Ramadijanti, N., Basuki, A.: Facial expression recognition system for analysis of facial expression changes when singing. In: International Electronics Symposium on Knowledge Creation and Intelligent Computing (IES-KCIC) (2018)
2. Ardhendu Behera, P.M.: Associating facial expressions and upper-body gestures with learning tasks for enhancing intelligent tutoring systems. Int. J. Artif. Intell. Educ. **30**, 236–270 (2020). https://doi.org/10.1007/s40593-020-00195-2
3. Ashwin, T.S., Guddeti, R.M.R.: Automatic detection of students' affective states in classroom environment using hybrid convolutional neural networks. Educ. Inf. Technol. **25**(2), 1387–1415 (2020)
4. Yang, B., Yao, Z., Lu, H., Zhou, Y., Xu, J.: In-classroom learning analytics based on student behavior, topic and teaching characteristic mining. Pattern Recogn. Lett. **129**, 224–231 (2020)
5. Kirana, K.C., Wibawanto, S., Herwanto, H.W.: Facial emotion recognition based on viola-jones algorithm. In: The Learning Environment. International Seminar on Application for Technology of Information and Communication, Malang, Indonecia. IEEE (2018)
6. Turan, C., Neergaard, K.D., Lam, K.M.: Facial expressions of comprehension (FEC). IEEE Trans. Affect. Comput. **13**(1), 335–346 (2019). https://doi.org/10.1109/TAFFC.2019.2954498
7. DataScientest (2023). https://datascientest.com/es/random-forest-bosque-aleatorio-definicion-y-funcionamiento
8. Ekman, P., Friesen, W.V.: Facial action codign system (FACS) (1978)
9. D'errico, F., Paciello, M., De Carolis, B., Vattanid, A., Palestra, G., Anzivino, G.: Cognitive emotions in e-learning processes and their potential relationship with students. Int. J. Emotional Educ., 89–111 (2018). ISSN 2073–7629
10. Gama Velasco, A.K.: Desarrollo de un Sistema interactivo basado en vision artificial para la rehabilitacion del tobillo. CENIDET, Morelos, Cuernavaca (2018)
11. Gunavathi, H.S., Siddappa, M.: Towards cognitive state detection using facial expression and hand-over-face gesture. In: IEEE International Conference on Recent Trends in Electronics, Information & Communication Technology. Bangalore, India. IEEE (2018)
12. DataScientest (2022). https://datascientest.com/es/que-es-el-algoritmo-knn
13. Obtenido de MediaPipe. https://developers.google.com/mediapipe/solutions/vision/face_landmarker
14. Anis, M., Baltrusaitis, T., Robinson, P.: CAM3D. Apollo - University of Cambridge Repository (2019).https://doi.org/10.17863/CAM.38196
15. PSY.TXT. https://psy.takelab.fer.hr/datasets/all/pandora/
16. Hu, G.L., et al.: Deep multi-task learning to recognise subtle facial expressions of mental states. In: Conference on Computer Vision, pp. 106–123 (2018)
17. Indhumathi, R., Geetha, A.: Emotional Interfaces for Effective E-Reading using Machine Learning Techniques. International Journal of Recent Technology and Engineering (IJRTE), pp. 4443–4449 (2019). ISSN: 2277–3878

18. Rao, K.P., Rao, M.C.S.: Recognition of learners' cognitive states using facial expressions in e-learning environments. J. Univ. Shanghai Sci. Technol. **22**(12), 93–103 (2020)
19. Pourmirzaei, M., Montazer, G.A., Mousavi, E.: Customizing an affective tutoring system based on facial expression and head pose estimation. Human Computer Interaction Cornell university (2021)

Can We Take Out CARLA from the Uncanny Valley? Analyzing Avatar Design of an Educational Conversational Agent

Pablo Isaac Macias-Huerta[1] , Carlos Natanael Lecona-Valdespino[2] ,
Guillermo Santamaría-Bonfil[3]([✉]) , and Fernando Marmolejo-Ramos[4]

[1] Escuela Superior de Cómputo, Mexico City, Mexico
pmaciash1800@alumno.ipn.mx
[2] Universidad Panamericana, Mexico City, Mexico
carlos.leconavaldespino@up.edu.mx
[3] BBVA Mexico, Data Portfolio Management, Mexico City, Mexico
guillermo.santamaria@bbva.com
[4] University of South Australia Online, Adelaide, Australia
fernando.marmolejo-ramos@unisa.edu.au

Abstract. The integration of conversational agents into virtual-reality educational environments holds immense promise. However, the uncanny valley phenomenon looms as a critical challenge, potentially impacting the effectiveness of these educational agents. This manuscript delves into the pivotal role of avatar design within virtual-reality educational systems and its implications for user experience during learning sessions. In this work, a comprehensive comparison of two distinct avatars used by a virtual-reality conversational educational agent was undertaken, exploring their impact on user acceptance and enjoyability. Through user feedback, the avatar design that mitigates the uncanny valley effect was determined, ultimately allowing to choose the best avatar to enhance human-computer interaction.

Keywords: Uncanny Valley · Avatar design · Conversational agent · Educational environments · Virtual reality · FACS · Educational agent

1 Introduction

The integration of virtual agents into virtual-reality (VR) educational environments is innovating modern education. Due to its potential to improve learners' immersion into dynamic and interactive virtual worlds, educational conversational agents have garnered significant attention for their ability to enhance engagement and knowledge retention [3]. Among the challenges for designing a proper avatar for an educational conversational agent stands the *uncanny valley*: a well-documented psychological phenomenon which states that, as a robot or virtual avatar appears more human-like, there is a point at which it elicits a strong feeling of eeriness or discomfort, rather than affinity [9,11]. The presence of the uncanny valley can significantly impact the overall learning experience [4,15]. Further, even when educational agents closely resemble humans but exhibit subtle imperfections or unrealistic behaviors, learners may find it challenging to establish rapport and engage effectively with these agents [2]. As a result, the promise of VR as an effective educational tool may be compromised: especially if displayed on a flat monitor instead of a head mounted display, which have shown to relieve the effects of the uncanny valley [8].

In general, embodied chatbots are considered to induce more eeriness over text chatbots [4]. In order to avoid a feeling of discomfort for the user, previous research emphasize key aspects to take into account when designing an avatar [13]. For instance, realism has a bigger impact in user perception than fidelity [1], this means that it is possible to achieve the same level of user engagement without the need of much rendering power and with lower costs. Further, cartoonish avatars are preferred over hyper-realistic ones given that they induced less eeriness and discomfort [14]. Even while in other research areas the impact of avatar design has been studied more deeply (e.g. [4]), the education field is not the case.

Other works have shown that similarity between the user and the avatar they're learning from significantly improves their learning rate [7]; although these works focus on comparing between already human-like avatars; not exploring the possibility of learning through completely non-human alternatives.

This manuscript explores the role of avatar design of an embodied educational conversational agent to discern between avatar designs the one that minimizes the uncanny valley effect. Specifically, we focus on the case of CARLA, an educational virtual agent designed to assist users in navigating a virtual-reality power systems environment, providing guidance on controls, and facilitating a wide range of learning activities [10]. Therefore, a comparison of two distinct avatars employed by the conversational educational agent is carried out. Experimentation revolves around the impact of the avatars appearance and the impact on users acceptance and enjoyability. Through empirical analysis and user feedback, it is expected to provide guidance in the selection of avatars that will best serve the educational goals and aspirations of learners in the virtual realm.

2   Materials and Methods

Once the conversational agent was finished, the design aim was to create a way for the user to not feel as though they were interacting with a disembodied voice. To resolve that issue, two virtual avatars were implemented.

2.1   Virtual Avatars

The two avatars created were made with two very different design philosophies: One aiming for a cartoon robot look (see Fig. 1a); and one with a more human-like approach (see Fig. 1b). They shall be referred to as "Simple CARLA" and

"Complex CARLA" respectively, from this point onward. Both were created using Blender[1], a free and open 3D software. Avatars were designed taking inspiration from characters in other media that have proven successful with the public. Simple Carla takes inspiration from characters like "Eve" from *WALL·E* and "BB-8" from *Star Wars*; while Complex Carla's design and implementation is based on the way different Real-Time Strategy (RTS) games would have characters interact with the user via an icon. The reason for such different design approaches was to find the type of avatar (Simple and in a cartoon style; or Complex and human) potential users would find more likable.

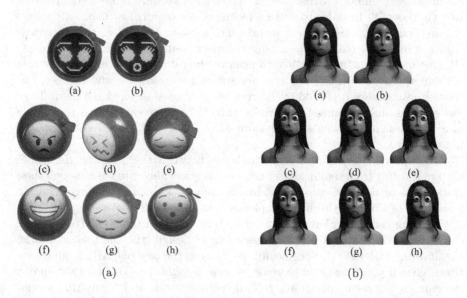

Fig. 1. Simple (a) and Complex (b) CARLA expressions and emotions: (a) Idle, (b) Talking, (c) Anger, (d) Disgust, (e) Fear, (f) Happiness, (g) Sadness, (h) Surprise.

2.2 Emotional Expression Design

Once the virtual avatars were created, however, the question of which emotions should it be able to communicate to the user became important. For that, we relied on Paul Ekman's Universal Emotions [5]: Anger, Disgust, Enjoyment, Fear, Sadness, Surprise. [6] Although Contempt is among that list, it was not implemented for this particular conversational agent, as Contempt's use in an educational environment is not clear. Once the emotions were selected, implementation for each of them was needed. For Simple CARLA, the use of emojis was preferred, given the fact that most user will already have certain familiarity with them. For Complex CARLA, FACS-based animation[2] was used. FACS (Facial Action Coding System) is a behavioral coding system for measuring observable facial action from muscle movements [12].

---

[1] https://www.blender.org/.
[2] https://github.com/dccsillag/unity-facs-facial-expression-animation.

3   Experiments and Results

Once both avatars could display the range of emotions the platform required, finding out which avatar was more liked became really important for the platform's development. For that, a questionnaire comparing the two avatars was designed[3]. Most answers were answered via a Likert Scale with discrete values ranging from 1 to 10. The questionnaire contained 5 sections:

1. An introduction to the project and an informed consent of what the user's data was being used for.
2. Personal data from the respondents. Although the respondents stayed anonymous, data such as level or education or age were important. A really important question for later analysis was how familiar the respondent was with video games.
3. An introduction to the user for both Simple CARLA and Complex CARLA, in their idle and talking states; and questions regarding how much the respondents liked each one, and how human they found them.
4. The respondents were shown each of Simple CARLA's emotion states and was asked questions regarding each one.
5. An analogous section to the previous, but regarding Complex CARLA's emotion states instead.

3.1   Results

Personal Data. Firstly, out of 79 total respondents, every single one accepted the terms stated in the informed consent for their results to be published.

Of the respondents, a vast majority (69.6%) were male, meaning there is the possibility of a bias for male-oriented preferences. Another possible bias is the fact that every respondent was Mexican. Age-wise, the respondents ranged from 15 to 67 years old, with a mode (20.3% of respondents) being 19 years old and an average of 24.8 years. Although this might indicate an age-based bias, this younger-aged trend follows vaguely the age distribution in Mexico[4]; so although the possibility of bias cannot be ruled out, it seems unlikely. A similar phenomenon occurs with the level of education: most (63.2%) respondents had below college level education. In Mexico, people over the age of 15 average just above secondary school[5].

Lastly, a vast majority of respondents (78.4%), on a discrete scale from 1–10, gave an answer of 6 or higher to their familiarity with video games (10 being "Very Familiar"): This means that results are most likely not affected by unfamiliarity with virtual avatars overall.

CARLA Avatars. In this section, two questions were asked for each avatar: How much did the respondent like it; and How human did they find it. Each question was answered via a Likert Scale with discrete values ranging from

---

[3] https://forms.gle/X7KfJmzmCB8zGE6m8.

[4] https://www.inegi.org.mx/programas/ccpv/2020/.

[5] https://www.cuentame.inegi.org.mx/poblacion/escolaridad.aspx.

1 to 10. In the case of perceived humanity, 1 was labeled as "Completely Non-human" and 10 as "Completely Human". In the case of likability, 1 was labeled as "Completely unlikable" and 10 as "Really Likable".

The answers for Simple CARLA (See Fig. 2) shows respondents found Simple CARLA very appealing ($Median = 8, 95\%$ CI $= [7.46, 8.53]$); while also showing they found the avatar very much nonhuman ($Median = 4, 95\%$ CI $= [3.46, 4.53]$).

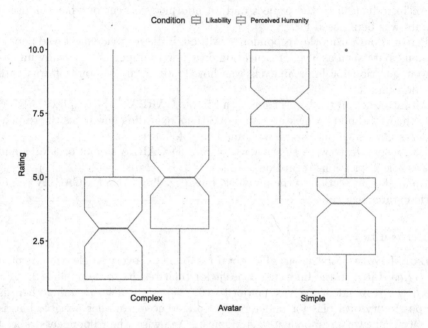

Fig. 2. Ratings of perceived humanity (red) and likability (blue) for the complex and simple avatars. Notches represent approximate 95% CIs. Quantile comparisons for paired samples indicated the two comparisons between simple and complex avatars' perceived humanity ratings and simple and complex avatars' likability ratings were significant (both $p < .005$). (Color figure online)

In contrast, the answers for Complex CARLA (See Fig. 2) shows respondents found her more human than Simple CARLA; however, they didn't find her very human overall ($Median = 5, 95\%$ CI $= [4.28, 5.71]$). Moreover, they found her very unappealing ($Median = 3, 95\%$ CI $= [2.28, 3.71]$).

This is the first indication Complex CARLA might fall under the uncanny valley: the avatar is perceived as almost human, but also very unlikable.

Emotion Recognition. This final section is divided into two almost identical sections, only difference being the first section is exclusive to Simple CARLA and the second one to Complex CARLA. The question structure was the following: The respondent was shown a picture of the respective CARLA avatar emotion state, followed by 3 questions:

1. "How much do you agree with the following statement? "[Avatar] seems [emotion]""
2. "How understandable is the expression of [Avatar] as a representation of [emotion]?"
3. "How much do you like how [emotion] is represented in [Avatar]?"

Where *[Avatar]* is either Complex or Simple CARLA, and *[emotion]* is the shown emotion. This structure was repeated for each of the 6 emotions. Each question was answered via a Likert Scale from 1 ("Completely Disagree", "Not Understandable", and "Completely Dislike", respectively) to 10 ("Completely Agree", "Really Understandable", and "Really Like", respectively).

Questions 1 and 2 refer to recognizability, while question 3 refers to likability. Therefore, for result showing purposes, we will merge the answers for questions 1 and 2. As the results for Simple CARLA (See Fig. 3), respondents found Simple CARLA's emotions easily readable ($Median = 10, 95\%$ CI $= [9.85, 10]$) and likable ($Median = 9, 95\%$ CI $= [8.78, 9.21]$). However, for Complex CARLA's emotions (See Fig. 3), respondents found them less recognizable ($Median = 8, 95\%$ CI $= [7.78, 8.21]$), and less likable ($Median = 7, 95\%$ CI $= [6.70, 7.29]$).

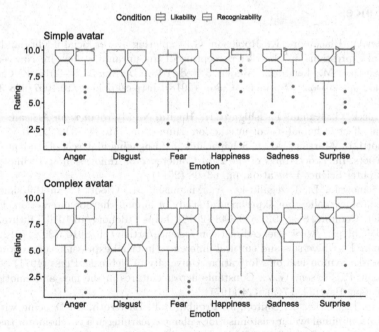

Fig. 3. Ratings of emotion recognizability (red) and likability (blue) for the complex and simple avatars in each of the emotions considered. Notches represent approximate 95% CIs. Quantile comparisons for paired samples indicated the two comparisons between simple and complex avatars' emotion recognizability ratings and simple and complex avatars' likability ratings were significant (both $p < .005$). (Color figure online)

4   Conclusions and Future Work

Although some of the results could be explained by bad animation, the unlikability in Complex CARLA both in idle and emotion states is not. This is most likely the effect of the uncanny valley on the respondents, as shown by the avatar's almost human features in comparison with it's likability score. With this, we conclude that, for educational purposes, keeping a less-human design approach yields better results with potential users. It is also possible, although requires further experimentation, that a human-like avatar that goes beyond the uncanny valley effect could yield good results.

For future works, experimenting with better animations could change the results of Complex CARLA emotions likability and recognizability results, although it is unlikely that it would change the idle results. Furthermore, implementation of a more human avatar has potential for better likability and readability among users.

Acknowledgments. Data files and R codes for the statistical analyses can be found at https://github.com/Rand0m-Guy/CARLA-Analyses.

References

1. Bailey, J., Blackmore, K., Robinson, G.: Exploring avatar facial fidelity and emotional expressions on observer perception of the uncanny valley. In: Naweed, A., Wardaszko, M., Leigh, E., Meijer, S. (eds.) ISAGA/SimTecT -2016. LNCS, vol. 10711, pp. 201–221. Springer, Cham (2018). https://doi.org/10.1007/978-3-319-78795-4_15
2. Bartneck, C., Kanda, T., Ishiguro, H., Hagita, N.: My robot is an African: Cross-racial effect in human-robot interaction. Hum.-Rob. Int. 341–348 (2009)
3. Benotti, L., Martínez, M.C., Schapachnik, F.: Engaging high school students using chatbots. In: Proceedings of the 2014 Conference on Innovation & Technology in Computer Science Education, pp. 63–68 (2014)
4. Ciechanowski, L., Przegalinska, A., Magnuski, M., Gloor, P.: In the shades of the uncanny valley: an experimental study of human-chatbot interaction. Future Gener. Comput. Syst. **92**, 539–548 (2019). https://doi.org/10.1016/j.future.2018.01.055, https://www.sciencedirect.com/science/article/pii/S0167739X17312268
5. Ekman, P.: Universals and cultural differences in facial expressions of emotion. In: Nebraska Symposium on Motivation. University of Nebraska Press (1971)
6. Ekman, P., Friesen, W.V.: Constants across cultures in the face and emotion. J. Pers. Soc. Psychol. **17**(2), 124 (1971)
7. Fitton, I., Clarke, C., Dalton, J., Proulx, M.J., Lutteroth, C.: Dancing with the avatars: minimal avatar customisation enhances learning in a psychomotor task. In: Proceedings of the 2023 CHI Conference on Human Factors in Computing Systems, CHI 2023. Association for Computing Machinery, New York (2023). https://doi.org/10.1145/3544548.3580944
8. Hepperle, D., Purps, C.F., Deuchler, J., Wölfel, M.: Aspects of visual avatar appearance: self-representation, display type, and uncanny valley. Vis. Comput. **38**(4), 1227–1244 (2022)

9. MacDorman, K.F., Ishiguro, H.: The uncanny advantage of using androids in cognitive and social science research. Interact. Stud. **7**(3), 297–337 (2006)
10. Macias-Huerta, P.I., Santamaría-Bonfil, G., Ibanñez, M.: CARLA: conversational agent in virtual reality with analytics. Res. Comput. Sci. **149**(12), 15–23 (2020)
11. Mori, M.: The uncanny valley. Energy **7**(4), 33–35 (1970)
12. Rosenberg, E.L., Ekman, P.: What the Face Reveals: Basic and Applied Studies of Spontaneous Expression Using the Facial Action Coding System (FACS). Oxford University Press, Oxford (2020)
13. Schwind, V., Wolf, K., Henze, N.: Avoiding the uncanny valley in virtual character design. Interactions **25**(5), 45–49 (2018). https://doi.org/10.1145/3236673
14. Shin, M., Kim, S.J., Biocca, F.: The uncanny valley: no need for any further judgments when an avatar looks eerie. Comput. Hum. Behav. **94**, 100–109 (2019). https://doi.org/10.1016/j.chb.2019.01.016, https://www.sciencedirect.com/science/article/pii/S0747563219300251
15. Slater, M., Wilbur, S.: A framework for immersive virtual environments (FIVE): speculations on the role of presence in virtual environments. Presence: Teleoper. Virtual Environ. **6**(6), 603–616 (1997)

HIS 2023

A GPT-Based Approach for Sentiment Analysis and Bakery Rating Prediction

Diego Magdaleno[1]([⊠]), Martin Montes[2], Blanca Estrada[1,3],
and Alberto Ochoa-Zezzatti[4,5]

[1] Universidad Autónoma de Aguascalientes, Aguascalientes, Mexico
diegomagdaleno@pm.me
[2] Universidad Politécnica de Aguascalientes, Aguascalientes, Mexico
[3] Tecnológico Universitario Aguascalientes, Aguascalientes, Mexico
[4] Doctorado en Tecnología, UACJ, Tokyo, Japan
[5] Centro de Alta Dirección en Ingeniería y Tecnología de La Universidad Anáhuac,
Naucalpan de Juárez, Mexico

Abstract. This paper presents a comprehensive approach to predicting the ratings of bakery establishments on diverse online platforms using natural language reviews. The study incorporates Large Language Models (LLMs) and Aspect-Based sentiment analysis to discern nuanced sentiment scores within specific categories. Utilizing advanced machine learning and regression methods, the paper introduces a robust predictive framework that employs LLMs, exemplified by GPT-3.5 Turbo, to accurately infer sentiment from natural language phrases. Framework's effectiveness is demonstrated in accurately forecasting bakery scores based on such sentiment analysis, obtaining a MAE of 0.27, demonstrating low error rates in comparison to other state-of-the-art models. The study also addresses challenges associated with sentiment analysis, including emojis and sarcasm, and underscores LLMs enhanced proficiency in handling intricate linguistic nuances. Taken together, this research highlights the potential of LLMs in constructing a precise predictive model for online platform ratings, offering valuable insights into consumer perceptions.

Keywords: Large Language Model · GPT · Intelligent System · Bakeries · Bread quality · Pattern Recognition · Online reviews

1 Introduction

In recent years, the use of online ratings for establishments has grown exponentially [1, 2]. Prominent platforms, including but not limited to TripAdvisor, FourSquare, and Google Places, have emerged as beneficiaries of this trend. It is noteworthy that the efficacy of these platforms is rooted in the aggregation of user-generated content, where individuals are encouraged to share their perspectives and provide star-based appraisals of specific locales. Sentiment Analysis, a foundational challenge in the realm of Natural Language Processing, has sparked significant research efforts focused on enhancing the accuracy and precision of this classification objective. However, despite attempts, the effectiveness of current approaches has revealed potential for enhancement [3].

© The Author(s), under exclusive license to Springer Nature Switzerland AG 2024
H. Calvo et al. (Eds.): MICAI 2023 Workshops, LNAI 14502, pp. 61–76, 2024.
https://doi.org/10.1007/978-3-031-51940-6_7

With the growth of pre-trained Transformers in various domains [4], such as Chat-GPT rapidly amassing millions of users in days [5], the use of these models in the realm of Sentiment Analysis became an inevitable progression. Evidence such as [6] demonstrates that models akin to GPT-3.5-Turbo exhibit superior performance vis-à-vis state-of-the-art counterparts in the domain, underpinned by the adeptness of Large Language Models (LLMs) in deciphering and analyzing linguistic nuances encompassing sarcasm, negation, common abbreviations, emoticons, and more.

In a formal context, aspect-based sentiment analysis involves assigning a specific sentiment, often categorized as positive, neutral, or negative, to a particular object mentioned within a text. Alternatively, one can employ a 5-point Likert scale [7], comprising descriptors such as Excellent, Good, Neutral, Bad, and Horrible.

While researching the abilities of LLMs capabilities in conducting sentiment analysis, a conceptual proposition crystallized: To harness LLMs as an additional layer within a Neural Architecture, to predict star-based ratings predicated on natural language input. Following a comparable foundational idea [8], though utilizing BERT, which stands for Bidirectional Encoder Representations from Transformers, a pre-trained language model that can be fine-tuned for various natural language understanding tasks, such as question answering, sentiment analysis, and named entity recognition, as the linguist processing layer, our study aims to propel this paradigm forward by harnessing the potential of LLMs in the realm of natural language processing. Notably, our hypothesis asserts that LLMs, armed with the ability to perceive the subject in context, showcase an elevated proficiency in performing aspect-based sentiment analysis, an intricacy particularly pertinent within this domain.

This paper addresses aspect-based sentiment analysis for various bakery attributes, including Flavor, Freshness, Customer Service, Price, and Variety, utilizing Google Places data from the Mexican state of Aguascalientes. To achieve this, our proposed approach involves the utilization of GPT-3.5-Turbo as the linguistic component, succeeded by a Multi-Layer Perceptron regressor.

The model developed in this study has demonstrated a commendable level of accuracy in forecasting star ratings through analysis of natural language reviews. These predictions hold significant potential value for both consumers and businesses. Consumers can benefit from the provision of an aggregated rating, even in cases where there is an insufficient number of star-based reviews available for a particular business. Simultaneously, businesses can gain valuable insights that can inform improvement strategies. Since this paper employs Aspect-Based Sentiment Analysis, the results yielded by this research may have broader applications, extending beyond star-rating predictions and offering valuable insights that can inform various domains.

2 Related Work

2.1 Aspect-Based Sentiment Analysis (ABSA)

Aspect-Based Sentiment Analysis (ABSA) represents a logical progression from general Sentiment Analysis, as emphasized by Liu [9]. In standard sentiment analysis, the primary components are the target and the sentiment. This involves understanding the viewpoint or emotion of a speaker or writer concerning a particular subject. Similarly, within the ABSA framework, the target is defined through either an aspect category or an aspect term. Similarly, sentiment encompasses a more comprehensive expression of opinions, which includes both the specific opinion term used and the broader sentiment orientation, such as sentiment polarity.

In the referenced survey [10], the authors outline four essential elements that constitute the primary focus of the Aspect-Based Sentiment Analysis (ABSA) research:

– **Aspect Category**: This refers to a specific feature or attribute of an entity categorized into a predetermined set specific to each domain. For example, in Bakeries, aspects could include "customer service" and "variety".
– **Aspect Term**: Denoted by the authors as 'a', the aspect term signifies the explicit target of sentiment within a given text. It pinpoints the element being appraised. For example, in the sentence "The bread here is decent", the aspect term is "bread".
– **Opinion Term**: Represented by the authors as 'o', the opinion term embodies the sentiment expressed by the opinion holder about the aspect term. In the statement "The bread here is amazing!", the opinion term is "amazing".
– **Sentiment Polarity**: Denoted as 'p' by the authors, sentiment polarity characterizes the sentiment's orientation toward an aspect category or term. Polarity generally falls into positive, negative, or neutral categories.

The progression of Aspect-Based Sentiment Analysis can be delineated through three pivotal phases: The initial era of feature-based models, followed by the advancement to pre-trained word embeddings, and ultimately culminating in the transformative age of pre-trained language models.

Feature-based models relied on hand-crafted features to represent the text and the aspects. The features can include lexical, syntactic, semantic, and sentiment information. For example, [11] and [12] utilize a rule-based approach, where the primary focus is on lexicons and dependency resolution and utilize manually defined patterns and extract aspects. These models also often employ traditional machine learning algorithms, such as support vector machines, naive Bayes, or decision trees. The main drawback of these models is that they require a lot of manual effort to design compelling features, and they cannot capture the complex and dynamic relations between words and aspects.

On the other hand, "Word Embeddings" are distributed representations of words that are learned from sizeable unlabeled corpora using neural network models, such as Word2Vec [13] or GloVe [14]. Word embeddings capture the semantic and syntactic similarities between words based on their co-occurrence patterns. For example, the word embeddings for "delicious" and "tasty" are close in the vector space. The main advantage of these models is that they can leverage large amounts of unlabeled data to learn to generalize word representations. The main limitations are that they do not consider task-specific information, such as aspect category or sentiment polarity. These key aspects of Aspect-Based Sentiment Analysis were highlighted by the survey authors in [10].

Pre-trained language models are advanced neural network models that are pre-trained on large-scale unlabeled corpora using self-supervised objectives such as masked language modeling or next-sentence prediction. These models can learn deep contextualized representations of words that capture both syntactic and semantic information. Some examples of these models are BERT [15], RoBERTa [16], or XLNet [17]. Commonly, these models are fined-tuned on downstream ABSA tasks using task-specific labels. The main benefit of these models is that they can achieve state-of-the-art performance on ABSA. The main challenge is that they require a lot of computation power to train and fine-tune.

2.2 Large Language Models

Large Language Models have become a fundamental aspect of NLP research due to their ability to harness extensive knowledge repositories, efficient learning techniques, and sophisticated comprehensions of linguistic subtleties. These models employ advanced statistical approaches to produce text resembling human language, facilitating a remarkable understanding of context and representation of intricate contexts.

A noteworthy contributor to the advancement of LLMs is OpenAI, a private research laboratory that specializes in developing AI in ways that benefit humanity, in 2020 announced its state-of-the-art Generated Pre-Trained Transformer (GPT), GPT-3, which is an auto-regressive language model. The GPT-3 model boasts an enormous parameter count of 175 billion [18]. It demonstrates proficiency in generating human-like text [19]. LLMs also show high intelligence, for example ChatGPT (Which utilizes GPT-3.5) has scored up to 155 in an IQ test [20].

The scope of large language models in processing text data is a broad and ever-growing field. The GPT-3 model has been used for sentiment analysis, as mentioned in [6]. A subsequent iteration of the GPT-3 model known as GPT-3.5-Turbo has demonstrated the capability to outperform state-of-the-art models in sentiment analysis.

Within the context of this research paper, the deliberate selection of GPT-3.5-Turbo as the model of choice is rooted in its compelling and empirically substantiated superiority over the previous generations of GPT models, most notably surpassing the capabilities of GPT-2. GPT-3.5-Turbo is characterized by a series of remarkable enhancements that render it exceptionally well suited for the multifaceted challenges posed by contemporary natural language processing tasks.

First and foremost, GPT-3.5-Turbo exhibits remarkable proficiency in few-shot learning, a critical attribute that empowers the model to swiftly adapt and generalize from limited examples, which is particularly valuable in scenarios where extensive labeled training data is scarce or nonexistent. Furthermore, this model excels in the domain of cross-lingual transfer, showcasing the ability to effectively transfer knowledge and understand linguistic subtleties across diverse languages, thereby broadening its applicability to a global spectrum of natural language data.

Of particular significance is GPT-3.5-Turbo's heightened aptitude in comprehending intricate linguistic nuances, an essential attribute for the processing of natural language reviews. This facet of the model's capabilities is of paramount importance as it ensures the extraction of nuanced sentiments and insights embedded within text, enhancing its ability to discern subtle contextual cues, idiomatic expressions, and the underlying sentiment of reviewers.

These qualities make GPT-3.5-Turbo an ideal choice for the tasks and objectives outlined in this paper, especially as it pertains to the analysis of natural language reviews. The significance of these model characteristics will be expounded upon in Sect. 3, where we delve deeper into the specific applications and implications of this model in the context of our research.

2.3 Multi-layer Perceptrons

Multi-Layer perceptrons (MLPs) represent a foundational category within artificial neural networks. They encompass a configuration of interconnected layers composed of "neurons", where these neurons operate through weighted connections. Each neuron undertakes a linear combination of its input signals, followed by the application of a nonlinear activation function, culminating in the generation of output. The acquisition of knowledge by MLPs emanates from data assimilation, achieved by iteratively adjusting the weights of interconnecting links employing methodologies like gradient descent or backpropagation.

It is noteworthy that MLPs can be mathematically proven to possess the attributes of Universal Function Approximators [21]. Due to their demonstrated versatility, MLPs found extensive applications across diverse domains within machine learning, encompassing tasks such as regression and classification [22]. While more advanced neural network architectures have emerged over time, including Convolutional Neural Networks (CNNs) and Recurrent Neural Networks (RNNs), these sophisticated structures predominantly find application in specialized domains; for instance, CNNs are well-suited for image classification tasks. Additionally, the utility of neural networks extends to domains such as machine translation, speech recognition, and video analysis.

The present research endeavors to harness the capabilities of MLPs. The intention behind this choice is to exemplify the efficacy of even a rudimentary regressor when faced with a high-quality data supply. We maintain our belief that our model, facilitated by the integration of Large Language Model embeddings, possesses the potential to excel in this regard.

3 Dataset

The corpus analyzed in this research comprises 3,494 bakery reviews, all authored in the Spanish language and collected from Google Maps. Notably, these reviews do not possess any distinctive linguistic or thematic characteristics. Each review comprises the organic expression of the reviewer's thoughts in natural language, accompanied by an associated star rating. We extracted these reviews from the Google Maps platform. To facilitate categorization and evaluation, we utilized a five-category classification scheme, employing a 5-point Likert Scale. The resulting categories are as follows:

1. Flavor (F)
2. Variety (V)
3. Freshness of Bread (FoB)
4. Customer Service (CS)
5. Price (P)

The 5-Likert scale ratings are as follows:

1. Horrible
2. Bad
3. Neutral
4. Good
5. Excellent

Running an initial data analysis, we noted that the data was highly unbalanced; that is, for polarity, some of their labels are more frequent than others. Figure 1 shows the rating distribution across all bakeries, while Fig. 2 shows the relationship between the features mentioned in the Likert Scale, in relation to the rating.

Additionally, some opinions contained unrelated notes about the Bakery ("Ayer me mordio un perro"), as well as emojis, repeated opinions ("El pan es muy bueno"), common English words (good, bad), abbreviations (pq, q), symbols ($, +), emoji-only reviews, grammatical errors ("Estos panes son muy mejor que los otros", "Pan horne-ando en el horno huele bien.") and spelling errors ("haiga", "beces"). Considering these factors, we carried out further data pre-processing to optimize the model's performance. The specific procedures undertaken are detailed in the following Sub-Sects. 4.1 and 4.3.

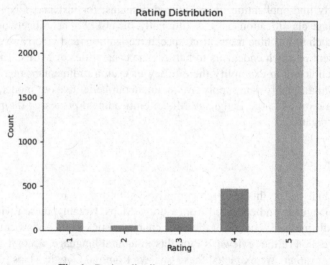

Fig. 1. Rating distribution across the dataset

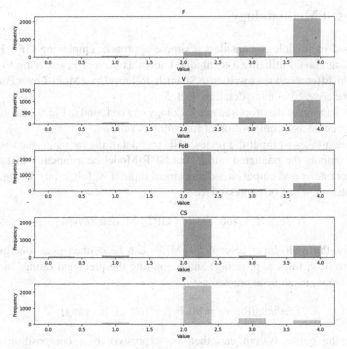

Fig. 2. Distribution of ratings in relation to features across the dataset

In the preceding section, we underscored the variability observed in the reviews. To further elucidate this phenomenon, we provide below illustrative examples of actual reviews, each paired with their respective tags and ratings.

> "rating": 5,
> "review": "El pan es excelente",
> "sentiment": "{'F': 'G', 'V': 'E', 'FoB': 'N', 'CS': 'N', 'P': 'N'}"

Ex. 1. Good review example, the text is coherent, there are no spelling or grammar mistakes, and the rating corresponds to the mentioned aspects.

> "rating": 5,
> "review": "Me mordio mi perro",
> "sentiment: "{'F': 'N', 'V': 'N', 'FoB': 'N', 'CS': 'N', 'P': 'N'}"

Ex. 2. Bad review example, the text is unrelated to the quality of the bread or the experience.

4 Proposed Methodology

The proposed methodology considers a simple approach, employing GPT-3.5-Turbo as the language layer, utilizing a technique known as prompt engineering to evaluate sentiments in different reviews written in Spanish, followed by a Multi-Layer Perceptron. These will be covered in more detail in Sect. 5.

A block diagram of the proposed methodology can be found in Fig. 3, and the system can be represented as a composition of functions as follows:

The system takes as input the review (NL text data), the rating (valued between 1 and 5) and outputs the predicted rating. Let GPT_Model be a function that takes the review as parameter and outputs a 5-dimensional tuple (f, v, fob, cs, p). Using standard function notation, it can be expressed as:

$$\text{Aspects}(f, v, fob, cs, p) = \text{GPT\_Model}(\text{Review}) \tag{1}$$

Similarly, the multi-layer perceptron (MLP) can be defined as a function taking 6 parameters (f, v, fob, cs, p, rating) and outputting the predicted rating. In standard function notation, it can be expressed as:

$$\text{PredictedRating} = \text{MLP}(f, v, fob, cs, p, rating) \tag{2}$$

Finally, the entire system can then be expressed as a composition of these components:

$$\text{System}(\text{Review, Rating}) = \text{MLP}(\text{GPT\_Model}(\text{Review}), \text{Rating}) \tag{3}$$

where System (Review, Rating) represents the final output of the system, which takes a natural language review and an associated rating as input and produces a predicted rating as output. This methodology will be discussed in more detail in Sect. 5.

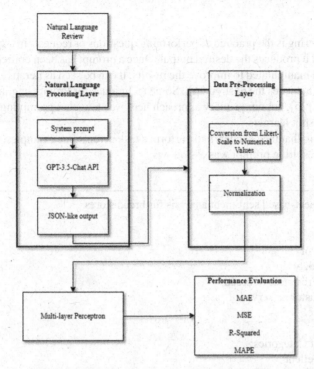

Fig. 3. Block Diagram of the proposed methodology

4.1 Data Pre-processing

In Sect. 3 some examples of the opinions in the dataset were remarked on the need for pre-processing data. Taking this into account, the steps for preprocessing were as follows.

- Reduction of duplicates.
- Manual removal of useless or unrelated reviews.
- Removal of emoji-only reviews

Text transformations were not applied, as LLMs show a particularly good ability to detect grammatical errors and abbreviations [23, 24].

After performing the pre-processing procedure, the number of reviews was reduced from 3,494 to 3,143.

4.2 Prompt

Prompt Engineering is the practice of performing questions or requests in a specific way to an LLM until it produces the desired output. Once a prompt has been crafted, the result is reviewed and manipulated to improve the results. It can be seen as iterative refinement until the model achieves the desired task. Some techniques for Prompt engineering have been discussed [25], but our primary approach here was few-shot prompting [26], since LLMs are few-shot learners.

[27] meaning that they can learn to perform a task given some examples. After some iterations, the resulting prompt was as follows:

Task: Aspect-based sentiment analysis for bread stores.

Aspects:
- F: Flavor
- V: Variety
- FoB: Freshness of bread
- ČS: Customer service
- P: Price

Sentiment Categories:
- E: Excellent
- G: Good
- N: Neutral
- B: Bad
- H: Horrible

Input Examples:
1. The bread here is horrible; the taste is disgusting.
2. Amazing place to hang out! the waitress is terrific!
3. I loved the bread. However, I think there could be more variety, but the current selection is okay.
4. Bread is nice, however, a little expensive.
Output Examples:
1. {'F': 'H', 'V': 'N', 'FoB': 'N', 'CS': 'N', 'P': 'N'}
2. {'F': 'N', 'V': 'N', 'FoB': 'N', 'CS': 'E', 'P': 'N'}
3. {'F': 'E', 'V': 'G', 'FoB': 'N', 'CS': 'N', 'P': 'N'}
4. {'F': 'G', 'V': 'N', 'FoB': 'N', 'CS': 'N', 'P': 'B'}

The results of this prompt will be discussed in Sect. 5.

4.3 Tabular Data Transformation

Multi-Layer Perceptrons (MLPs) inherently require numerical inputs for processing. To facilitate the integration of textual reviews as inputs, a pre-processing procedure was employed.

First, since star ratings follow a quintessential 5-point rating system, for each value, one was subtracted to obtain 0-indexed values. Following this practice, "Min-Max Normalization" or "Feature-Scaling" was performed, this corresponds to the method of squishing values in the range from 0 to 1. After the mentioned procedure, for example, a review without any processing was:

> "sentiment": {'F': 'G', 'V': 'N', 'FoB': 'N', 'CS': 'N', 'P': 'B'},
> "rating": 4

After feature scaling was performed, the result was as follows:

> "sentiment": 'F': 0.8, 'V': 0.4, 'FoB': 0.4, 'CS': 0.4, 'P': 0.2,
> "rating": 4

4.4 Modeling and Training

The dataset was randomly partitioned into three distinct subsets, namely Training, Validation, and Testing, each assigned with 80%, 10%, and 10% of the data, respectively.

The formulation of the model's architecture was achieved by levering PyTorch, a prominent and extensively utilized framework within the field of Machine Learning [28].

The Multi-Layer Perceptron (MLP) configuration employed herein is characterized by an input layer comprising 5 neurons, complemented by a hidden layer housing 6 neurons, said number was attained by taking the mean between the input and output layer, as suggested by Heaton in [29]. The activation of this hidden layer is effectuated by the application of the Rectified Linear Unit (ReLU) activation function, the said activation function was chosen because it is widely used and has shown better computational performance and better convergence [30]. This arrangement culminates in a single output layer, aligning with the customary practice for regression tasks.

The network's learning rate was determined through PyTorch Lightning's [31] tuner module, a process information by foundational research [32]. After an exhaustive iterative exploration involving 100 trials, the tuner ascertained an optional learning rate of 0.06.

The model underwent a 25-epoch training regimen, a duration chosen based on the discernment that performance ceased to exhibit significant improvements beyond this point. To confirm this observation, a series of tests involving 50, 100, 150, and 200 epochs were conducted. However, none of these extended training regimens produced appreciable performance enhancements compared to the 25-epoch regimen.

A summary of the hyperparameters described in this section can be found in Table 1.

Table 1. Hyperparameters employed during training

Hyperparameter	Value
Input Layer Size	5
Hidden Layer Size	6
Output Layer Size	1
Activation Function	ReLU
Learning Rate	0.06
Number of Epochs	25

4.5 Metrics

Since our research utilized regression, we found it compelling to utilize standard metrics to measure the performance of the model.

The primary evaluative measure employed in this study was the Mean Squared Error (MSE), a well-established metric within the realm of regression analysis. The MSE quantifies the average of the squared disparities between the predicted values generated by the model and the corresponding actual values. By doing so, it provides an indication of the alignment level between the model's predictions and the empirical data points. The calculation formulation for this metric is presented in Eq. 4.

$$\text{MSE} = \frac{1}{n} \sum_{i=1}^{n} (y_i - \hat{y}_i)^2 \tag{4}$$

In conjunction with MSE, an additional prevalent metric, R-Squared (R2), was used for the assessment. The R-Square metric serves as a statistical gauge of the extent to which variance in the dependent variable can be elucidated by variations in the independent variable. Ranging between 0 and 1, higher R-Squared values signify an improved congruence between the model and the dataset. Hensler [33] introduced a heuristic categorization for acceptable R-square values, designating 0.75, 0.50 and 0.25 as thresholds for substantial, moderate, and weak explanatory power, respectively. The formula underlying the computation of this metric is delineated in Eq. 5.

$$R^2 = 1 - \frac{\sum_{i=1}^{n} (y_i - \hat{y}_i)^2}{\sum_{i=1}^{n} (y_i - \overline{y})^2} \tag{5}$$

Lastly, the Mean Absolute Percentage Error (MAPE), a prevalent metric in the realm of regression analysis, was adopted as the final evaluative criterion.

MAPE gauges the average absolute percentage discrepancy between projected values and actual data points. Equation 6 details the mathematical framework to calculate this metric.

$$\text{MAPE} = \frac{1}{n} \sum_{i=1}^{n} \left| \frac{y_i - \hat{y}_i}{y_i} \right| \times 100 \tag{6}$$

5 Results

Following the training phase, the model was tested on previously unseen data sourced from the designated testing set. The protocol employed for partitioning the dataset is explained in Sect. 4. After evaluation, the results presented in Table 2 were derived.

Table 2. Results obtained through testing the model

	MAE	MSE	R-Squared	MAPE
Model	0.27	0.26	0.69	0.10%

To guarantee the precision and relevance of our model, it is imperative that we conduct a comparative analysis with other relevant works. To gain a more holistic perspective, we can refer to the data presented in 3. It is pertinent to note that the datasets utilized in the comparative studies are dissimilar from those employed in our investigation.

	Test MAE
UMU-Team	0.258
UC3M	0.260
CIMAT2020_Beto	0.267
BreadAllYouNeed (This paper)	0.271
CIMAT_Bo-TextAutoAugment	0.315
Raschka Research - CORN	0.412

To gauge our accuracy, we determined our Mean Absolute Error (MAE) baseline. This was achieved by identifying the median of all available label values and then calculating the average absolute difference between the computed average prediction and the actual label values. Thanks to our refined model, we saw an impressive 41.30% reduction in MAE, indicating a significant improvement in our prediction accuracy.

6 Conclusion

6.1 Summary

We have shown that our theory that large language models can be used for Aspect-Based sentiment analysis has been proven, the research shined insight into the use of these types of models to produce high quality, sentiment analysis data, which can accurately aid in the creation of new technologies.

6.2 Future Improvements and Research

Despite our success, we believe that our model faced significant challenges that limited its abilities and challenges related to LLMs.

- **Techniques for class imbalances**: Our dataset exhibited a significant class imbalance, which can adversely impact model training and performance. To mitigate this problem, commonly employed methods such as oversampling and undersampling were applied. Despite these efforts, they did not yield substantial improvements in model performance. We posit that further investigation is warranted in this domain. Subsequent research directions could encompass the development of an augmentation framework designed to effectively mitigate class imbalances.
- **Exploration of ordinal regression**: In the deliberation of our modeling approach, the choice between classification and regression became a crucial decision. The decision to utilize regression over classification was motivated by the unsatisfactory performance in preliminary classification experiments. However, during the research process, we recognized the inherited ordinal nature of star-based ratings, prompting an exploration of the "Ordinal Regression" technique. Prominent implementations for neural networks, such as CORAL [34] and CORN [35], leverage customized loss functions to facilitate the integration of ordinal regressors within machine learning models. Although we endeavored to incorporate these techniques, discernible enhancements in model performance were not achieved. This underscores the need for further inquiry, particularly across diverse datasets, to unlock the full potential of ordinal regression in our context.
- **Reducing computational costs**: The computation demands posed by GPT models, a pillar in our research, are known to be substantial. The financial implications of utilizing OpenAI's API were identified as a significant challenge. To address this concern in future work, we propose an avenue that involves leveraging open-source LLMs. These models, coupled with frameworks such as Petals [36], hold promise to circumvent the computational cost constraints associated with proprietary APIs.
- **Assessment of Cross-Linguistic Applicability:** GPT-3.5-Turbo has demonstrated its competence in multiple languages [27, 37]. In pursuit of broader validation, further investigations are planned across a diverse array of linguistic and geographic contexts. This will entail the acquisition and analysis of datasets sourced from various languages and regions, thereby probing the model's applicability beyond its original training domain.

6.3 Future Work

In continuation of our ongoing research within the bread industry, our upcoming work will concentrate on the development of an innovative approach to evaluate wheat quality utilizing the Sorensen Similarity Index across diverse geographic locations within the state of Aguascalientes, Mexico. Wheat, as a key crop in bread production, has a significant influence over both agricultural producers and consumers. Consequently, the establishment of a reliable and efficient methodology for assessing wheat quality is of paramount importance in ensuring the prosperity of wheat cultivation and subsequent marketing efforts.

Our novel approach will leverage advanced machine learning techniques and Big Data analysis to explore novel avenues for assessing wheat quality. This cutting-edge methodology will empower consumers with the capability to ascertain the quality of the raw material used in bread production even before the baking process commences. This represents a significant leap forward in providing transparency and quality assurance to consumers in the bread industry.

References

1. TripAdvisor: Transparency Report (2023)
2. Leader, I: How reviews on Google Maps work (2022)
3. Brun, C., Nikoulina, V.: Aspect based sentiment analysis into the wild. In: Proceedings of the 9th Workshop on Computational Approaches to Subjectivity, Sentiment and Social Media Analysis. Association for Computational Linguistics, Brussels, Belgium, pp 116–122 (2018)
4. Kaddour, J., Harris, J., Mozes, M., Bradley, H., Raileanu, R., McHardy, R.: Challenges and applications of large language models (2023)
5. Hu, K.: ChatGPT sets record for fastest-growing user base - analyst note (2023)
6. Kheiri, K., Karimi, H.: SentimentGPT: exploiting GPT for advanced sentiment analysis and its departure from current machine learning. arXiv (2023)
7. Tian, L., Lai, C., Moore, J.: Polarity and intensity: the two aspects of sentiment analysis. In: Proceedings of Grand Challenge and Workshop on Human Multimodal Language (Challenge-HML), pp 40–47. Association for Computational Linguistics (2018)
8. Santiibáñez, E., Castillo, Y.A., Moctezuma, D.A., Muñiz, V.H.: BERT and data augmentation for sentiment analysis in tripadvisor reviews. In: Ceur Workshop Proceedings, vol. 3202 (2022)
9. Liu, B.: Sentiment Analysis and Opinion Mining. Morgan & Claypool Publishers, San Rafael (2012)
10. Zhang, W., Li, X., Deng, Y., Bing, L., Lam, W.: A survey on aspect-based sentiment analysis: tasks, methods, and challenges (2022)
11. Qiu, G., Liu, B., Bu, J., Chen, C.: Opinion word expansion and target extraction through double propagation. Comput. Linguist. **37**, 9–27 (2011). https://doi.org/10.1162/coli_a_00034
12. Liu, Q., Liu, B., Zhang, Y., Kim, D.S., Gao, Z.: Improving opinion aspect extraction using semantic similarity and aspect associations In: Proceedings of the AAAI Conference on Artificial Intelligence, vol. 30 (2016). https://doi.org/10.1609/aaai.v30i1.10373
13. Rong, X.: word2vec parameter learning explained (2016)
14. Pennington, J., Socher, R., Manning, C.: Globe: global vectors for word representation. In: Proceedings of the 2014 Conference on Empirical Methods in Natural Language Processing (EMNLP). Association for Computational Linguistics, Doha, Qatar, pp 1532–1543 (2014)
15. Devlin, J., Chang, M.-W., Lee, K., Toutanova, K.: BERT: pre-training of deep bidirectional transformers for language understanding (2019)
16. Liu, Y., et al.: RoBERTa: a robustly optimized BERT pretraining approach (2019)
17. Yang, Z., Dai, Z., Yang, Y., Carbonell, J., Salakhutdinov, R., Le, Q.V.: XLNet: generalized autoregressive pretraining for language understanding (2020)
18. Alarcon, N.: OpenAI presents GPT-3, a 175 billion parameters language model | NVIDIA Technical Blog (2023)
19. Elkins, K., Chun, J.: Can GPT-3 pass a writer's turing test? J. Cult. Anal. **5** (2020). https://doi.org/10.22148/001c.17212
20. Roivainen, E.: I Gave ChatGPT an IQ Test. In: Here&rsquos What I Discovered

21. Hornik, K., Stinchcombe, M., White, H.: Multilayer feedforward networks are universal approximators. Neural Netw. **2**, 359–366 (1989). https://doi.org/10.1016/0893-6080(89)900 20-8
22. Murtagh, F.: Multilayer perceptrons for classification and regression. Neurocomputing **2**, 183–197 (1991). https://doi.org/10.1016/0925-2312(91)90023-5
23. Fan, Y., Jiang, F., Li, P., Li, H.: GrammarGPT: exploring open-source LLMs for native Chinese grammatical error correction with supervised fine-tuning (2023)
24. Buruk, O.O.: Academic writing with GPT-3.5: reflections on practices, efficacy and transparency (2023)
25. White, J., et al.: A prompt pattern catalog to enhance prompt engineering with ChatGPT. arXiv (2023)
26. Ma, H., et al.: Fairness-guided few-shot prompting for large language models. arXiv (2023)
27. Brown, T.B., et al.: Language models are few-shot learners. arXiv (2020)
28. Paszke, A., et al.: PyTorch: an imperative style, high-performance deep learning library. In: Advances in Neural Information Processing Systems, vol. 32, pp 8024–8035. Curran Associates, Inc. (2019)
29. Heaton, J.; Introduction to Neural Networks with Java, 2nd edn. Heaton Research, Incorporated, St. Louis, Mo (2008)
30. Waoo, A.A., Soni, B.K.: Performance analysis of sigmoid and relu activation functions in deep neural network. In: Sheth, A., Sinhal, A., Shrivastava, A., Pandey, A.K. (eds.) Intelligent Systems. AIS, pp. 39–52. Springer, Singapore (2021). https://doi.org/10.1007/978-981-16-2248-9_5
31. Falcon, W.: The PyTorch Lightning team. PyTorch Lightning (2019)
32. Smith, L.N.: Cyclical learning rates for training neural networks. arXiv (2017)
33. Rights, J.D., Sterba, S.K.: A framework of R-squared measures for single-level and multilevel regression mixture models. Psychol. Methods **23**, 434–457 (2018). https://doi.org/10.1037/met0000139
34. Cao, W., Mirjalili, V., Raschka, S.: Rank consistent ordinal regression for neural networks with application to age estimation. Pattern Recogn. Lett. **140**, 325–331 (2020). https://doi.org/10.1016/j.patrec.2020.11.008
35. Shi, X., Cao, W., Raschka, S.: Deep neural networks for rank-consistent ordinal regression based on conditional probabilities (2021)
36. Borzunov, A., et al.: Petals: collaborative inference and fine-tuning of large models. arXiv preprint arXiv:220901188 (2022)
37. Armengol-Estapé, J., Bonet, O.D.G., Melero, M.: On the multilingual capabilities of very large-scale English language models (2021)

Brake Maintenance Diagnostic
with Fuzzy-Bayesian Expert System

Misael Perez Hernández[1]([⊠]), Martín Montes Rivera[1], Ricardo Perez Hernández[2],
and Roberto Macias Escobar[1]

[1] Research and Postgraduate Studies Department in Universidad Politécnica de Aguascalientes,
Aguascalientes, Mexico
mc220003@alumnos.upa.edu.mx, martin.montes@upa.edu.mx
[2] Colegio Bosques de Aguascalientes, Aguascalientes, Mexico
jrperez@bosques.edu.mx

Abstract. Brakes, one of a vehicle's most crucial safety systems, are necessary
to ensure safety. They are our primary protection mechanism while driving a car
on the road. A brake failure can end up causing an accident and putting lives at
risk, which is why it is essential to check all its elements periodically. The car
must go to service in case of brake issues: unusual noises, abnormal movements
or sensations, inability to stop quickly, and warning lights. Sometimes, drivers do
not associate them with brake failure and wait to take the car to check service.
However, if they could determine that an issue relates to brake problems, they could
immediately seek assistance. State of the art shows that expert systems and domain
expertise revolutionize maintenance, reshaping diagnostics, decision-making, and
predictive strategies by blending advanced AI techniques, data analysis, and real-
time monitoring. On the one hand, fuzzy logic is a branch of artificial intelligence
and mathematics used to model and manage uncertainty and imprecision in data
and expert systems. On the other hand, Bayesian reasoning allows determining
beliefs about a hypothesis based on facts. In this work, we propose developing
a Fuzzy-Bayesian expert system for assisting the drivers in the maintenance of
car brake systems encompassing goal setting, knowledge acquisition, interface
design, and testing. Our proposal, programmed in Python, uses UPAFuzzySystems
to describe fuzzy rules and Twilio to allow SMS integration in a user interface,
empowering users to make informed brake system decisions from their mobile
and obtain information about the status of their vehicle's brake system.

Keywords: expert systems · brake maintenance · fuzzy systems · forward
chaining · backward chaining

1 Introduction

The primary purpose of the brake is to perform a unique and crucial function: to help
reduce speed to completely stop the vehicle or prevent possible collisions with obstacles
on the road. Brakes perform the fundamental function of counteracting the friction
between the wheels and the pavement, thus preventing the wheels from slipping [1].

© The Author(s), under exclusive license to Springer Nature Switzerland AG 2024
H. Calvo et al. (Eds.): MICAI 2023 Workshops, LNAI 14502, pp. 77–101, 2024.
https://doi.org/10.1007/978-3-031-51940-6_8

This is why it is one of the most important systems in vehicles. Being the first safety system in cars, it is imperative to keep it in good condition to prevent collisions and accidents [2]. The system itself sends signs of a malfunction before it fails completely or requires a preventive change. This type of signals is presented through sound, (such as squeaks), sensations (both in the pedal and in driving), as well as visual aspects (irregularities in the brake disc, light on the dashboard) [3].

These failures sometimes share symptoms that can make the diagnosis for the repair of the failure delayed, in addition, several factors such as distance traveled and time elapsed since last maintenance will affect which component has the highest probability of failing. The procedure to begin an automotive diagnosis is based on the mechanic's memory and experience, which could be biased due to the limitation of the number of vehicles with that fault that the automotive technician has repaired. To solve this problem, an expert system was designed which emulates the decision making of the automotive technician to give a diagnosis with the most probable failure depending on the client's input variables [4].

The expert system will help customers obtain a diagnosis at any time they want and eliminating the bias of the technician's experience by using fuzzy logic that helps solve uncertainty problems where each fuzzy set is determining for a temperature range where it varies. The membership value of said set moving it further or further away from that set, for example the temperature, where there would be several fuzzy sets such as "cold", "warm" and "hot", in which the membership value decreases for a set and increasing for another, the range of values of belonged is a closed interval from 0 to 1 or mathematically written as [0, 1] [5].

On the other hand, for probabilities the Bayes theorem will be used since it is a mathematical formula that describes how we can update our beliefs about the probability of a given event occurring as we obtain new relevant evidence. It is based on two key elements: a priori probability (our initial belief) and conditional probability (the probability of the evidence given the hypothesis), which will help improve precision given that several conditional events, such as several failures at the same time, squeal, braking feel, etc., to determine which part is most likely to fail [6].

For the user interface, tkinter was used as it is a standard Python library used to create graphical user interfaces (GUIs). Provides tools and widgets for designing windows, buttons, text boxes, menus, and other user interface elements in desktop applications. Tkinter is based on the Tcl/Tk toolset and is a popular way to create simple, functional user interfaces in Python. Tkinter is easy to learn and suitable for small and medium-sized projects that require a basic GUI [7].

In addition, an option to send the diagnosis via SMS through Twilio was incorporated since it is a cloud communications platform that allows developers to integrate communication functions, such as text messages, voice and video calls, into their applications. And websites. With Twilio, developers can send automated text messages, make programmatic phone calls, create chatbots, and enable two-factor authentication, among many other features. The platform is widely used in a variety of industries, including customer services, marketing, telemedicine, and more [8].

Furthermore, this paper delves into the specific steps involved in developing an expert system for brake system fault diagnosis. These steps encompass defining the system's objectives, acquiring relevant knowledge from experts, representing the knowledge in a suitable format, designing a user-friendly interface, implementing an inference engine, and rigorously testing and validating the system's performance.

1.1 Objective

This paper focuses on the development of an expert system for fault diagnosis in brake systems, a critical component of automotive vehicles..The objective of this research is to leverage the power of expert systems and Bayesian reasoning to accurately diagnose brake system faults and provide effective recommendations for repairs or maintenance.

Overall, the development and implementation of an expert system for brake system fault diagnosis using Bayesian reasoning, fuzzy systems, inference Mamdani system, in addition to the visual interface with tkinter and communication via SMS with TWILIO hold great potential in improving automotive maintenance practices.

2   State of the Art

The development of expert systems for vehicle brake maintenance represents a significant advancement in the automotive industry. These systems combine the power of artificial intelligence (AI) and domain expertise to assist mechanics, technicians, and vehicle owners in diagnosing, maintaining, and repairing brake systems [9].

Trends and advancements have emerged in this field:

Integration of AI and Brake Maintenance: Expert systems leverage AI techniques such as machine learning, rule-based reasoning, and data analysis to interpret sensor data, historical maintenance records, and real-time vehicle performance metrics. This integration enhances the accuracy and speed of brake system diagnostics [10].

Data-Driven Decision Making: The availability of extensive data from modern vehicles has paved the way for data-driven decision-making in brake maintenance. Expert systems can analyze sensor data, wear patterns, temperature fluctuations, and more to provide informed recommendations for maintenance or repair [11].

Real-time Monitoring: Advanced sensor technologies have enabled real-time monitoring of brake components, ensuring continuous assessment of brake health while the vehicle is in operation. Expert systems can process this data in real-time to detect anomalies or potential issues [12].

Predictive Maintenance: Predictive maintenance models have gained prominence, allowing expert systems to predict when specific brake components are likely to fail based on historical data and usage patterns. This proactive approach reduces downtime and prevents unexpected failures [13].

User-Friendly Interfaces: User interfaces for these expert systems have evolved to become more intuitive and user-friendly. Mechanics and technicians can interact with the system through graphical interfaces, voice commands, and mobile applications, enhancing accessibility and usability [14].

The integration of AI, real-time monitoring, predictive analytics, and user-friendly interfaces has revolutionized brake maintenance practices. As vehicles continue to evolve, these systems are poised to play a crucial role in ensuring safe and efficient brake system operation [9].

3   Theorical Framework

3.1   Expert Systems

Expert systems (SE) are computer programs that aim to solve a specific problem and use Artificial Intelligence (AI) to simulate the reasoning of a human being. They are called expert systems because these programs mimic the decision-making of a professional in the field [15].

Every expert system consists of two main parts: the knowledge base; and reasoning, or inference engine.

The knowledge base of expert systems contains actual and heuristic knowledge. Effective knowledge is task domain knowledge that is widely shared, typically found in textbooks

Heuristic knowledge is the least rigorous, most experimental, most critical knowledge of functioning. In contrast to factual knowledge, heuristic knowledge is rarely discussed and is largely individualistic. It is knowledge of good practice, good judgment, and admissible reasoning in the field. It is the knowledge that is the basis of the "art of good inference" [16].

The development of an expert system in the field of programming involves the following steps:

- Definition of the objective: Determines the specific objective of the expert system in the field of programming. For example, it might be helping programmers solve programming problems, providing software design recommendations, or diagnosing bugs in code.
- Knowledge acquisition: Identifies and collects relevant knowledge from programming experts. You can conduct interviews, review technical documentation, analyze existing code, among other methods. Make sure you capture the knowledge in a format suitable for further processing by the expert system.
- Knowledge Representation: Choose a way of representing knowledge that best suits your programming domain. Some common options include production rules, decision trees, semantic networks, or fuzzy logic. The choice will depend on the type of knowledge and the characteristics of the problem you are addressing.
- User interface design: Create an interface that allows users to interact with the expert system. It can be a command line interface, a GUI, or a web-based interface. Make sure that the interface is intuitive and facilitates communication between the user and the system.
- Implementation of the inference engine: Develops the inference engine that will be used by the expert system to process the knowledge and provide answers or solutions. You can implement rule-based reasoning algorithms, search algorithms, or even machine learning techniques, depending on the complexity of the problem and the resources available.
- Testing and validation: Perform extensive testing of the expert system to ensure that it is working correctly and producing accurate and reliable results. Validates the results of the expert system by comparing them with the knowledge of human experts or by using test data sets [17, 18].

3.2 Faults and Diagnosis for the Brake System

Brake systems play a crucial role in ensuring the safety and performance of vehicles, and timely detection and resolution of faults are essential for maintaining optimal functionality [19].

The most common failures in the brake system are the following: in addition to adding the most common way to diagnose and correct said failure of each one, these failures must also take into account various factors such as the brand and material of the parts, time. And vehicle route.

Proper maintenance and prompt repair of brake system problems are essential to ensure safe driving on the road and prevent severe accidents [20].

Noises When Braking
The wear of component eventually generates noise in the braking system, which causes the metal section to come into contact with the brake disc and emit a high-pitched sound.

- Causes: They may be due to wear of the brake pads or metallic contact between the pads and the discs.
- Diagnosis: The problem occurs if hearing squeaks, squeals, or growls when braking.
- Repair: Replace worn brake pads and, if necessary, grind or replace the discs [20].

Spongy Brake Pedal
This impression tends to arise when the brake pedals appear closer than usual. The brake pedal should have a firm or sturdy feel in typical situations. It should resist pressure, allowing it to be applied progressively rather than instantly.

- Causes: It may be due to an air or brake fluid leak in the hydraulic system.
- Diagnosis: If the pedal feels spongy or sinks to the floor, it is a sign of this problem.
- Repair: Repair leaks in the brake lines, bleed the system to remove air, and replace the brake fluid if necessary [20].

Steering Wheel Vibration When Braking
The brake disc tends to experience deformations since it is a constantly rotating piece that comes into contact with the brake pads with each advance. Additionally, applying the brakes could produce uneven pressure, causing vibrations in the steering wheel, brake pedal, or even the entire vehicle.

- Causes: Detected if warped or unevenly worn brake discs.
- Diagnosis: Feeling vibrations in the steering wheel when braking is a clear sign of damaged brake discs.
- Repair: Grind or replace the brake discs and ensure the calipers work correctly [20].

Brake Fluid Leak
Hydraulic lines are robust components in the brake system but are subject to corrosion, wear damage, or punctures. If these parts deteriorate, brake fluid leaks are likely. Additionally, if the piston seals, which have the function of containing the brake fluid, are damaged, this can cause leaks.

- Causes: May be caused by deteriorated seals in the brake system or damaged lines.

- Diagnosis: Seeing brake fluid on the ground or under the vehicle is an obvious sign of a leak.
- Repair: Replace faulty seals or damaged lines and bleed the brake system [20].

Irregular Brake Pad Wear

The two brake pads within the same caliper may have different levels of wear, but this disparity should not be significant. If there is a marked discrepancy in wear levels between the two pads, it is essential to perform an inspection on the brake system to determine if any component, such as the cylinder piston or caliper bolt, is experiencing any type of restriction in their mobility.

- Causes: May be due to stuck calipers, hydraulic system problems, or incorrect brake alignment.
- Diagnosis: Visually inspect the brake pads for uneven wear.
- Repair: Fix the underlying cause, such as repairing or replacing calipers, bleeding the system or correcting brake alignment, and replacing worn pads [20].

Dashboard Brake System Warning Lights

If the ABS warning light illuminates during everyday driving, it indicates a problem with the proper functioning of the ABS system. Although the brakes should continue to operate normally, there is a strong possibility that the ABS system will not activate in emergency braking situations.

- Causes: This may be due to problems with the ABS (Anti-lock Braking System) system or low brake fluid level.
- Diagnosis: The warning light on the dashboard will come on if there is a problem in the brake system.
- Repair: Diagnose the underlying cause using a diagnostic scanner and make necessary repairs [20].

3.3 Experta

To develop the expert system, Python and the library called expert are used to develop it. This library helps to generate an expert system [21].

Fundamentals: An expert system is a program capable of combining a set of facts with a set of rules and executing actions based on the rules that match the facts.

Facts: Facts are the basic unit of information in Expert. They are used to reason about the problem and are represented by Python classes.

Rules: Rules in Expert are defined as functions decorated with @Rule. The rules have two components: the LHS (left side) and the RHS (right side). The LHS describes the conditions that must be met for the rule to run, and the RHS contains the actions that will be performed when the rule is triggered.

DefFacts: It is a decorator used to declare a set of initial facts that are needed for the system to work correctly. Methods decorated with @DefFacts must be generators that generate fact instances.

KnowledgeEngine: It is the main class where the execution of the expert system occurs. You must create subclasses of this class and use the @Rule decoration in the methods to define the rules. The knowledge engine execution cycle consists of selecting an active rule and executing the actions defined in its RHS.

Fact manipulation: Methods such as declare, retract, modify, and duplicate are provided to manipulate facts in the knowledge engine. These methods allow you to add, delete, and modify facts in working memory.

Engine Execution Cycle: The Knowledge Engine execution cycle involves selecting an active rule from the rule set, executing the corresponding actions, and updating the active rule set based on changes in the facts [21].

3.4 Bayesian Reasoning

Bayesian reasoning is a method of logical inference that is based on Bayes' theorem, a mathematical formula that relates the probability of a hypothesis to the available evidence. Bayesian reasoning allows you to update beliefs about a hypothesis as new information is obtained, so that you can estimate the posterior probability of the hypothesis given the evidence [22].

Bayesian reasoning is a logical, probabilistic approach to inference and decision making, based on Bayes' theorem. This approach uses probabilities to quantify uncertainty and update them as new information is obtained.

Bayesian reasoning is based on the following elements:

1. Priori: It is the initial probability or belief about an event or a hypothesis before obtaining new information. It is denoted as $P(H)$, where H represents the hypothesis.
2. Likelihood: It is the probability of observing the data or evidence given a hypothesis. It is denoted as $P(D|H)$, where D represents the data.
3. Posteriori: It is the revised or updated probability of a hypothesis after considering the data. It is calculated using Bayes' theorem and is denoted as $P(H|D)$.

Bayes' theorem states that the posterior probability of a hypothesis is proportional to the product of the prior probability and the likelihood:

$$P(H|D) = (P(H) * P(D|H))/P(D) \tag{1}$$

where $P(D)$ is the marginal probability of the data, which can be calculated by adding the probabilities of all possible hypotheses multiplied by their respective likelihoods.

Bayesian reasoning involves continually updating the posterior probability as new information is obtained. This is accomplished by incorporating new data into the posterior probability calculation, leading to a revision and refinement of initial beliefs [23].

3.5 UPAFuzzySystems

The UPAFuzzySystems library used in this research serves as an instrument for generating inference systems using fuzzy logic. Additionally, it facilitates the execution of simulations and control assignments involving fuzzy controllers, transfer functions, and state-space models, encompassing both discrete and continuous domains. This library effectively fills a notable void in the realm of open-source resources by seamlessly incorporating these capabilities within a Python-based framework [24].

The capabilities of this library are manifold. It permits the formulation of fuzzy universes tailored to diverse scenarios. Furthermore, it streamlines the process of defining rules within a Fuzzy Inference System (FIS), creating connections between premises, connectives, and consequences. Moreover, the library stands out in its capacity to replicate complex problem scenarios by computing outcomes based on input arrays, establishing a mapping between input premises and corresponding consequences within a continuous realm [24].

Furthermore, the UPAFuzzySystems library demonstrates adeptness in governing and emulating the conduct of a DC motor plant through the utilization of an array of fuzzy controllers. These encompass single-input and dual-input Mamdani and Fuzzy Logic System (FLS) controllers, along with Takagi-Sugeno controllers available in both single-input and dual-input configurations. This comprehensive suite of controllers efficiently rectifies errors, accomplishing reduction of errors by less than 1% under standard circumstances, and by less than 4% even when subjected to random disturbances of 10% uniform intensity [24].

The library empowers users to delve into pivotal attributes of control systems, encompassing aspects like overshoot, steady time, time rising, and time peak. The manifestation of these attributes is contingent upon the controller chosen, due to discernible modifications within control structures, which encompass premises, consequences, connectives, implication, fuzzification, and defuzzification techniques. Furthermore, the library is equipped to incorporate derivatives that enable projection of changes in error behavior, as well as integrals that facilitate incremental error reduction [24].

4 Methodology

For the development of the following expert system, the methodology described below was followed:

Identification and Definition of Relevant Facts: The first step involves identifying and defining the facts that are essential for the problem domain. These facts represent the information or data that the expert system will reason with. For example, in the context of brake maintenance, relevant facts may include the condition of brake pads, the level of brake fluid, etc. Each fact is defined with its possible values and attributes. Code1 represents an example of the identified facts.

```
1    class Spedal(Fact):
2        "Sensacion del pedal"
3        pass
4    class Cfrenar(Fact):
5        "Chillido al frenar"
6        pass
7    class Rfrenar(Fact):
8        "Vacio al frenar"
9        pass
10   class Nivel(Fact):
11       "Nivel de liquido de freno"
12       pass
13   class Recorrido(Fact):
14       "Distancia sin mantenimiento"
15       pass
16   class Rnivel(Fact):
17       "Rellenaste el nivel"
18       pass
19   class NivelBajo(Fact):
20       "Nivel Bajo"
21       pass
22   class ok(Fact):
23       "Todo bien"
24       pass
25   class MalaB(Fact):
26       "Balatas malas"
27       pass
28   class MU(Fact):
29       "Mantenimiento Urgente"
30       pass
```
Code1. Classes Definition.

Design of Inference Rules: Inference rules capture the knowledge and relationships between the facts in the expert system. These rules define the logic and conditions under which certain actions or recommendations are made. The rules are designed based on the expertise and experience of human experts in the field. The activation and execution conditions for each rule are established to ensure that they are triggered when the necessary conditions are met. Code2 illustrates an example of such inference rules.

```
1    @Rule(NOT(Rpedal(r_ped=W())))
2      def ask_rpedal(self):
3        self.declare(Rpedal(r_ped=ask_info("¿El pedal recorre más de lo normal? (si/no) ")))
4
5      @Rule(NOT(Spedal(s_ped=W())))
6      def ask_spedal(self):
7        self.declare(Spedal(s_ped=ask_info("¿Qué sensación tiene el pedal? (mas duro/mas blando/normal) ")))
8
9      @Rule(NOT(Dfrenar(d_frenar=W())))
10     def ask_Dfrenar(self):
11       self.declare(Dfrenar(d_frenar=ask_info("¿Tarda más en frenar? (si/no) ")))
12
13     @Rule(NOT(Cfrenar(c_frenar=W())))
14     def ask_cfrenar(self):
15       self.declare(Cfrenar(c_frenar=ask_info("Limpia las balatas y el disco, ¿Se escucha un chillido cuando frenas?
16   (si/no) ")))
17
18     @Rule(NOT(Vibracion(vibe=W())))
19     def ask_vibrar(self):
20       self.declare(Vibracion(vibe=ask_info("¿El vehiculo vibra al frenar? (si/no) ")))
21
22     @Rule(NOT(Rfrenar(r_frenar=W())))
23     def ask_Rfrenar(self):
24       self.declare(Rfrenar(r_frenar=ask_info("¿Se escucha un ruido de vacio o como una fuga de aire cuando frenas?
25   (si/no) ")))
26
27     @Rule(NOT(Nivel(nivel_bajo=W())))
28     def ask_Nivel_de_liquido(self):
29       self.declare(Nivel(nivel_bajo=ask_info("¿El nivel del liquido de frenos esta bajo? (si/no) ")))
30
31     @Rule(NOT(Tpastillas(tiempoc=W())))
32     def ask_tiemposincambio(self):
33       self.declare(Tpastillas(tiempoc=ask_info("¿Ha pasado mucho tiempo sin hacerle cambio de balatas? (si/no) ")))
```

Code2. Rules for Forward.

Implementation of the Inference Engine: The inference engine is responsible for controlling the reasoning process of the expert system. It processes the facts and applies the inference rules to make informed decisions or provide recommendations. The engine evaluates the conditions of the rules, matches them with the available facts, and triggers the appropriate rules. The inference engine is implemented using a programming language, such as Python, and suitable libraries or frameworks, such as the "Experta" library mentioned earlier. The engine interacts with the knowledge base, applies the rules, and produces the desired outputs.

Fig. 1. Example of request and engine created.

4.1 Bayesian Rules

The objective of the Bayesian system is initially selected, with the hypothesis being that brake pads need to be replaced, and the evidence being the vehicle's screeching sound when braking. For the probabilistic calculations, a maintenance record was obtained, which includes documented instances of failure symptoms and the corresponding actions taken for repairs. The maintenance record was provided by "Julian's Mechanical Workshop," a well-established automotive maintenance facility with over 20 years of experience in providing general vehicle maintenance services (Table 1).

Table 1. Brake system maintenance record.

Vehiculo No.	El pedal del freno se siente esponjoso/suave.	Se escucha un chirrido al frenar.	Vehiculo tarda en detenerse.	El pedal del freno se hunde hasta el fondo.		Accion
1	no	si	si	no		Cambio de balatas
2	si	si	si	no		Cambio de balatas
3	no	no	si	no		Cambio de cilindro
4	no	no	si	no		Cambio de cilindro
5	no	si	no	no		Exceso de suciedad
6	no	no	no	no		Sin accion
7	si	si	si	no		Cambio de balatas
8	no	no	si	no		Reparacion de Booster
9	si	si	si	no		Cambio de balatas
10	no	no	no	no		Sin accion
11	no	no	no	si		Nivel bajo de liquido
12	si	no	no	si		Reparacion de fuga
13	si	no	no	si		Reparacion de fuga
14	no	si	no	no		Exceso de suciedad
15	si	no	no	no		Cambio de balatas
16	no	si	no	no		Exceso de suciedad
17	no	no	si	no		Exceso de suciedad
18	si	no	si	no		Cambio de balatas
19	si	no	no	si		Reparacion de fuga
20	si	no	si	no		Cambio de cilindro

In the present study, Bayesian reasoning was employed to calculate the probability of the event by dividing it into two components: *LS* (Likelihood of Sufficiency) and *LN* (Likelihood of Necessity). *LS* (Likelihood of Sufficiency): In the context of risk analysis or safety assessment, "Likelihood of Sufficiency" refers to the probability that a quantity or level of resources, safety measures, or any other factor is sufficient to meet certain criteria or standards.

LN (Likelihood of Necessity): could be interpreted as the probability that necessity occurs in certain actions or processes. Necessity generally involves failing to fulfill a duty or responsibility. These components are determined using the following equations [25, 26]:

$$LS = P(E|H)/P(E|\neg H) \tag{2}$$

$$LN = P(\neg E|H)/P(\neg E|\neg H) \tag{3}$$

where:
$P(E|H)$ = Probability that the evidence given the hypothesis.

$P(E|\neg H)$ = Probability that the evidence does not give the hypothesis.

$P(\neg E|H)$ = Probability that the evidence is not present given the hypothesis.

$P(\neg E|\neg H)$ = Probability that the evidence is not present given the non-hypothesis.

By substituting the values into the formula, the expression can be represented as follows:

$$LS = P(E|H)/P(E|\neg H) = 0.66/0.214 = 30.8 \qquad (4)$$

$$LN = P(\neg E|H)/P(\neg E|\neg H) = 0.33/0.785 = 0.42 \qquad (5)$$

To conclude the probabilistic calculation, the following equations need to be applied, which involve probability assignment and normalization in order to obtain a specific probability.

$$O(H) = P(H)/(1 - PH) \qquad (6)$$

$$O(H|E) = LS * O(H) and O(H|\neg E) = LN * O(H) \qquad (7)$$

Applying Bayesian reasoning in the Python-based expert system results in the implementation illustrated in the following figure.

```
1    class BY(KnowledgeEngine):
2
3        @DefFacts()
4        def inicializar_hechos(self):
5            yield cambiopastillas()
6            yield cambiodiscos()
7            yield estado()
8
9        @Rule(NOT(estado(state=W())))
10       def AsKCHILLA(self):
11           self.chilla_ahora=estado(state=ask_info("Chilla al frenar el vehiculo?(si/no)"))
12           self.declare(self.chilla_ahora)
13
14
15
16       @Rule(estado(state="si")| estado(state="no"))
17       def probabilidaddecambiopastillas(self):
18           LS=3.08
19           LN=0.42
20           OS=0.5
21           OS=OS/(1-OS)
22           if self.chilla_ahora['state']=='si':
23               OS_T=OS*LS
24               OS=OS_T/(1+OS_T)
25
26           if self.chilla_ahora['state']=='no':
27               OS_N=OS*LN
28               OS=OS_N/(1+OS_N)
29           self.cambiopastillas=cambio(OSr=OS)
30           self.declare(self.cambiopastillas)
31           respuesta=("Probabilidad de cambio de pastillas de freno %f" %self.cambiopastillas['OSr'])
32           listas_resp.append(respuesta)
33           show_info(respuesta)
34           if self.chilla_ahora['state']=='si':
35               respuesta=("Limpia los discos y las balatas")
36               listas_resp.append(respuesta)
37               show_info(respuesta)
```

Code3. Bayesian reasoning in expert system

4.2 Fuzzy Sets

It is important to note that, in the context of brake maintenance, the kilometers traveled play a crucial role in determining the need for a brake change. On the other hand, the elapsed time is considered as a maintenance standard, even if the vehicle has not traveled a considerable distance. The importance of maintaining, at least once a year, an adequate maintenance of the brake system is highlighted since this constitutes one of the fundamental components to guarantee the safety in the operation of the vehicle.

In this context, the development of an expert system for the maintenance of brakes in vehicles, proceeds in the first instance to the creation of fuzzy sets of inputs. These fuzzy sets represent two universes: one related to the number of kilometers traveled since the last brake maintenance and another concerning the amount of time elapsed in months since said last maintenance, as can be visually appreciated in Figs. 1 and 2, respectively.

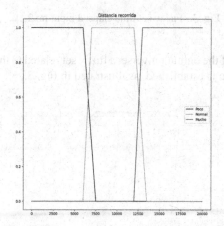

Fig. 2. Fuzzy set kilometers driven input no.1

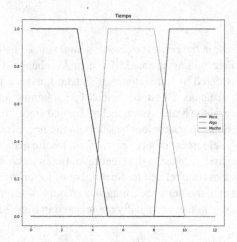

Fig. 3. Fuzzy set time elapsed input no.2.

For the generation of the output universe, a fuzzy set related to the remaining months for the next maintenance is established, as illustrated in Fig. 3.

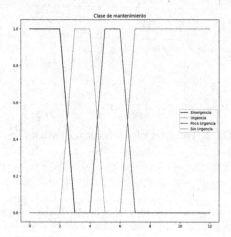

Fig. 4. Fuzzy set remaining months output

4.3 FIS

In the inference process, it was possible to observe that as more time elapses since the last maintenance and as the vehicle travels a greater number of kilometers, the number of months remaining for the next maintenance decreases. That is, these two factors are inversely related to the time remaining to carry out brake maintenance again.

This inverse relationship identified in the rules of inference highlights the importance of periodic and proper maintenance of the brake system. As the time and distance traveled increase, it is essential to be aware of the need for maintenance, since the safety and optimal performance of the vehicle are closely linked to proper care of the braking system. As shown in Fig. 4 (Fig. 5).

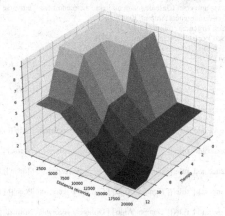

Fig. 5. FIS For the relationship between the inputs with the output.

Everything described above is shown in code4.

```
1    #Fuzzy Logic
2    tiempo_universe = np.arange(0,12.1,0.1)
3    TiempoSinMantenimiento=UPAfs.fuzzy_universe('Tiempo',tiempo_universe,'continuous')
4    TiempoSinMantenimiento.add_fuzzyset('Poco','trapmf',[0,0,3,5])
5    TiempoSinMantenimiento.add_fuzzyset('Algo','trapmf',[4,5,8,10])
6    TiempoSinMantenimiento.add_fuzzyset('Mucho','trapmf',[8,9,12,12])
7    TiempoSinMantenimiento.view_fuzzy()
8
9    #ask_info2 Recorrido
10   distancia_universe = np.arange(0,20000.1,0.1)
11   DistanciaRecorrida=UPAfs.fuzzy_universe('Distancia recorrida',distancia_universe,'continuous')
12   DistanciaRecorrida.add_fuzzyset('Poco','trapmf',[0,0,6000,7500])
13   DistanciaRecorrida.add_fuzzyset('Normal','trapmf',[6000,7000,12000,13500])
14   DistanciaRecorrida.add_fuzzyset('Mucho','trapmf',[12000,13000,20000,20000])
15   DistanciaRecorrida.view_fuzzy()
16
17   #Output
18   mantenimiento_universe = np.arange(0,12.1,0.1)
19   NiveldeUrgencia=UPAfs.fuzzy_universe('Clase de mantenimiento',mantenimiento_universe,'continuous')
20   NiveldeUrgencia.add_fuzzyset('Emergencia','trapmf',[0,0,2,3])
21   NiveldeUrgencia.add_fuzzyset('Urgencia','trapmf',[2,3,4,5])
22   NiveldeUrgencia.add_fuzzyset('Poca Urgencia','trapmf',[4,5,6,7])
23   NiveldeUrgencia.add_fuzzyset('Sin Urgencia','trapmf',[6,7,12,12])
24   NiveldeUrgencia.view_fuzzy()
25
26   #FIS
27   MantenimientoFrenos_Inference = UPAfs.inference_system('Mantenimineto de Frenos')
28   MantenimientoFrenos_Inference.add_premise(TiempoSinMantenimiento)
29   MantenimientoFrenos_Inference.add_premise(DistanciaRecorrida)
30   MantenimientoFrenos_Inference.add_consequence(NiveldeUrgencia)
31   MantenimientoFrenos_Inference.add_rule([['Tiempo','Poco'],['Distancia recorrida','Poco']],['and'],[['Clase de manteni-
     miento','Sin Urgencia']])
32   MantenimientoFrenos_Inference.add_rule([['Tiempo','Poco'],['Distancia recorrida','Normal']],['and'],[['Clase de mante-
     nimiento','Sin Urgencia']])
33   MantenimientoFrenos_Inference.add_rule([['Tiempo','Poco'],['Distancia recorrida','Mucho']],['and'],[['Clase de manteni-
     miento','Urgencia']])
34   MantenimientoFrenos_Inference.add_rule([['Tiempo','Algo'],['Distancia recorrida','Poco']],['and'],[['Clase de manteni-
     miento','Sin Urgencia']])
35   MantenimientoFrenos_Inference.add_rule([['Tiempo','Algo'],['Distancia recorrida','Normal']],['and'],[['Clase de mante-
     nimiento','Poca Urgencia']])
36   MantenimientoFrenos_Inference.add_rule([['Tiempo','Algo'],['Distancia recorrida','Mucho']],['and'],[['Clase de manteni-
     miento','Urgencia']])
37   MantenimientoFrenos_Inference.add_rule([['Tiempo','Mucho'],['Distancia recorrida','Poco']],['and'],[['Clase de manteni-
     miento','Poca Urgencia']])
38   MantenimientoFrenos_Inference.add_rule([['Tiempo','Mucho'],['Distancia recorrida','Normal']],['and'],[['Clase de man-
     tenimiento','Urgencia']])
39   MantenimientoFrenos_Inference.add_rule([['Tiempo','Mucho'],['Distancia recorrida','Mucho']],['and'],[['Clase de man-
     tenimiento','Emergencia']])
40   MantenimientoFrenos_Inference.configure('Mamdani')
41   MantenimientoFrenos_Inference.build()
```

Code4. Code of fuzzy systems with the inference system in python.

The final and crucial stage in this development of the expert system consists in the creation of the graphical interface by using the renowned Tkinter library. As experts in the field of programming, we understand the importance of an attractive, intuitive and functional graphical interface to improve the user experience and ensure application efficiency.

By using the power and versatility of Tkinter, we can design and develop a graphical interface that perfectly suits the needs and objectives of the project. Widely recognized in the developer community, this library offers a variety of widgets and tools that make it easy to create windows, buttons, labels, text boxes, and many other essential visual elements for a robust user interface.

In the process of developing the graphical interface, we focus on ensuring consistency with the visual identity and design standards of the project. Our experience in interface design allows us to create attractive color schemes, efficient element distribution and a harmonious layout that promotes a high-quality and pleasant user experience.

Additionally, by working with Tkinter, we have the ability to implement interactive and dynamic functionality such as user event response, input validation, and application logic management. This real-time responsiveness adds significant value to the application, allowing for fluid and friendly interaction with the user.

```
1    from tkinter import *
2    from tkinter import ttk
3    def center_window(window):
4        window.update_idletasks()  # Actualizar la ventana antes de centrarla
5        screen_width = window.winfo_screenwidth()
6        screen_height = window.winfo_screenheight()
7        x = (screen_width // 2) - (window.winfo_width() // 2)
8        y = (screen_height // 2) - (window.winfo_height() // 2)
9        window.geometry(f"+{x}+{y}")
10   def sendInfo():
11       global information, root, info
12       info = information.get()
13       print(info)
14       root.destroy()
15   def showInfo():
16       global root
17       root.destroy()
18   def closeAllWindows():
19       global root
20       root.destroy()
21   def ask_info(question):
22       global information, info, root
23       root = Tk()
24       root.title("Ask information")
25       Ask_information = StringVar()
26       Ask_information.set(question)
27       Ask_information_label = ttk.Label(root, textvariable=Ask_information)
28       Ask_information_label.pack(pady=5)
29       information = StringVar()
30       information_entry = ttk.Entry(root, width=30, textvariable=information)
31       information_entry.pack(pady=5)
32       Send_information_button = ttk.Button(root, text="Siguiente", command=sendInfo)
33       Send_information_button.pack()
34       # Agregar el botón "Cerrar"
35       Close_button = ttk.Button(root, text="Cerrar", command=closeAllWindows)
36       Close_button.pack()
37       root.update_idletasks()  # Actualizar la ventana antes de obtener las dimensiones
38       root.geometry(f"{root.winfo_width()}x{root.winfo_height()}")
39       center_window(root)
40       root.mainloop()
41       return info
42   def show_info(info):
43       global root
44       root = Tk()
45       root.title("Show information")
46
47       show_information = StringVar()
48       show_information.set(info)
49       show_information_label = ttk.Label(root, textvariable=show_information)
50       show_information_label.pack(pady=5)
51       Send_information_button = ttk.Button(root, text="Siguiente", command=showInfo)
52       Send_information_button.pack()
53       # Agregar el botón "Cerrar"
54       Close_button = ttk.Button(root, text="Cerrar", command=closeAllWindows)
55       Close_button.pack()
56       root.update_idletasks()  # Actualizar la ventana antes de obtener las dimensiones
57       root.geometry(f"{root.winfo_width()}x{root.winfo_height()}")
58       center_window(root)
59       root.mainloop()
```

Code5. tkinter implementation in python code.

Finally, we proceed to implement the SMS interaction with TWILIO to confirm the user's phone number and receive responses from the expert system, allowing efficient access and management of information related to the status of the vehicle's braking system.

The following figure shows the implementation in python code of the interaction code as well as the functions to receive and send SMS.

```
1   from twilio.rest import Client
2   from twilio.twiml.messaging_response import MessagingResponse
3   import time
4   account_sid = 'AC200585a739c26d58bda368d19e478b45'
5   auth_token = '9634eed0e78e8317595cedbffeb15525'
6   client = Client(account_sid, auth_token)
7   def send_sms(ask, numero_a_enviar):
8       message = client.messages.create(
9           from_='+14706137168',
10          body=ask,
11          to= numero_a_enviar
12      )
13      print(message.sid)
14  def recive_sms(numero_a_recibir):
15      time.sleep(20)  # Espera 20 segundos antes de leer los mensajes
16      messages = client.messages.list(from_=numero_a_recibir, limit=5)  # Ajusta el límite según tus necesidades
17      messages_list = list(messages)  # Almacenar los mensajes en una lista
18      if messages_list:
19          ultimo_mensaje = messages_list[0]  # El último mensaje será el primer elemento de la lista
20          respose = ultimo_mensaje.body
21          return str(respose)
22      else:
23          respose = "No se encontraron mensajes."
24          return str(respose)
```

Code6. Code of the callable functions of TWILIO to receive and send messages.

```
1   #Mandar Resultados por SMS
2   Pregunta_Numero = ask_info("Desea recibir las respuestas por SMS")
3   if Pregunta_Numero.lower() == 'si':
4       numero = ask_info("Ingrese su número telefónico")
5       numero = "+52" + numero
6       send_sms('Confirme su número telefónico respondiendo con un: Si',
7   numero)
8       numero_confirmado = recive_sms(numero)
9       if numero_confirmado.lower() == 'si':
10          for respuesta in listas_resp:
11              send_sms(respuesta, numero)
```

Code7. Functions called to send responses after executing the expert system.

5   Result

The results shown below are the results of one input tests, the first uses where the user gives as input that the pedal travels more than normal and that its travel since the last change has been more than normal, as it does not present Another symptom of failure, the probabilistic conclusion reaches natural wear and only brake fluid is added to compensate for the wear of the pad with the fluid inside the piston (Fig. 6).

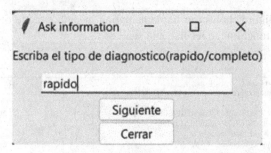

Fig. 6. Choice of diagnosis type.

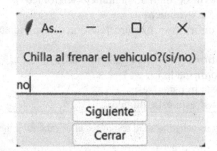

Fig. 7. Distance Traveled Entry.

In this entry the client commented that he has traveled approximately 10,000 km (Fig. 7).

Then the client specifies that he has no problems with noise (Fig. 8).

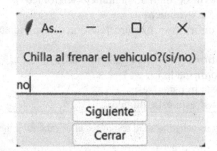

Fig. 8. Input, customer without sound problems

The client mentions that the brake fluid level is below the maximum level, which indicates that the brake pistons are traveling more than normal (Fig. 9).

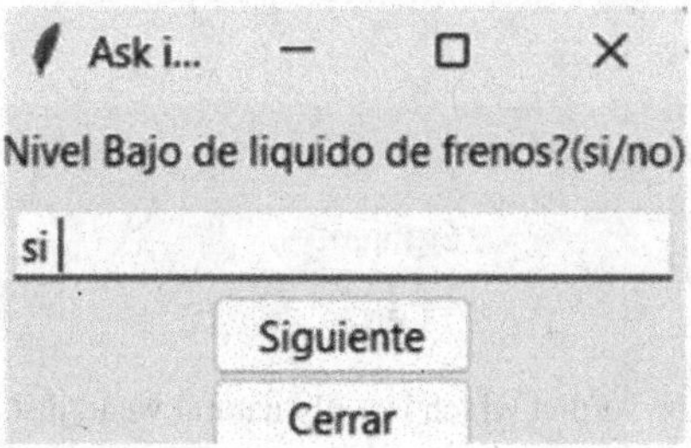

Fig. 9. Positive response of low brake fluid level.

The expert system returns responses based on the database and probabilistic calculations, which shows that it is normal wear since it does not present any other failure symptom other than the low fluid level (Figs. 10, 11 and 12).

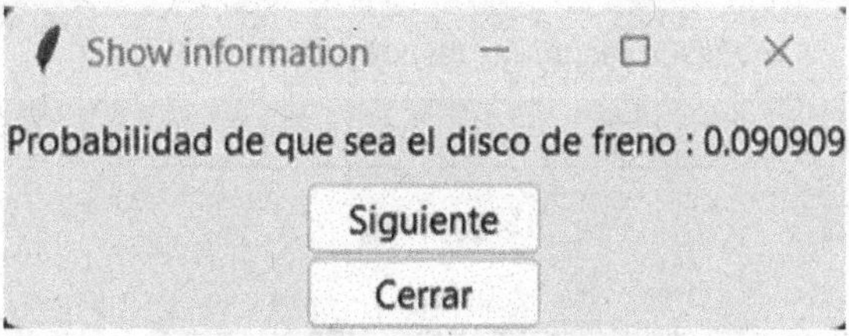

Fig. 10. Shows the probability of changing the brake disc.

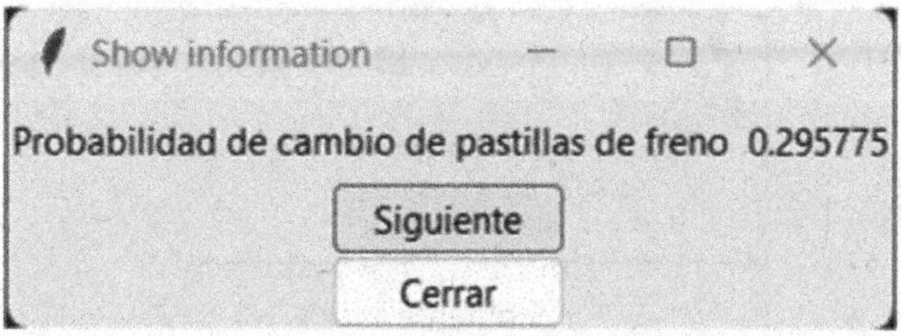

Fig. 11. Shows the probability of changing the brake pads.

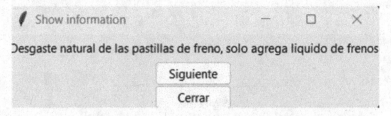

Fig. 12. It shows the verdict which is: only natural wear, just add brake fluid.

Finally, it asks the client if they want the answers by SMS, to which if the client does require it, it will ask them to enter their phone number, then the client has to confirm the number by responding via SMS to obtain the diagnostic answers (Figs. 13, 14 and 15).

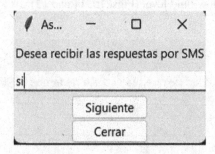

Fig. 13. Affirmative response to the request to send responses via SMS.

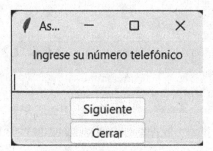

Fig. 14. Box to enter the phone number to receive the answers.

Fig. 15. Confirmation of cell phone number and receipt of answers and diagnosis.

6 Conclusion

In summary, expert systems have proven to be valuable tools in the automotive industry. By combining the specialized knowledge of auto mechanic experts with the power of computational processing, these systems offer accurate diagnosis, component condition assessment, and repair guidelines. They simulate human decision-making, incorporating both actual and heuristic knowledge into their knowledge bases, and utilize inference engines to provide informed recommendations.

Furthermore, the integration of Bayesian reasoning in brake system fault diagnosis shows promising results. The use of probabilistic calculations, based on previous maintenance records, improves accuracy in assessing probabilities of recurring failures and identifying potential solutions, such as brake pad replacement.

Another innovative approach involves combining fuzzy logic and expert systems for vehicle brake maintenance. This method aims to enhance vehicle safety and performance by utilizing intelligent maintenance systems that can adapt to various situations and handle uncertainty in data. It lays the groundwork for developing advanced maintenance systems capable of adjusting to changing conditions and dealing with imprecision in data.

Moreover, the utilization of TWILIO for SMS communication enhances the system's capabilities by providing users with real-time results and maintenance recommendations directly on their mobile devices. This combination of technology ensures a seamless and efficient user experience, allowing users to make informed decisions regarding the maintenance and safety of their brake systems.

References

1. Chapi-Chamorro, E.F., Fraga-Portilla, J.A., Caiza-Quispe, L.: Existing influence on the viscosity of fluids in the anti-lock braking system (ABS). Polo Conocimiento **7**, 619–629 (2022)
2. Borawski, A., Mieczkowski, G., Szpica, D.: Composites in vehicles brake systems-selected issues and areas of development. Materials **16**, 2264 (2023). https://doi.org/10.3390/MA16062264
3. Guerra, S.A.C., Correa, L.A.S., Maigua, D.P.P.: Eficiencia del sistema de frenos en vehículos eléctricos. Open J. Syst. (2022)
4. Bousdekis, A., Lepenioti, K., Apostolou, D., Mentzas, G.: A review of data-driven decision-making methods for industry 4.0 maintenance applications. Electronics (Basel) **10**, 828 (2021). https://doi.org/10.3390/electronics10070828
5. Amirkhani, A., Molaie, M.: Fuzzy controllers of antilock braking system: a review. Int. J. Fuzzy Syst. **25**, 222–244 (2023). https://doi.org/10.1007/S40815-022-01376-Y/METRICS
6. Knaiber, M., Alawieh, L.: Bayesian inference using an adaptive neuro-fuzzy inference system. Fuzzy Sets Syst. **459**, 43–66 (2023). https://doi.org/10.1016/J.FSS.2022.07.001
7. Naik, K.N., Patil, A.R., Patil, K.N., et al.: A python-based grade converter application. In: Proceedings of the 2023 2nd International Conference on Electronics and Renewable Systems, ICEARS 2023, pp. 180–184 (2023). https://doi.org/10.1109/ICEARS56392.2023.10084961
8. Timko, D., Rahman, M.L.: Commercial anti-smishing tools and their comparative effectiveness against modern threats. In: WiSec 2023 - Proceedings of the 16th ACM Conference on Security and Privacy in Wireless and Mobile Networks, pp. 1–12 (2023). https://doi.org/10.1145/3558482.3590173
9. Singh, A.: Evaluating user-friendly dashboards for driverless vehicles: evaluation of in-car infotainment in transition (2023). https://doi.org/10.25394/PGS.23750994.V1
10. Daniyan, I., Mpofu, K., Muvunzi, R., Uchegbu, I.D.: Implementation of artificial intelligence for maintenance operation in the rail industry. Procedia CIRP **109**, 449–453 (2022). https://doi.org/10.1016/J.PROCIR.2022.05.277
11. Bousdekis, A., Lepenioti, K., Apostolou, D., Mentzas, G.: A review of data-driven decision-making methods for industry 4.0 maintenance applications. Electronics **10**, 828 (2021). https://doi.org/10.3390/ELECTRONICS10070828
12. Alamelu Manghai, T.M., Jegadeeshwaran, R., Sakthivel, G.: Real time condition monitoring of hydraulic brake system using naive bayes and bayes net algorithms. IOP Conf. Ser. Mater. Sci. Eng. **624**, 012028 (2019). https://doi.org/10.1088/1757-899X/624/1/012028
13. Arena, F., Collotta, M., Luca, L., et al.: Predictive maintenance in the automotive sector: a literature review. Math. Comput. Appl. **27**, 2 (2021). https://doi.org/10.3390/MCA27010002
14. Le, T.T., Le, M.V.: Development of user-friendly kernel-based Gaussian process regression model for prediction of load-bearing capacity of square concrete-filled steel tubular members. Mater. Struct./Mater. Constr. **54**, 1–24 (2021). https://doi.org/10.1617/S11527-021-01646-5/METRICS
15. ¿Qué es un sistema experto? Usos y aplicaciones en la IA. https://www.unir.net/ingenieria/revista/sistema-experto/. Accessed 24 May 2023
16. Tecnológica Nacional, U., Regional Rosario Autor, F., Juan Manuel, P.: Sistemas Expertos Sistemas Expertos Sistemas Expertos Sistemas Expertos (Expert System) (Expert System) (Expert System) (Expert System) Orientación I: Informática aplicada a la Ingeniería de Procesos 1 Ingeniería Química

17. Horvitz, E.J., Breese, J.S., Henrion, M.: Decision theory in expert systems and artificial intelligence* (1988)
18. Horvitz, E.J., Breese, J.S., Henrion, M.: Decision theory in expert systems and artificial intelligence. Int. J. Approximate Reasoning **2**, 247–302 (1988). https://doi.org/10.1016/0888-613X(88)90120-X
19. Guzmán, J.J.C., Téllez, E.M., Macias, M.G.: Un software analítico de vehículos y un sonido de alerta la salvación de muchas vidas humanas. J. Sci. Res. **7**, 612–633 (2022)
20. Avliyokulov, J.S., Pulatovich, M.S., Rakhmatov, M.I.: Main failures of the vehicle brake system, maintenance and repair. Cent. Asian J. Math. Theory Comput. Sci. **4**, 63–69 (2023). https://doi.org/10.17605/OSF.IO/SMAUF
21. The Basics—experta unknown documentation. https://experta.readthedocs.io/en/latest/thebasics.html. Accessed 24 May 2023
22. Gigerenzer, G., Hoffrage, U.: How to improve Bayesian reasoning without instruction: frequency formats. Psychol. Rev. **102**, 684–704 (1995). https://doi.org/10.1037/0033-295X.102.4.684
23. Ayal, S., Beyth-Marom, R.: The effects of mental steps and compatibility on Bayesian reasoning. Judgm. Decis. Mak. **9**, 226–242 (1930). https://doi.org/10.1017/S1930297500005775
24. Montes Rivera, M., Olvera-Gonzalez, E., Escalante-Garcia, N.: UPAFuzzySystems: a python library for control and simulation with fuzzy inference systems. Machines **11**, 572 (2023). https://doi.org/10.3390/machines11050572
25. Mandel, D.R.: The psychology of Bayesian reasoning (2014). https://doi.org/10.3389/fpsyg.2014.01144
26. Vista de SEDFE: Un Sistema Experto para el Diagnóstico Fitosanitario del Espárrago usando Redes Bayesianas. https://dspace.palermo.edu/ojs/index.php/cyt/article/view/785/687. Accessed 29 June 2023

Use of IoT-Based Telemetry via Voice Commands to Improve the Gaudiability Rate of a Generation Z Pet Habitation Experience

Alberto Ochoa-Zezzatti[1,2], Jose De los Santos[1(✉)], Maylin Hernandez[1], Ángel Ortiz[1], Joshuar Reyes[1], Saúl González[1], and Luis Vidal[1]

[1] UACJ, Universidad Autónoma de Ciudad Juárez, Chihuahua, Mexico
al212177@alumnos.uacj.mx
[2] CADIT, Universidad Anahuac, Naucalpan, Mexico

Abstract. This research seeks to explore the usefulness of IoT-based telemetry for the management of an aquarium, which will contain different fish species in environments ranging from 16 to 22 °C. This is with an emphasis on the perspective of Generation Z, as well as measuring the enjoyment index as a key reference to determine the quality of the experience sought with this work. To manage the habitat optimally, various factors must be considered, such as water quality, lighting, temperature, among other fundamental characteristics for the well-being of a fish. With that said, the goal is to achieve this objective through the creation of a system capable of self-regulating through the proper use of input and output transducers, these being sensors and actuators to give the project the ability to monitor along with good autonomy to remedy an unfavorable environment in case of encountering one. This is to promote superior experiences for today's youth. Similarly, voice commands were implemented to improve the enjoyment index by providing greater convenience to the user. In conclusion, the implementation of IoT-based telemetry along with the addition of voice command seeks to represent a significant advance in the improvement of both a captive habitat, such as an aquarium, and the experience obtained. This research contributes to the field of study of the interaction between humans and animals with the implementation of innovative technologies to improve the quality of life of pets.

1 Introduction

Telemetry is an important tool for monitoring fish behavior in a fish tank. In addition, the use of Internet of Things (IoT) technology can further enhance monitoring and decision making in fish tank management. Fish tanks are a popular means of keeping and caring for a variety of fish species. Proper monitoring of the fish tank is important to ensure the health and well-being of the fish, as well as to maintain a stable environmental balance. Telemetry is a valuable tool for monitoring key fish tank parameters such as water

The original version of this chapter has been revised. this paper important information and one figure was missing. This was corrected. A correction to this chapter can be found at
https://doi.org/10.1007/978-3-031-51940-6_35

temperature, pH and oxygen levels. In addition, the use of IoT technology can improve monitoring efficiency and the ability to make real-time adjustments in two separate components as in: Telemetry in the fish tank refers to the transmission of data from a measuring device to a remote receiver. In the case of a fish tank, measurement devices may include temperature, pH and dissolved oxygen sensors, among others. These sensors may be located directly in the fish tank or in a filtering system connected to the tank. The data collected by the sensors are transmitted to a remote receiver, which can be a mobile device, a computer or a cloud server. Telemetry in the fish tank can be especially useful for monitoring water temperature. Water temperature can affect fish health and behavior. For example, if the water temperature is too high, fish can become stressed and sick. If the water temperature is too low, fish may become inactive and stop feeding. Telemetry can help detect fluctuations in water temperature and take action to correct any problems. Internet of Things in the fish tank IoT technology refers to connecting everyday devices and objects to the Internet. In the case of the fish tank, IoT technology can include the use of Internet-connected devices, such as sensors, cameras, and automatic feeding devices. These devices can collect data about the fish and the fish tank environment and transmit this data to a server in the cloud for analysis and processing. IoT technology in the fish tank can provide valuable information about fish behavior. For example, sensors can monitor fish activity, including their swimming speed and feeding frequency. This information can help fish tank keepers determine if fish are healthy and active. Automatic feeding devices can also help ensure that fish are receiving the right amount of food, even when caretakers are not present. Benefits of telemetry and IoT technology in the fish tank Telemetry and IoT technology can provide a variety of benefits for managing a fish tank. Some of these benefits include detection of contaminants primarily in algae and food debris. The benefits of telemetry and IoT technology in the fish tank are diverse and can be especially relevant to Generation Z, which has grown up in an increasingly technological and connected world. Some of the most prominent benefits include the following: Constant monitoring of their habitat: Telemetry and IoT technology allows for constant monitoring of fish tank parameters and fish behavior, which can help detect any problems early and take immediate action to correct them. This can be especially relevant for Generation Z, as they have been raised in a world where information and feedback are increasingly immediate. Automation: IoT technology enables the automation of certain tasks, such as feeding and changing the water in the fish tank. This can free up time and resources for other aspects of fish tank management and make it more efficient. Generation Z values efficiency and optimization of time and resources. Remote access - Telemetry and IoT technology allows fish tank data to be accessed and adjustments to be made from anywhere at any time via mobile devices or computers. This can be especially useful for Generation Z, who value mobility and flexibility in their lifestyle. In the Fig. 1, we consider different components and stages to our model. Reduced human error: automating tasks in the fish tank using IoT technology can reduce the possibility of human error in fish tank management. This can improve the accuracy and efficiency of fish tank management and provide peace of mind to caregivers. Generation Z values accuracy and precision in tasks and activities. Learning through assisted reinforcement of tacit knowledge: IoT technology can provide a wealth of data and statistics about the fish tank and fish, which can be used to learn more about fish biology and behavior [1].

This can be especially appealing to Generation Z, which has a keen interest in learning and exploring new knowledge. In summary, telemetry and IoT technology can provide a variety of benefits for managing a fish tank, from constant monitoring to automation and reducing human error. These benefits may be especially relevant to Generation Z, which has grown up in an increasingly technological and connected world and values efficiency, accuracy and constant learning.

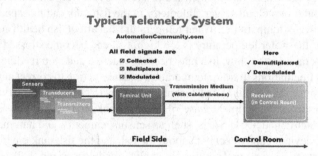

Fig. 1. Components and stages in a Telemetry Model.

2   Gaudability Analysis: An Approach from the Immersive Enjoyment of the Experience

Measuring the Gaudability Index is an important aspect to consider so that Generation Z can enjoy their pets in the fishbowl even more. This index refers to a person's ability to find pleasure in a given activity or experience, in this case, the care and maintenance of a fish tank with different species of fish. IoT technology and telemetry can have a significant impact on the gaudability index by providing a more immersive and engaging experience for the Z generation. Some ways this can be achieved are as follows: Intuitive user interface should be easy to use and navigate so that fish tank keepers can easily access data and adjust settings. Generation Z is accustomed to intuitive user interfaces and expects a similar experience in fish tank management. Access to data and statistics can provide a wealth of data and statistics about the fish tank and fish. Generation Z values access to relevant information and statistics, which can increase their interest and engagement with fish tank management, as is shown in Fig. 2. Notifications and alerts can send real-time notifications and alerts about any problems or changes in the fish tank. This can increase the fish tank caretakers' sense of control and accountability, which can improve their gaudability index [2]. Integration with social networks can provide a more social and shared experience for Generation Z. For example, fish tank keepers can share data, photos and videos of the tank and fish on their social media profiles, which can increase their emotional connection with their pets. Personalization: IoT technology can enable personalization of the fish tank management experience. For example, fish tank keepers can customize the alerts and notifications they receive, or adjust fish tank settings based on the specific needs of each fish species. This can increase the caretakers' sense of control and responsibility and improve their gaudability index. In conclusion, IoT technology and telemetry can have a significant impact on the gaudability index of Generation Z fish tank management. To achieve this, it is important to consider aspects such as an intuitive user interface, access to data and statistics, real-time notifications and alerts, integration with social networks and personalization of the experience. Doing so can enhance Generation Z's emotional connection and engagement with their pets in the fish tank.

3   Telemetry Components Associated with the Project for This Research

In order to determine the most critical components for using IoT-based telemetry in the continuous improvement of a fish tank habitat, several factors need to be considered. First, it is important to understand the specific needs of the different fish species inhabiting the fish tank. Each species has unique requirements in terms of temperature, pH, oxygen level and other parameters. Once the specific needs of each species are understood, the key components needed to effectively use IoT-based telemetry to improve the fish tank habitat can be identified. These components may include the following: Monitoring sensors are essential for measuring and monitoring critical fish tank habitat parameters such as temperature, pH, oxygen concentration, salinity, and others. Monitoring sensors must be accurate and reliable to ensure that accurate data is collected and informed decisions can be made about fish tank habitat management [3, 4]. Some of these sensors include the following: It is impressive how life has thrived from a simple specimen to what we are today as a society, with our values, with our technologies, with each of our unique characteristics. This has led to the creation of distinctive technologies to make life easier for us as well as for those beings who accompany us on this earthly plane. Our project focused on telemetry and the Internet of Things opens doors for us to create devices capable of developing truly functional tools for our daily lives. As mentioned before, the goal was to create an aquarium capable of maintaining the care of 27 fish of different species in a coexistence environment of 47 L, the type of fish intended to be used in this project would be those whose biological characteristics are similar to that of the goldfish, Bronze Corydoras, Tiger Barb, Angel Fish, Shubunkin Goldfish, Suckerfish, Rainbowfish, Harlequin Rasbora, among others where it is known that a healthy coexistence exists among specimens. In the quest to create an optimal environment for the fish in the aquarium, a series of specialized sensors has been selected, each playing a crucial role. Each of these sensors is responsible for monitoring a specific variable, such as the water's pH, temperature, dissolved oxygen level, electrical conductivity, and water turbidity.

The pH sensor (**PH-4502C**) is fundamental in ensuring that the water in the aquarium maintains the appropriate level of acidity, as pH fluctuations can significantly impact the fish's health. This high-precision sensor connects to the ESP32 microcontroller, providing real-time measurements and enabling continuous monitoring of the water's pH [10], as is shown in Fig. 2.

The temperature sensor (**DS181B20**) ensures that the water temperature stays within the optimal range for the fish species in the aquarium. Maintaining the right temperature

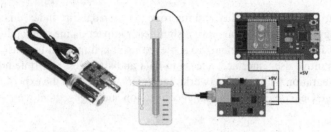

Fig. 2. PH-4502 sensor and its connection simulation.

is essential for the fish's well-being, and this sensor allows for accurate monitoring and the detection of any deviations from the ideal temperature [11], as in Fig. 3.

Fig. 3. DS181B20 sensor and its connection simulation.

The dissolved oxygen sensor (**Gravity DO meter V1.0**) is another essential component for the aquarium, as it measures the concentration of dissolved oxygen in the water. Adequate levels of dissolved oxygen are crucial for the fish's health, and this sensor allows for monitoring and ensuring that the levels remain within a safe range [12], as in Fig. 4.

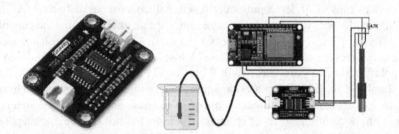

Fig. 4. Gravity DO meter V1.0 sensor and its connection simulation.

The electrical conductivity sensor (**Gravity: analog electrical conductivity sensor**) is used to measure the number of salts or minerals dissolved in the water, which can affect the ionic balance and salinity in the aquarium. Maintaining the right electrical conductivity is essential to ensure the fish are in an optimal environment. [13], as in Fig. 5.

The turbidity sensor (**SEN0189**) is important for assessing the water's clarity in the aquarium.

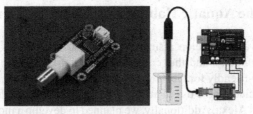

Fig. 5. Gravity: analog electrical conductivity sensor and its connection simulation.

Turbidity can affect the visibility of the fish and the efficiency of filtration systems, so this sensor helps identify and address any water turbidity issues [14], as in Fig. 6.

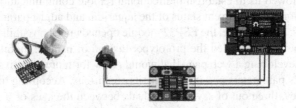

Fig. 6. SEN0189 sensor and its connection simulation.

Lastly, the water level sensor (**T1592**) has been included as an additional safety measure. It detects any water leakage in the aquarium, which could be harmful to the fish. Early detection of a leak is essential to take preventive measures and protect the fish [15], as in Fig. 7.

Fig. 7. T1592 sensor and its connection simulation.

Collectively, these sensors enable comprehensive and continuous monitoring of the aquarium's environment, ensuring optimal care for the fish and a healthy artificial habitat. IoT technology and telemetry play a vital role in this project, allowing real-time data collection and informed decision-making to enhance the aquarium's habitat. The use of specialized sensors exemplifies how technology can improve life for both us and the beings that share our world.

4   Enhancing the Aquatic Habitat Experience for Generation Z

Initially, our goal was to develop a smart aquarium prototype that could be adjusted both directly and remotely, and that could also be controlled using voice commands, making it more user-friendly for Generation Z. To tackle this project, multiple solutions were considered, including the use of an Arduino with a WiFi module to enable remote communication with Alexa. Additionally, we planned to develop a mobile application to facilitate the remote monitoring of the prototype's parameters. As we progressed in the development of the prototype, potential challenges became apparent. Adapting an application for different devices and optimizing it presented significant hurdles. Ultimately, the decision was made to utilize an ESP32 module due to its cost-effectiveness and versatility, which includes integrated Bluetooth and WiFi technology. The built-in WiFi of the ESP32 allowed us to establish bidirectional remote communication with Alexa, enabling us to monitor the current status of the aquarium and adjust parameters through voice commands. Furthermore, the ESP32 module opened up the possibility of changing the way we remotely controlled the prototype. Instead of using a mobile application, we considered developing a web page that would allow for more effective modification of the aquarium's parameters and monitoring its conditions. According to Petco's most recent user survey, three out of seven individuals between the ages of 8 and 24 have a fish tank or wish to purchase one in the near term, with more than 7 fish on average. To improve the gaudability of IoT-based telemetry for Generation Z, it is important to consider the following: User interface design should be easy to use and navigate, with a clear and consistent layout of information. The information should be easy to understand and visually appealing to keep the user's attention. Personalization: IoT-based telemetry should allow users to customize the experience to their needs and preferences. This can include the ability to adjust monitoring sensor settings, receive customized alerts, and set schedules for task automation. Gamification: Gamification can be an effective way to increase the gaudability of IoT-based telemetry for Generation Z. Gamification involves incorporating game elements into the experience, such as rewards, achievements and challenges, which can make the experience more engaging and motivating for the user. Integration with mobile devices: Integration with mobile devices is essential to increase the gaudability of IoT-based telemetry for Generation Z. Users must be able to access and control fish tank telemetry from their mobile devices, allowing them to stay connected at all times. Visual feedback: IoT-based telemetry should provide clear and effective visual feedback to the user. This can include easy-to-understand graphs and charts showing changes in fish tank habitat parameters over time. In summary, to improve the gaudability of IoT-based telemetry for Generation Z in fish tank habitat management, it is important to consider user interface design, personalization, gamification, integration with mobile devices, and visual feedback. Implementing these factors can make the experience more engaging and satisfying for the user, which in turn can improve fish tank habitat management and fish health.

4.1 Inclusion of Voice Commands for Telemetry in an Aquatic Habitat Project

The inclusion of voice commands in an IoT-based telemetry project for fish tank habitat management can significantly improve the user experience, especially for Generation Z, who are accustomed to interacting with devices by voice. Voice commands can make interaction with telemetry more natural and seamless, which in turn can increase the gaudability of the experience [6]. To include voice commands in an IoT-based telemetry project for fish tank habitat management, a voice recognition system would need to be incorporated into the user interface. This voice recognition system should be able to detect the user's voice commands and convert them into actions in the telemetry system. To achieve this, the following steps could be followed: Selection of a voice recognition platform: there are several platforms available for voice recognition, such as Google Assistant, Amazon Alexa or Microsoft Cortana. The platform selection will depend on the compatibility with the telemetry system and the specific speech recognition capabilities needed. Integration of the voice recognition system into the user interface: Once the voice recognition platform has been selected, it needs to be integrated into the telemetry user interface. This may require the creation of a custom interface for voice commands or the integration of voice commands into the existing interface. Design of specific voice commands: Voice commands must be specific and easy for users to remember [8].

Voice commands should be designed for the most common actions the user wants to perform, such as changing the water temperature, turning the light on or off, feeding the fish, among others. Testing and adjustments: Once the voice commands have been integrated into the telemetry, tests should be performed to ensure that the voice recognition system is working properly and that the voice commands are designed effectively [9].

If necessary, adjustments should be made to the user interface and voice commands to improve the user experience. The inclusion of voice commands in an IoT-based telemetry project for fish tank habitat management can significantly improve the user experience, especially for Generation Z. To achieve this, it is necessary to select a voice recognition platform, integrate it into the user interface, design specific voice commands, and perform testing and tuning to improve the user experience.

4.2 The Development of the Prototype

The prototype consists of three main components: a WiFi server enabling communication between devices, a web page, and the internal code of the ESP32 for sensor control. The "WiFi" library was used to create a server on a wireless network, acting as a communication bridge between the ESP32, the web page, and Alexa. For optimal operation, all devices need to be connected to the same WiFi network, which may limit user freedom, but this issue is planned to be addressed in future versions. A webpage was developed in HTML that allows users to monitor parameters such as pH, water temperature, food supply, and water level in the fish tank. In addition to creating the server and running the webpage from the ESP32, the internal code also controls sensors, including the pH sensor and water level sensor, as in Fig. 8.

Regarding the food dispenser, an autonomous mechanism was designed with a clock and a plastic container that dispenses food at regular intervals. For future versions, consideration is given to allowing users greater control over pet feeding scheduling, which

Fig. 8. Test version of the web site, both for computer and cell phone.

can be communicated wirelessly with the help of a servo motor and a plastic container, connecting to Alexa and the webpage. The webpage was designed with aesthetics and adaptability to different devices in mind. HTML and CSS were used for its development, allowing for a visually pleasing presentation and a good user experience on a variety of devices. In summary, the prototype consists of a WiFi server, a webpage, and the internal code of the ESP32 for sensor control and a pet food dispenser. HTML and CSS were used to design the webpage, which provides visually pleasing monitoring of fish tank parameters and is compatible with various devices. Flexibility is planned to be improved in future versions, as in Fig. 9.

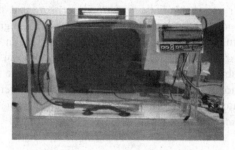

Fig. 9. Prototype already made.

5   Analysis of Project Results

The analysis of results for the project of using IoT-based telemetry using voice commands to improve the Gaudiability Index of the inhabitation experience for a fish tank with 27 fish, one of each species, and a water volume of 47 L, with a water filter and a water pump: In this project, we sought to improve the inhabitation experience for fish in a fish tank using IoT-based telemetry and voice commands. The goal was to increase the Gaudiability index, i.e., to achieve an environment that promotes their well-being and satisfaction. This approach is especially relevant considering the needs of Generation Z, which seeks to interact with their pets in a more technological and innovative way. The fish tank selected for the study had a water volume of 47 L and included 27 fish, each belonging to a different species. An IoT-based telemetry system was implemented that collected data on water temperature, oxygen level and pH using specialized sensors [5]. In addition, a water filter and water pump were installed to maintain the quality of the aquatic environment. When monitoring a fish tank, it is recommended to consider the following seven important variables, along with their respective units of measurement:

Water temperature: Water temperature is a crucial factor in fish welfare. It can be measured in degrees Celsius (°C) or Fahrenheit (°F). Water pH: The pH of the water indicates its acidity or alkalinity. It is measured on a scale of 0 to 14, where 7 is neutral. There is no specific unit of measurement for pH, as it is represented as a numerical value. Dissolved oxygen level: Dissolved oxygen in water is essential for fish respiration. It can be measured in milligrams per liter (mg/L) or parts per million (ppm). The prototype was tested to neutralize the pH in the fish's environment, artificially lowering the pH level in the water drastically. Considering the current situation, Alexa was asked to stabilize the environment by means of an alkaline buffer, thus raising the pH as shown below, in Fig. 10:

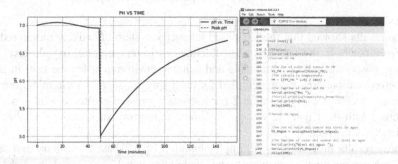

Fig. 10. Ph control test via telemetry with Alexa along with its code.

Ammonia level: Ammonia is a by-product of fish waste and can be toxic in high concentrations. It is measured in milligrams per liter (mg/L) or parts per million (ppm).

Nitrates: Nitrates are another by-product of waste and their accumulation can affect water quality. It is measured in milligrams per liter (mg/L) or parts per million (ppm).

Electrical conductivity: Electrical conductivity provides information on the concentration of minerals and other solutes in water. It is measured in microsiemens per centimeter (μS/cm) or millisiemens per centimeter (mS/cm).

Turbidity: Turbidity refers to the clarity of water and is related to the presence of suspended particles. It is measured in nephelometric turbidity units (NTU) or turbidity units (TU).

These variables provide a comprehensive view of water quality and the fish tank environment. Regular monitoring of these variables will help maintain a proper balance and ensure the well-being of the fish in the tank. During the project, continuous data on water parameters were collected using telemetry sensors.

These data were recorded at regular intervals to analyze fluctuations and trends over time [7]. Observations on fish behavior and Generation Z participants' interactions with their pets were also recorded. Telemetry data were analyzed using statistical techniques to identify patterns and correlations between water parameters. Graphs and visualizations were used to represent changes in temperature, oxygen level, and pH over time. In addition, qualitative analyses of observations of fish behavior and participant responses to interaction with voice commands were conducted, as in Fig. 11.

Fig. 11. Monitoring control of parameters in an Aquarium.

The results showed that the IoT-based telemetry system effectively monitored the water parameters in the tank. Minimal fluctuations in temperature, oxygen level and pH were observed, indicating a stable and suitable environment for the fish. In addition, behavioral patterns were identified for each species, such as swimming, feeding and resting preferences. Generation Z participants expressed greater satisfaction when interacting with their fish via voice commands, suggesting an improved inhabitation experience for the fish.

These results indicate that the implementation of IoT-based telemetry and voice commands can positively contribute to the inhabitation experience for fish in a fish tank. Constant monitoring of water parameters allows maintaining a stable and healthy environment for fish, avoiding abrupt changes that could affect their well-being. In addition, the behavioral patterns observed in each species provide valuable information about their specific needs and allow the environmental conditions to be adjusted accordingly. Interaction via voice commands also triggers responses from the fish, demonstrating further stimulation and enrichment of their environment. Generation Z participants expressed greater satisfaction in being able to interact with their pets in a more intuitive and technological way. This suggests that IoT technology and voice commands provide a more engaging living experience attuned to the preferences of this generation, as in Fig. 12.

It is important to note some limitations of this study. Since a sample of 27 fish, one of each species, was used, the results may not be generalizable to all fish species or to fish tanks of different sizes. Furthermore, although positive responses were observed in fish and Generation Z participants, it is important to further investigate and optimize the implementation of IoT-based telemetry and voice commands to maximize the benefits and minimize any potential negative impacts.

For all of the above, the use of IoT-based telemetry using voice commands in a fish tank with 27 fish of different species and a water volume of 47 L, along with a water filter and water pump, has been shown to improve the inhabitation experience for

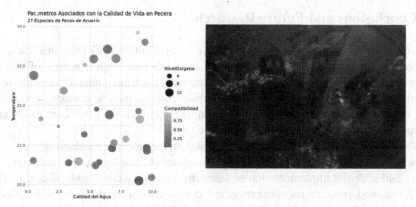

Fig. 12. Final Project associated with an optimal and ideal Aquarium to diverse pet fish species.

the fish and meet the needs of Generation Z. Constant monitoring of water parameters, identification of behavioral patterns and interaction via voice commands have contributed to a more stable, enriched and attractive environment for the fish, and allowed for greater interaction between participants and their pets. These findings support the idea that IoT technology and voice commands can play a significant role in improving the pet-living experience and owner satisfaction, especially among Generation Z. Continuing to investigate and refine this technology in the context of pet habitats may lead to additional advances in human-animal welfare and interaction. In the next Fig. 13 is possible show these levels of reaction in this project.

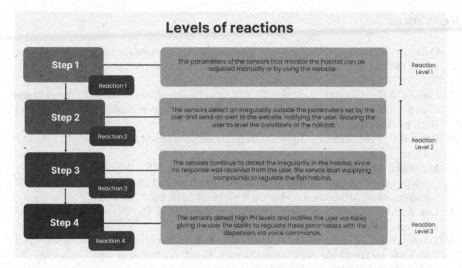

Fig. 13. Diverse levels of reaction in this project associated with habitat control of Aquarium.

6 Conclusions and Future Research

The following are the conclusions and possible areas of future work from this research: The implementation of IoT-based telemetry in a fish tank with a school of fish of seven different species can significantly improve fish tank habitat management and user experience. Key components for telemetry implementation include sensors, monitoring devices, and automation systems. In addition, the inclusion of voice commands can improve the user experience and increase gaudability.

To measure improvement in fish tank habitat management, a gaudability index can be used. It is recommended to measure user satisfaction, fish health, and water quality before and after the implementation of telemetry and voice commands. Based on the results obtained, percentages of improvement in each area can be determined to evaluate the success of the project. Extend the study to fish tanks with larger fish populations and/or more sensitive species.

Develop an early warning system to detect problems in the fish tank habitat and take action automatically. Integrate telemetry with mobile applications for real-time monitoring and remote management. Explore the possibility of using artificial intelligence to improve the accuracy of fish tank parameter measurement. Include self-cleaning systems to ensure water quality on a continuous basis.

Regarding improvement in fish tank habitat management, significant improvement in user satisfaction, fish health, and water quality is expected after implementation of telemetry and voice commands. The percentages of improvement will depend on initial conditions and the specific needs of each fish tank, but are expected to be significant and evidence the importance of IoT-based telemetry in aquatic habitat management.

References

1. Quiñonez, Y., Lizarraga, C., Aguayo, R., Arredondo, D.: Communication architecture based on IoT technology to control and monitor pets feeding. J. Univers. Comput. Sci. **27**(2), 190–207 (2021)
2. Christos, S.C., Nektarios, G., Fotios, G., Nikolaos, D., Panagiotis, P., Areti, P.: Development of an IoT early warning platform for augmented decision support in oil & gas. In: MOCAST 2021, pp. 1–4 (2021)
3. Pöhls, H.C., Petschkuhn, B.: Towards compactly encoded signed IoT messages. In: CAMAD 2017, pp. 1–6 (2017)
4. Mossinger, M., Petschkuhn, B., Bauer, J., Staudemeyer, R.C., Wójcik, M., Pöhls, H.C.: Towards quantifying the cost of a secure IoT: overhead and energy consumption of ECC signatures on an ARM-based device. In: WoWMoM 2016, pp. 1–6 (2016)
5. Lee, J., Kim, D.K.: Smart telemetry monitoring technique for TV transmitter using RF watermark signal. In: ICCE 2019, pp. 1–2 (2019)
6. Tavares, S.A.C., Cavalcanti, R.J.B.V.M., Silva, D.R.C., Nogueira, M.B., Rodrigues, M.C.: Telemetry for domestic water consumption based on IoT and open standards. In: MetroInd4.0&IoT 2018, pp. 1–6 (2018)
7. Thurmer, C.R., et al.: Toward underwater wireless telemetry for inland waterways using low frequency electromagnetic communication. In: WiSEE 2018, pp. 110–112 (2018)
8. Wang, L., Wang, S., Ran, Y.: Data sharing and data set application of watershed allied telemetry experimental research. IEEE Geosci. Remote Sens. Lett. **11**(11), 2020–2024 (2014)

9. Liu, T., Luo, W., Yan, B.: The application of the sensor model language in the HeiHe watershed allied telemetry experimental research. In: GRMSE (1) 2013, pp. 486–49 (2013)
10. Gamiño, A.M., Ruiz, M.S., Acosta, D.: Real-time monitoring of water conditions for fish farming. UAEH (Universidad autonoma del estado de Hiddalgo), pp. 17–19 (2020)
11. Violante, D.: Why is it important to measure the temperature in aquariums? In: HANNA Instruments (2021). https://hannainst.com.mx/blog/por-que-es-importante-medir-la-temper atura-en-los-acuarios/
12. Jordan Press, Noah Press. Gravity analog dissolved oxygen meter. In: AtlasScientific (2021). https://files.atlas-scientific.com/Gravity-DO-datasheet.pdf
13. DFRobot Gravity. Analog electrical conductivity meter V2. In: DFROBOT (2018)
14. Matilde Quintana: Por qué tienes un acuario turbio (y cómo solucionarlo)? In: Carpaskoi (2022–2023). https://carpaskoi.com/acuario-turbio/
15. usar um sensor de nivel de agua. In: Cortocircuito (2021)

Analysis of Convolutional Neural Network Models for Classifying the Quality of Dried Chili Peppers (Capsicum Annuum L)

David Navarro-Solís[1] (iD), Carlos Guerrero-Méndez[2](✉) (iD),
Tonatiuh Saucedo-Anaya[2] (iD), Daniela Lopez-Betancur[2] (iD), Luis Silva[2] (iD),
Antonio Robles-Guerrero[2] (iD), and Salvador Gómez-Jiménez[1] (iD)

[1] Universidad Autónoma de Zacatecas, Unidad Académica de Ingeniería, Zacatecas, México
[2] Universidad Autónoma de Zacatecas, Unidad Académica de Ciencia y Tecnología de la Luz y la Materia (LUMAT), Zacatecas, México
{guerrero_mendez,tsaucedo}@uaz.edu.mx

Abstract. In this paper, an analysis of convolutional neural network (CNN) models to classify the quality of dried chili pepper is described. The classifier models can determine the categories of a set of images that could be encountered in a sorting machine, such as "Extra", "First class", and "Second class" which correspond to different qualities of dried chili peppers. Additionally, two more classes were added as "Trash" and "Empty" which corresponds to cases that could occur in a sorting machine. To determine the best model for image classification, a set of state-of-the-art architectures were compared from the Torchvision library, including ResNet, ResNeXt, Wide ResNet, EfficientNet, and RegNet. The models were trained using feature extraction on the transfer learning approach, and were evaluated using cross-validation method and various advanced metrics such as Precision, Recall, Specificity, F1-score, Geometric mean, and Index of Balanced Accuracy. The results of the cross-validation process indicate that ResNet-152 is the best CNN model for implementation in a sorting machine, with a mean validation accuracy of 95.04%. By using this model, agricultural producers can ensure that their products are sorted according to international standards.

Keywords: Deep learning in agricultural products · Dried chili peppers classification · Visual algorithm in sorting machines

1 Introduction

Dried chili pepper is a widely-used spice in numerous dishes worldwide. It is mainly used in Asian cuisine, but it is also used in Indian and Middle Eastern cuisine, as well as in Mexican food. Given its relevance in soups, sauces, and other dishes, peppers play a pivotal role in the global agricultural economy. As a result of high demand, the pepper market is one of the fastest-growing food markets worldwide. In 2017, the global pepper production was estimated to be around 36,092,631 million tons, with China producing the highest quantity worldwide (17,821,238 tons), followed by Mexico (3,296,875 tons) [1,

H. Calvo et al. (Eds.): MICAI 2023 Workshops, LNAI 14502, pp. 116–131, 2024.
https://doi.org/10.1007/978-3-031-51940-6_10

2]. The quality of dried chilies depends on several factors, such as their size (fragmented or not) and uniform color. Discoloration or brown spots are signs of poor quality. The Mexican Official Norm (NMX-FF-107/1-SCFI-2014) classifies dried chili pepper into three quality levels, namely Extra, First, and Second Class [3]. However, the quality of chili peppers is often graded and sold based on the personal experience of the buyer and seller, leading to disagreements or inequality in the negotiation of this agricultural product.

The Food industry is constantly evolving. One of the biggest changes in this area has been the introduction of automation, which has been used to improve and guarantee the quality and efficiency of food processing operations. Automation in food sorting has been implemented to improve quality, increase efficiency, and reduce operational costs in food companies. Since sorting dried chili peppers is a challenging, labor-intensive, and time-consuming task, the advancement of sorting technologies can be a highly attractive area of interest for many chili pepper marketers.

One of the most promising emerging technologies is Artificial Intelligence (AI). AI is a branch of computing that focuses on creating digital devices capable of performing tasks that require human-like intelligence. One subfield of AI that has recently gained significant success is Deep Learning (DL). DL can be applied to a variety of disciplines, including medicine [4–6], agriculture [7, 8] food processing [9, 10], physics [10, 11], and many others [12–14].

Recently, several visual inspection systems have been developed for precision agriculture and the food industry with the aim of reducing the time and cost associated with manual inspection. Typically, sorting machines or visual inspection systems utilize computer vision and image analysis to detect product anomalies without human intervention. State-of-the-art sorting machines employ machine learning algorithms to automatically detect abnormal products or to identify specific defects with high accuracy, and they are nearly comparable in performance to experienced manual inspectors.

AI and DL algorithms have been used to develop sorting machines to classify dried chili peppers. Sorting machines can differentiate chili samples according to the product size, color, shape, or other properties. The development of these machines is of significant interest to agricultural producers as they can more effectively categorize their agricultural products.

Due to the crucial importance of the classification of peppers, some novel studies, methods and techniques have been developed. In a study conducted by [15], they classified images of red chili pepper according to two categories, worth in chili and no worth in chili (feasible and not feasible). They recorded 80 images of the two quality types of chilies using a smartphone. This database was splitted into 70 images for training and 10 for validation. The authors designed a simple CNN model that achieved a classification accuracy of 80% on the validation dataset. However, when a small dataset is used to train a CNN model, it is common for the model to experience underfitting or overfitting.

Saad et al. [16] developed a system to detect chili (*Capsicum frustecens*) and its flower. They recorded five hundred images of chili plants (with multiple target objects) to train and validate their algorithm. The authors utilized the Faster Regions with ResNet-50 (CNN model) as a feature extractor for the object detection algorithm. Saad et. al. Achieved a detection confidence level of 65%.

Other research by Herdiyeni [17] used the You Only Look Once (YOLO) version 3 object detection algorithm to detect and classify chili peppers. The authors created two classes (A and B) based on a set of parameters of the chili samples. They used a commercial smartphone to record the images. The dataset has 100 photos with 5 peppers per image. The image dataset was split into 80% for training and 20% for validation. They used 10000 iterations to train the detector. Its object detector algorithm achieved 100% accuracy in object detection and 99.4% accuracy in classification tasks. The authors also studied classification under conditions of overlapping red chili peppers and achieved an accuracy of 75.6%.

In addition, Cruz-Dominguez [18] created a classification method for dried chilies by implementing a simple artificial neural network. The authors calculated and used the histogram of the chili image as input of a multilayer perceptron to develop the classification task and obtain the class of the chili as output. In the classification results, they obtained 82.7% accuracy.

Inspired by these related works, the authors of this research are motivated to evaluate (using cross-validation and advance performance metrics) the performance of set of state-of-the-art architectures from Torchvision (open-source computer vision package for the Torch machine learning library) for chili pepper quality using CNN models trained on a large image dataset.

In this paper an analysis of a set of CNN models that can be implement in a dried chili pepper sorting machine is described. The analysis and comparison of the models is performed according to a set of statistical metrics commonly used in DL.

2 Methods, Techniques, and Instruments

This section describes the use of CNN models for the image classification task of dried chili pepper quality grades. It also details the equipment, methods, and performance metrics employed in this study.

2.1 CNN Models

Convolutional neural networks (CNNs) are a type of artificial neural networks that are inspired by the structure of the human visual system, and they are commonly implemented in computer vision tasks [19]. These networks are composed of many blocks that work together to process images (See Fig. 1). Traditionally, CNN architectures consisted of stacked convolutional layers, while the newest layers or architectures look for new and novel ways of constructing convolutional layers in order to improve learning efficiency. Some models are good at learning the features of cars, while others are better at learning the features of dogs, and so on. Therefore, it is important to compare different CNN models to determine the most optimal CNN architecture to use for the quality classification of dried chili peppers.

In order to compare the latest and best models in image classification tasks, the authors implemented the most accurate models that were reported on the Torchvision website, an open-source computer vision package for the Torch machine learning library that has several CNN models available. The next part of this section introduces the ResNet, ResNeXt, Wide ResNet, EfficientNet, and RegNet CNN models that were used.

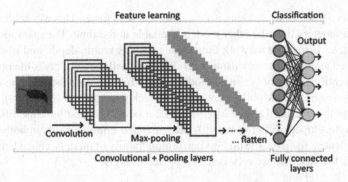

Fig. 1. Architecture of a general CNN to classify dried chili pepper images.

1. *ResNet (ResNet-152)*

   ResNet, short for Residual Network, is a CNN architecture that introduces the skip connection (or shortcut connection). This allows for the development of deeper networks with a large number of layers, avoiding the vanishing/exploding gradient problem [20]. ResNet was the winner of the 2015 ImageNet Large Scale Visual Recognition Challenge (ILSVRC), which evaluates algorithms for large-scale image classification [21]. In this research, we implemented the ResNet-152 version, which means that the model architecture is 152 layers deep.

2. *ResNext (ResNext-101)*

   The ResNeXt model, proposed by Xie et al. [22], also implements the skip connection from the previous block to the next block (like ResNet) and aggregates a set of transformations. This new dimension (hyperparameter) is called "cardinality," which refers to the size of the set of transformations or independent paths. The idea is to stack the same transformation blocks inside the residual block. Experiments have shown that accuracy can be improved more effectively by increasing the cardinality than by deepening or widening the model. ResNeXt was proposed by Facebook AI Research in 2017 and was designed for image classification tasks. It won second place in the 2016 ILSVRC challenge. The model used in this research is the ResNeXt-101-32x8d, which means it has 101 depth layers, 32 cardinality, and 8 basewidth.

3. *Wide ResNet (Wide Resnet-101-2)*

   Wide ResNet is a widened version of the ResNet model and was introduced by Zagoruyko and Komodakis [23] to address the problem of diminishing feature reuse and long training time caused by the increasing number of stacked layers in residual networks. The creators of Wide ResNet reduce the depth and increase the width of residual networks by using wide residual blocks (wide-dropout residual blocks), which are a type of residual block that utilize two Conv 3x3 layers with dropout. Technically, the model is the same as ResNet-101 (100 layers deep) except for the bottleneck number of channels, which is twice as large in each block.

4. *EfficientNet (EfficientNet-B0 and EfficientNet-B7)*

   EfficientNet is a family of image classification models introduced by Tan and Le [24]. Upon their introduction, EfficientNet models outperformed most other models in image classification. The EfficientNet-B7 version achieved higher accuracy on the

ImageNet database, while being 8.4x smaller and 6.1x faster. This was achieved using fewer parameters than the other models available at the time. The main idea behind EfficientNet is to scale network dimensions such as width, depth, and resolution in a balanced and uniform way using a "compound coefficient". This is the opposite of the conventional way of scaling a model using arbitrary scale factors. The Efficient-Net architecture uses Mobile Inverted Bottleneck Convolution (MBConv), which is similar to the one used in the MobileNetV2 model. The first baseline EfficientNet network developed (EfficientNet-B0) demonstrated improvements in both accuracy and efficiency. To create the highest performing family of models, called EfficientNet (B1-B7), the baseline network is scaled up accordingly.

5. *RegNet (X_32gf and Y_32gf)*

   In 2020, Facebook AI researchers Radosavovic et al. [25] published "Designing Network Design Spaces," which introduced a new network design paradigm known as RegNet. They presented a novel low-dimensional design space that produces simple, fast, and versatile networks. This network design combines the advantages of manual design and Neural Architecture Search (NAS), which overcomes the limitations of traditional network design. The process involves parametrizing the population of networks within the design space design. According to the authors, the main purpose of this project is to advance the understanding of network design and discover design principles that generalize across settings. RegNet also uses only one type of network block out of the many different architectures already available, such as the bottleneck block. In experiments, RegNet models outperform EfficientNet models and are up to 5x faster on GPUs. RegNet has two variants: RegNetX, which uses the residual block of the classic ResNet, and RegNetY, which takes advantage of squeeze-and-excite blocks. Although RegNet is not considered a CNN model, it is a design space. In this study, we analyze and implement the pre-trained model using ImageNet, as published on the Pytorch website. Likewise, we will analyze the versions of RegNet X and Y (32 gigaflops) in this research.

2.2 Transfer Learning

Training a CNN model is a highly computationally intensive task. Moreover, training a CNN model from scratch for a classification task requires a large number of labeled images. An alternative to reduce training time and computational resources is to use the Transfer Learning (TL) technique. TL takes advantage of the knowledge that a CNN model has learned in a previous training process and retrains only a portion of the model to perform the new image classification task. Additionally, the final layer is reshaped to have the same number of outputs as the number of classes for this process. Figure 2 illustrates the TL process in a pre-trained model using ImageNet and the same model reused to classify dried chili peppers.

   In this research, the authors implemented the TL approach known as feature extraction, in which all the network layers are frozen except for the final fully connected layer. The last layer is then reshaped and replaced with a new one that has random weights and is retrained using the new database. Table 1 lists the total parameters of the CNN models and the parameters to train after the network freeze. It also lists the size on disk used by the model.

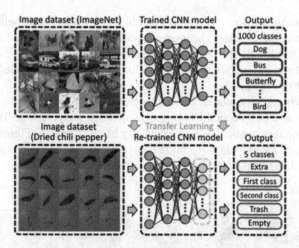

Fig. 2. Architecture of a general CNN to classify dried chili pepper images.

Table 1. Size and total of trainable parameters in the CNN model.

CNN model	CNN with TL (Trainable parameters)	CNN pre-trained with ImageNet (Total parameters)	Size on disk (MB)
ResNet-152	10245	60192808	230
ResNeXt-101-32x8d	10245	88791336	340
Wide ResNet-101-2	10245	126886696	243
EfficientNet-B0	6405	5288548	20.5
EfficientNet-B7	12805	66347960	255
RegNet_x_32gf	12605	107811560	412
RegNet_y_32gf	18565	145046770	554

2.3 Data Acquisition

According to the Mexican Official Norm NMX-FF-107/1-SCFI-2014 [3], the quality of dried chili pepper can be categorized into Extra, First Class, and Second Class. This classification depends mainly on the brightness, color uniformity (with no brown spots), size, and integrity of the dried chili. In order to develop the sorting machine, two additional categories were added: Trash, which refers to any other object that is not a chili pepper, and Empty, which indicates that the conveyor belt is empty and the sorting machine can continue. To train the sorting machine, a set of images of each category was recorded.

To generate the image database, several dried chili peppers were manually classified by an experienced seller of the product. Each sample was placed in front of a suspended camera at a distance of 50 cm to capture the images. The dried chili samples and a white background (same for all samples) were illuminated with an RGB LED lamp positioned next to the camera. The camera and lighting distance were selected to avoid casting shadows on the samples. The image acquisition process was conducted in a dark room using the Toshiba HV-F31F camera, which has a resolution of 1024×768 (H $\times$ V) pixels. However, the images of the dried chili peppers were recorded in an image dimension of 640×480 pixels.

The dataset comprises a total of 2,866 images, which are categorized into 564 Extra, 529 First Class, and 611 Second Class images of dried peppers. The Trash category consists of 541 images, while the Empty category has 621 images. Figure 3 depicts an example from each sample class. To mitigate overfitting risks and enhance the size of the training set, data augmentation was implemented.

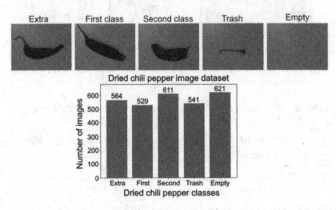

Fig. 3. Number of images of each class

1. *Data Augmentation*

Data Augmentation is the process of obtaining new image samples by applying mathematical operations to raw data (images). The result of this process is new data that can be used to improve the model training.

In this research, three numerical transformations were implemented to generate new images in the training dataset. To create new images, the authors used the torchvision.transforms module from the Pytorch library. The first transformation is RandomRotation, in which the image is randomly rotated by an angle within a range of 0° to 90° degrees. The second transformation is RandomHorizontalFlip, in which the raw image is randomly flipped horizontally. Finally, the last transformation is RandomVerticalFlip, in which the raw image is randomly flipped vertically to create a flipped image. Figure 4 shows examples of the image transformations applied to some images in the dataset.

2.4 Training Parameters

Some important parameters involved in the training process are hyperparameters, which are adjustable parameters in CNN models that determine their behavior and how they will affect performance in the task of interest. They are the "tunable" parts of a model

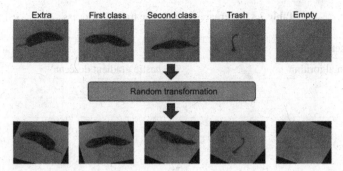

Fig. 4. The original input images and augmented output images, which were generated using the data augmentation process.

during the training process. In addition to their influence on the classification task performance, hyperparameters affect the computing power and time required to train a CNN model. The process of training a CNN model is complex and involves a wide variety of hyperparameters.

In DP, the process of finding the "best" or "optimal" parameters for the performance of a CNN model is called optimization. The classic optimization method is Stochastic Gradient Descent (SGD), a simple procedure that involves iteratively finding the values that will result in the lowest possible error (loss) based on the training dataset. This method has become the gold standard for training CNN models. Although newer and more powerful optimization algorithms exist, SGD provides consistency in the overall training process and results.

One of the most important hyperparameters is the "learning rate". It is responsible for adjusting the rate at which the model calculates the gradient of the loss function. Therefore, the learning rate controls how much the model changes its predictions as it updates its results based on model error. A high learning rate makes the model change its parameters quickly, while a low learning rate makes the model change its parameters slowly. The best option is to select a learning rate value that makes the error decrease correctly (not too quickly) and finds the minimal error in the fewest number of epochs. An "epoch" is another relevant hyperparameter, referring to the number of times the entire training dataset is passed through the CNN model. However, a model is trained using batches. In the context of a single training epoch, "batch size" refers to the amount of data passed to be processed by the CNN and update model parameters at a time until an epoch is complete. Larger batches allow for more computational parallelism and can often lead to better performance. However, larger batches also require more memory and can cause latency when passed into the training function. Finally, the hyperparameter "momentum" is employed to accelerate gradient descent by taking into account a fraction of the previous gradients to update to the current one.

The selection of hyperparameters was established without benefiting any specific model. The hyperparameters used are listed in Table 2.

The developed algorithms were executed using Python 3, and the CNN models were trained using PyTorch (version 1.9.1). An open-source, Python deep learning framework, developed by Facebook that is used by AI researchers and engineers to build and train

Table 2. Hyperparameters in the training process.

Hyperparameter	Value
Optimization algorithm	Stochastic gradient descent
Epochs	50
Batch size	32
Momentum	0.9
Learning rate	0.001
Seed	42
Folds	5

machine learning models. To train and evaluate the CNN models, we used the Torchvision package, which provides a collection of pre-trained models and is also used to build high-quality computer vision applications. In this research, we used Torchvision version 0.10.1. Performance metrics were verified using Imbalanced-learn [26], an open-source Python library that provides tools for classification with imbalanced datasets.

2.5 Performance Evaluation

1. *Cross Validation*

In other words, the accuracy results of a model might only be high when we use a specific dataset and not on other datasets. A more appropriate solution is to create multiple subsets to train and validate a model using a single dataset, and then calculate an average accuracy value. In this research, the authors generated five folds using a single dataset to establish the best model for an image classification task of dried peppers. Figure 5 shows a representation of the cross-validation process using five folds.

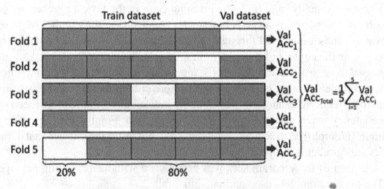

Fig. 5. K-fold Cross-Validation, with k = 5.

2. Confusion Matrix

In the field of machine learning, a confusion matrix is a table used to evaluate the performance of a model. This table is also used to visually summarize the results of the training process. The columns of the table represent the true values, while the rows show the predicted values of the model. The diagonal elements of the matrix indicate the number of cases that were correctly classified, while the off-diagonal elements denote the number of cases that were misclassified. The primary purpose of a confusion matrix is to determine the best set of hyperparameters for a CNN model. Figure 6 shows a multiclass confusion matrix.

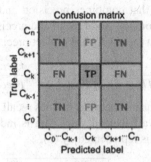

Fig. 6. K-fold Cross-Validation, with k = 5. Representation of a multiclass (n classes) confusion matrix. The class of interest is k. The observation terms are TP (True Positives, blue), TN (True Negatives, purple), FP (False Positive, orange), and FN (False Negatives, brown).

Four relevant terms can be extracted from a confusion matrix. The first is called True Positives (TP), which indicates the number of correctly classified cases. That is, the model predicted that they belong to a class, and they actually belong to the same. Another term is called True Negatives (TN), which refers to the number of elements that have been predicted not to belong to a class and actually do not belong. False Positives (FP) are the number of elements when the model predicts that a case belongs to a class when it actually does not. The last case is False Negatives (FN), which occurs when the model predicts that an element does not belong to a class when in fact it does.

Using the four terms extracted from the confusion matrix, a set of important performance metrics can be calculated. These metrics are known as accuracy, precision, recall, specificity, F1-score, G-mean, and Index of Balanced Accuracy or simply IBA.

The accuracy of the model is the fraction of the total samples that were correctly classified by the model. Equation (1) can be used to calculate accuracy.

$$Accuracy = \frac{TP + TN}{(TP + TN + FP + FN)}. \tag{1}$$

Precision is the ability of a model to correctly classify positive elements. It is represented by Eq. (2).

$$Precision = \frac{TP}{(TP + FP)}. \tag{2}$$

Recall (also called sensitivity) indicates the fraction of positive cases that the model correctly identified as positive. Equation (3) can be used to calculate the recall metric.

$$Recall = \frac{TP}{(TP + FN)}.$$ (3)

Specificity is a metric that indicates the fraction of negative cases that the model correctly predicted as negative. It is defined by Eq. (4).

$$Specificity = \frac{TN}{(TN + FP)}.$$ (4)

The F1-score is a metric that combines precision and recall into a single score. Mathematically, it is the harmonic mean of recall and precision and is expressed as (5). An F1-score of 1 indicates that the model has perfect precision and recall.

$$F1 - score = \frac{2TP}{(2TP + FP + FN)} = 2 \times \frac{Precision \times Recall}{Precision + Recall}.$$ (5)

The geometric mean (G_mean) metric combines recall and specificity into a single metric. This metric produces a balanced value that is independent of the number of positive and negative cases. It is represented by (6).

$$G\_mean = \sqrt{\frac{TP}{TP + FN} \times \frac{TN}{TN + FP}}.$$ (6)

The Index of Balanced Accuracy (IBA) is a metric used to evaluate the performance of a classifier on imbalanced datasets by giving more weight to the positive class (which is generally considered the most important class). In this research, a weighting factor of 0.1 is used [27]. The IBA is represented by (7).

$$IBA = \left(1 + 0.1\left(\frac{TP}{TP + FN} - \frac{TN}{TN + FP}\right)\right)\frac{TP}{TP + FN} \times \frac{TN}{TN + FP}$$ (7)

The metrics presented in this section were used to evaluate and compare the performance of the models in the classification task.

3 Results and Discussion

A cross-validation process was performed on each CNN model using five folds, and the final validation score for each model was used to calculate the mean validation score. The high mean validation score indicates the best model for the image classification task, based on the cross-validation process. Each fold used 80% of the image dataset for training and 20% for validation. The results show that ResNet-152 is the optimal CNN model to be implemented in a dried chili sorting machine. Table 3 lists the validation accuracy of each fold and the mean fold score for each model.

Based on the results presented in Table 3, all models achieved at least 90% accuracy on the validation dataset. However, ResNet-152 achieved the highest mean fold accuracy at 95.04%, closely followed by RegNet_y_32gf at 94.87% and RegNet_x_32gf at 94.66%.

Table 3. Validation accuracy (%) in each fold for every CNN model.

CNN model	Fold 1	Fold 2	Fold 3	Fold 4	Fold 5	Mean Fold
ResNet-152	94.77	95.46	94.76	94.77	95.46	95.04
RegNet_y_32gf	94.77	95.99	94.42	94.24	94.94	94.87
RegNet_x_32gf	95.30	94.59	95.81	93.72	93.89	94.66
EfficientNet-B0	94.43	94.42	93.19	95.11	94.59	94.35
ResNeXt-101-32x8d	92.68	93.37	92.67	92.50	92.67	92.78
Wide ResNet-101-2	93.03	93.72	91.27	90.92	92.84	92.36
EfficientNet-B7	90.59	90.40	90.40	90.23	90.75	90.47

The individual fold values indicate that RegNet_y_32gf achieved the highest accuracy values in Fold 1 and Fold 3. According to Fold 2, the highest value was achieved by RegNet_x_32gf with 98.99%. It should be noted that Fold 2 has the highest value among all folds. EfficientNet-B0 reached the highest accuracy value in Fold 4 with 95.11%. Finally, ResNet-152 achieved the highest value in Fold 5 with 95.46%.

ResNet-152, which has 152 residual layers, demonstrated better performance than its advanced counterparts such as ResNeXt and Wide ResNet, both of which have 101 residual layers. This suggests that, at least in dry chili classification, the performance of the classifier depends mainly on the type and number of images and not on advanced models. Additionally, for this research, a deeper CNN model with simple residual layers resulted in higher accuracy. A similar trend is observed in EfficientNet models, where a simple architecture such as EfficientNet-B0 achieved better results than a more advanced or complex model like EfficientNet-B7. As for RegNet models, their accuracy values demonstrate good overall performance (with RegNet_y_32gf achieving better mean fold accuracy than RegNet_x_32gf). However, the mean fold accuracy indicates a better balance of fold accuracy values for ResNet-152.

Finally, a set of mean performance metrics was calculated using the performance metrics at the epochs with the highest accuracy (one for each fold). The mean performance metrics calculated for each model are listed in Table 4.

The mean scores of the performance metrics listed in Table 4 indicate how well the models predict classes and how balanced or stable their performance is. Therefore, ResNet-152 achieves the best-balanced performance, as it obtains the highest values in all mean values of the performance metrics. It should be noted that RegNet_x_32gf and RegNet_y_32gf also reach an acceptable performance value, while EfficientNet-B7 achieves the worst performance values. Another important aspect to consider when developing a classifier algorithm is the time required to train the CNN model. Before developing an image classifier, we need to assess whether we have the time and computational power to train a CNN model. Table 5 lists the training time required to train the CNN models in this research, sorted by minimum time.

Table 5 shows that the fastest CNN models to train are EfficientNet-B0, ResNet-152, and Wide ResNet-101-2, with mean times of 23.4, 28.76, and 31.81 min per fold, respectively. On the other hand, RegNet models require more training time due to their larger

Table 4. Mean scores of performance metrics (%).

CNN model	Precision	Recall	Specificity	F1-score	G_mean	IBA
ResNet-152	95.04	95.04	98.75	94.98	96.84	93.55
RegNet_y_32gf	94.88	94.87	98.74	94.83	96.75	93.37
RegNet_x_32gf	94.69	94.66	98.69	94.62	96.62	93.11
EfficientNet-B0	94.49	94.35	98.61	94.25	96.37	92.71
ResNeXt-101-32x8d	92.96	92.78	98.27	92.81	95.44	90.78
Wide ResNet-101-2	92.38	92.36	98.11	92.29	95.13	90.18
EfficientNet-B7	90.40	90.47	97.65	90.37	93.91	87.87

Table 5. Training time (min).

CNN model	Fold 1	Fold 2	Fold 3	Fold 4	Fold 5	Mean Fold
ResNet-152	23.53	23.30	23.38	23.48	23.32	23.4
RegNet_y_32gf	29.80	28.53	28.30	28.53	28.63	28.76
RegNet_x_32gf	31.27	31.82	31.88	32.12	31.97	31.81
EfficientNet-B0	33.85	32.83	32.98	33.10	33.15	33.18
ResNeXt-101-32x8d	32.97	33.27	33.92	33.38	33.67	33.44
Wide ResNet-101-2	40.82	40.67	41.32	41.33	41.13	41.05
EfficientNet-B7	43.07	42.63	43.61	44.07	43.68	43.41

number of parameters. As reported in this research, the training times of RegNet models are consistent with the observation that the number of parameters directly influences training time. This is because models with more parameters require more computational resources to calculate the weights and biases of the neurons. Therefore, models with fewer parameters require less time to train, but also require less computing power in the hardware used in the sorting machine. Table 1 provides information on the number of parameters of each model, which can be helpful in choosing the optimal hardware device for implementation. While the RegNet model family achieves high accuracy, their size on disk and number of parameters makes them challenging to implement on devices with low computing power.

During the cross-validation process, a CNN model is trained a number of times equal to the number of folds. For the development of the image classifier, the best trained model of ResNet-152 could be implemented, which was trained at fold 5. The model learning process at fold 5 is shown in Fig. 7. The highest validation accuracy score was obtained at epoch 42, and the confusion matrix at this epoch is shown in Fig. 8. Therefore, if we implement the model (ResNet-152) of fold 5 in a sorting machine, we can expect the performance listed in Table 6 for each class.

Additionally, if we implement a CNN model to classify dried chili peppers, the trained model is expected to achieve mean precision, recall, and specificity scores of 95.52%, 95.46%, and 98.80%, respectively.

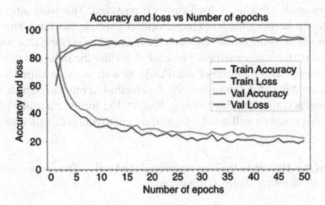

Fig. 7. Training process of the ResNet-152 model in the Fold 5.

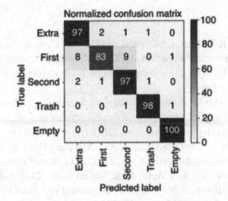

Fig. 8. Confusion matrix of the best epoch in the Fold 5.

Table 6. Performance metrics (%) for epoch 42.

Class	Precision	Recall	Specificity	F1-score	G_mean	IBA
Extra	92.62	96.58	98.03	94.56	97.30	94.54
First	96.20	82.61	99.38	88.89	90.61	80.72
Second	92.70	96.95	97.74	94.78	97.34	94.68
Trash	97.83	97.83	99.58	97.83	98.70	97.25
Empty	98.60	100.00	99.54	99.30	99.77	99.58

4 Conclusions

In this research, the performance of state-of-the-art pre-trained convolutional neural network models (ResNet, ResNeXt, Wide ResNet, EfficientNet, and RegNet) for classifying images of quality grades of dried chilies is analyzed. This work aims to establish the best or optimal network model to implement in a sorting machine. Therefore, each model was trained using the cross-validation method, and its performance was evaluated using advanced performance metrics. The goal of finding the most suitable model was made considering several qualities of dried chili, as well as two additional cases commonly encountered in sorting machines. At the conclusion of this research, the authors achieved a mean accuracy of 95.04% using ResNet-152 for the classification task. It is expected that this research will make a significant contribution to agricultural and food processing.

Contribution of the Authors in the Development of the Work.   The authors declare that they contributed equally to the realization of this research.

Interest Conflict.   The authors declare that there is no conflict of interest.

References

1. Kittler, P.G., Sucher, K.P., Nelms, M.: Food and Culture. Cengage Learning (2016)
2. Russo, V.: Peppers: Botany, Production and Uses. CAB International (2012)
3. NMX-FF-107/1-SCFI-2014. https://www.dof.gob.mx/nota_detalle.php?codigo=5379404& fecha=23/01/2015#gsc.tab=0
4. Lopez-Betancur, D., Bosco Durán, R., Guerrero-Mendez, C., Zambrano Rodríguez, R., Saucedo Anaya, T.: Comparación de arquitecturas de redes neuronales convolucionales para el diagnóstico de COVID-19. Comput. Sist. **25**(3), 601–615 (2021)
5. Ortiz-Rodriguez, J.M., Guerrero-Mendez, C., Martinez-Blanco, M., Castro-Tapia, S., Moreno-Lucio, M., Jaramillo-Martinez, R., Garcia, J.: Breast cancer detection by means of artificial neural networks. Adv. Appl. Artif. Neural Netw. **28**, 161–179 (2018)
6. Sarvamangala, D., Kulkarni, R.V.: Convolutional neural networks in medical image understanding: a survey. Evol. Intell. **15**(1), 1–22 (2022)
7. Maeda-Gutiérrez, V., et al.: Comparison of convolutional neural network architectures for classification of tomato plant diseases. Appl. Sci. **10**(4), 1245 (2020)
8. Too, E.C., Yujian, L., Njuki, S., Yingchun, L.: A comparative study of fine-tuning deep learning models for plant disease identification. Comput. Electron. Agric. **161**, 272–279 (2019)
9. Naranjo-Torres, J., Mora, M., Hernández-García, R., Barrientos, R.J., Fredes, C., Valenzuela, A.: A review of convolutional neural network applied to fruit image processing. Appl. Sci. **10**(10), 3443 (2020)
10. Xu, X., et al.: 11 TOPS photonic convolutional accelerator for optical neural networks. Nature **589**(7840), 44–51 (2021)
11. Guerrero-Mendez, C., Saucedo-Anaya, T., Moreno, I., Araiza-Esquivel, M., Olvera-Olvera, C., Lopez-Betancur, D.: Digital holographic interferometry without phase unwrapping by a convolutional neural network for concentration measurements in liquid samples. Appl. Sci. **10**(14), 4974 (2020)

12. Li, Z., Liu, F., Yang, W., Peng, S., Zhou, J.: A survey of convolutional neural networks: analysis, applications, and prospects (2021)
13. Lopez-Betancur, D., et al.: Convolutional neural network for measurement of suspended solids and turbidity. Appl. Sci. **12**(12), 6079 (2022)
14. Wang, W., Yang, Y., Wang, X., Wang, W., Li, J.: Development of convolutional neural network and its application in image classification: a survey. Opt. Eng. **58**(4), 040901–040901 (2019)
15. Purwaningsih, T., Anjani, I.A., Utami, P.B.: Convolutional neural networks implementation for chili classification. In: 2018 International Symposium on Advanced Intelligent Informatics (SAIN), pp. 190–194. IEEE (2018)
16. Saad, W., Karim, S., Razak, M., Radzi, S., Yussof, Z.: Classification and detection of chili and its flower using deep learning approach. J. Phys.: Conf. Ser. 012055 (2020)
17. Herdiyeni, Y., Haristu, A., Hardhienata, M.: Chilli quality classification using deep learning. In: 2020 International Conference on Computer Science and Its Application in Agriculture (ICOSICA), pp. 1–5. IEEE (2020)
18. Cruz-Domínguez, O., et al.: A novel method for dried chili pepper classification using artificial intelligence. J. Agric. Food Res. **3**, 100099 (2021)
19. Li, Q., Cai, W., Wang, X., Zhou, Y., Feng, D.D., Chen, M.: Medical image classification with convolutional neural network. In: 2014 13th International Conference on Control Automation Robotics & Vision (ICARCV), pp. 844–848. IEEE (2014)
20. He, K., Zhang, X., Ren, S., Sun, J.: Deep residual learning for image recognition. In: Proceedings of the IEEE Conference on Computer Vision and Pattern Recognition, pp. 770–778 (2016)
21. Russakovsky, O., et al.: Imagenet large scale visual recognition challenge. Int. J. Comput. Vision **115**, 211–252 (2015)
22. Xie, S., Girshick, R., Dollár, P., Tu, Z., He, K.: Aggregated residual transformations for deep neural networks. In: Proceedings of the IEEE Conference on Computer Vision and Pattern Recognition, pp. 1492–1500 (2017)
23. Zagoruyko, S., Komodakis, N.: Wide residual networks (2016)
24. Tan, M., Le, Q.: EfficientNet: rethinking model scaling for convolutional neural networks. In: International Conference on Machine Learning, pp. 6105–6114. PMLR (2019)
25. Radosavovic, I., Kosaraju, R.P., Girshick, R., He, K., Dollár, P.: Designing network design spaces. In: Proceedings of the IEEE/CVF Conference on Computer Vision and Pattern Recognition, pp. 10428–10436 (2020)
26. Lemaître, G., Nogueira, F., Aridas, C.K.: Imbalanced-learn: a python toolbox to tackle the curse of imbalanced datasets in machine learning. J. Mach. Learn. Res. **18**(1), 559–563 (2017)
27. García, V., Sánchez, J.S., Mollineda, R.A.: On the effectiveness of preprocessing methods when dealing with different levels of class imbalance. Knowl.-Based Syst. **25**(1), 13–21 (2012)

Fuzzy-Bayesian Expert System for Assistance in Bike Mechanical Issues

Roberto Macías Escobar[1](✉), Martín Montes Rivera[1], and Daniel Macias Escobar[2]

[1] Research and Postgraduate Studies Department in Universidad Politécnica de Aguascalientes, Aguascalientes, Mexico
mc220005@alumnos.upa.edu.mx, martin.montes@upa.edu.mx
[2] Design and Engineering in Esolar S.A. de C.V., Aguascalientes, Mexico

Abstract. Cycling is a popular recreational activity and mode of transportation that offers numerous health benefits and environmental advantages. Individuals of all skill levels can benefit from cycling, from beginners to seasoned professionals. However, as with any sport or activity, cyclists face inherent risks and challenges, ranging from safety concerns to performance optimization. Maintenance and mechanical fixing are standard activities in cyclists but could become so complex if it is the first time the user will perform them. Moreover, they could also be dangerous for professional or amateur users. However, it could be easier to fix mechanical issues in bikes if the user has the tools information, the repair instructions, and the probability of fixing the problem so that the user better let an expert fix it. Alternatively, Expert systems have exhibited skills assisting in different science fields, allowing users to query knowledge with forward and backward chaining to obtain information and answers to diagnose a problem and how to solve it. Additionally, Bayesian theory and fuzzy logic allow working with conditional probabilities and imprecise knowledge in expert systems. In this research, we propose a Fuzzy-Bayesian expert system that helps amateur and professional cyclists diagnose and decide whether to fix the issue themselves or search for an expert. Our proposal, developed in Python, uses UPAFuzzySystems to describe fuzzy rules and Twilio to allow SMS communication to send the report to the user, enabling him to maintain the information at hand while repairing the bike.

Keywords: Expert System · Fuzzy Logic · Bayesian Theory · Diagnose Mechanical Issues

1 Introduction

A bicycle or a bike is a maneuverable vehicle with two wheels operated by the rider's foot-powered pedaling. The propulsion of bikes implies the rotation of pedals connected to cranks and a chain wheel. The power transfer occurs through a continuous loop that links the chain to a sprocket on the rear wheel. Riding a bicycle is a skill that can be quickly acquired, and it requires minimal exertion, making it an accessible mode of transportation [1–3].

© The Author(s), under exclusive license to Springer Nature Switzerland AG 2024
H. Calvo et al. (Eds.): MICAI 2023 Workshops, LNAI 14502, pp. 132–151, 2024.
https://doi.org/10.1007/978-3-031-51940-6_11

Cycling has emerged as a health-promoting activity. The World Health Organization (WHO) identifies that incorporating a daily routine of at least 30 min of exercise helps to decrease the risk of overweight and obesity and coronary heart disease, hypertension, and diabetes. The WHO's recommendations for maintaining health entail 150 min of moderate or 75 min of vigorous physical activity per week. Additionally, WHO urges urban planners to create environments conducive to physical activity and active transportation [4].

Government and large corporate entities strategize to develop less costly transportation infrastructure and modernize public utility vehicles to reduce traffic congestion [5]. Riding a bicycle can benefit these policies, requiring only moderate physical effort.

Besides the benefits, some risks occur to the urban cyclist. We can see that a traffic accident can happen anytime if the vehicles are not in the best condition. Most of these accidents are related to the wheels' problems with brakes. Professionally, the mechanic should think about all the possible failures and be extremely careful with his work because the speed is higher than in the urban way [6].

In this work, we propose a Fuzzy-Bayesian expert system that helps amateur and professional cyclists diagnose and decide whether to fix the issue themselves or search for an expert. Our Expert System offers a guide for cyclists to repair their vehicles in the necessary procedures. The combination of Bayesian and fuzzy logic allows us to manage uncertainty, learn and improve over time, provide transparent reasoning, and take advantage of human knowledge and technology.

1.1 Common Bicycle Mechanical Issues

- Flat Tires: One of the most common issues occurs due to punctures from sharp objects like nails, glass shards, or thorns. Cyclists should carry spare tubes and inflation devices to mitigate the inconvenience of a flat tire.
- Chain Issues: These issues disrupt gear shifting and pedaling efficiency, necessitating prompt maintenance.
- Brake Problems: Brake pads wear down and may become misaligned or contaminated by dirt or oil.
- Gearing Problems: Problems with the derailleurs can lead to difficulty smoothly shifting gears.
- Wheel Misaligned: Bicycle wheels misaligned cause wobbling or an uneven ride.
- Bottom Bracket Issues: creaking or grinding noises during pedaling.
- Headset Problems: can undermine steering and overall control of the bicycle, posing safety concerns.
- Saddle and Seat Support Issues: impact riding position and control.
- Pedal Problems: reduce pedaling efficiency and control.
- Spoke Breakage: can result in wheel instability and requires immediate attention.
- Cables: can corrode over time, affecting the bicycle's ability to shift and brake smoothly.
- Tire Wear: affect grip and rolling resistance [7, 8].

1.2 Expert Systems

Knowledge-based or expert systems are software applications designed to exhibit a level of intelligent performance similar to human experts and are written as IF-THEN rules, some examples of ES are Dendral for chemistry analysis, MYCIN for infection analysis and Deep Blue for playing chest in a good level. Expert systems' essential parts include a knowledge base, a search or inference engine, a knowledge acquisition system, and a user interface or communication system [9].

- Knowledge Base
- Inference Engine: Its primary function is to employ inference rules on the stored knowledge base, thereby generating conclusions [10].
- User Interface [11].

Developing an Expert System corresponds to knowledge engineers— individuals ensure that the computer system possesses the necessary information to address a specific problem. The knowledge engineer must use one or more suitable methods for representing this essential knowledge as symbolic patterns within the computer's memory [12].

1.3 Bayesian Theory

Bayesian statistics is primarily concerned with conditional probability; event A occurs given event B. The concept of conditional probability finds wide application in medical testing, particularly when there is the possibility of false positives and negatives. A false positive is when a medical test indicates a positive result for a disease when, in fact, the patient does not have the disease. In simpler terms, it represents the probability of testing positive when the individual is free of the disease, applying vice versa. Both indicators have significant importance when making informed medical decisions [13].

Mathematically, when a scientist, during the course of an investigation, assigns a probability distribution to a hypothesis H—referred to as the prior probability of H as Pr(H)—and assigns probabilities to the observed evidence E under the assumption of H being true (PrH(E)) and under the assumption of H being false (Pr–H(E)), Bayes's theorem provides a means to compute the probability of the hypothesis H given the observed evidence E using the Eq. (1) [14]:

$$\Pr_E(H) = \frac{\Pr(H) \cdot \Pr_H(E)}{\Pr(H) \cdot \Pr_H(E) + \Pr(\neg H) \cdot \Pr_{\neg H}(E)} \tag{1}$$

1.4 Bayesian Inference

Bayesian inference employs the posterior distribution to derive various summaries for model parameters. These summaries include point estimates like posterior means, medians, percentiles, and interval estimates referred to as credible intervals. Furthermore, all statistical tests regarding model parameters are probability statements based on the estimated posterior distribution.

The Bayesian analysis allows the use of prior information in the analysis, provides an intuitive interpretation of credible intervals as predefined ranges where a parameter is known to belong with a specified probability, and enables the assignment of an actual probability to any hypothesis of interest [15].

Bayesian analysis offers an approach to join prior information with data, all within a robust decision-theoretic framework. This methodology provides exact inferences contingent upon the available data, eliminating the reliance on asymptotic approximations. In small sample scenarios, Bayesian analysis operates similarly to large sample settings, maintaining consistency.

There are some negative points when using Bayesian analysis:

- It does not offer a definitive method for selecting a prior distribution. Determining the appropriate prior is subjective. Bayesian inferences need the ability to translate subjective prior beliefs into a mathematically formulated prior. Careful consideration is essential to avoid generating strange results.
- The priors influence posterior distributions—a problem when we show the result to the experts who may question the validity of the selected prior [16].

1.5 Fuzzy Inference System

Fuzzy inference it's a tool that helps us figure out the right values for a group based on some rules. It's not just a yes or no thing; it's more like a sliding scale of truth. Fuzzy inference helps us connect input and output in a way that helps us make decisions and spot patterns, using Mamdani inference system.

- Fuzzification stage: In this stage, numerical inputs are converted into fuzzy sets.
- Rule base: Each rule specifies a relationship between the fuzzy sets of the inputs and the fuzzy set of the output.
- Combination of rules: This is done by union operations, intersection or using specific fuzzy logic operators, such as maximum or minimum.
- Defuzzification phase: The output fuzzy distribution is converted to a numerical value. It means finding the center of mass of the fuzzy distribution or using other aggregation methods [17].

1.6 Fuzzy Logic

Lotfi Zadeh introduced fuzzy logic during the 1960s. Fuzzy logic represents a computational paradigm that relies on "degrees of truth" instead of the conventional "true or false" (1 or 0). Fuzzy logic emulates human cognition and is extensively employed in AI systems across various fields like vehicle intelligence, consumer electronics, medicine, software development, chemistry, and aerospace [18].

1.7 Tkinter

Tkinter, is a library used to create Graphical User Interface (GUI) applications in Python and it's embedded in his system so it's open access and you get it once you install Python, prevalent for its simplicity and user-friendly nature [19].

Key Applications of Tkinter

- Generating Windows and Dialog Boxes: Create windows and dialog boxes that enable user interaction within your program.
- Constructing a GUI for Desktop Applications: Employed elements like buttons, menus, and various interactive components and create customized widgets tailored to specific needs.
- Rapid GUI Prototyping: Enabling you to experiment with and refine various design concepts before committing to a final implementation [20].

1.8 Twilio

Twilio is a service that allows adding phone calls, messaging, video, and two-factor authentication with a library adapted to Python [21].

2 Method

The tools for developing the proposed rule-based expert system are Visual Studio Code, Python Version 3.8.10, and the Expert Library. These tools allow building the required knowledge rules for implementing the mechanical issues expert system. The 2.1 section describes those rules and shows a diagram of their behavior.

2.1 Diagram

How does the expert system work.

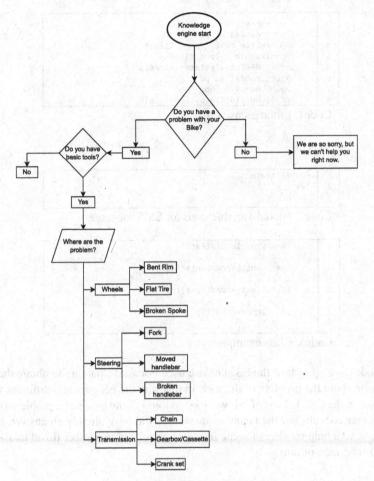

Fig. 1. Flow function diagram.

2.2 Knowledge Engine

To work with this ES the next libraries are required, Experta is the library to used expert systems, also have the way to define de classes as we can see in Code3. Are some examples of how to use the classes. For the text send it in the SMS we need a variable that collect all the information generated by the user while he is using the expert system as we can see in Code2.

```
1    from experta import *
2    import pandas as pd
3    from twilio.rest import Client
4    from IPython import display
5    import UPAFuzzySystems as UPAfs
6    import pandas as pd
7    import numpy as np
8    import matplotlib.pyplot as plt
```

Code1. Libraries used

```
1    global texto
2    texto =""
```

Code2. Global variable used for SMS message

```
1    class Experiencia(Fact):
2        pass
3    class Cant_herramienta(Fact):
4        pass
5    class Efectividad(Fact):
6        pass
7    class Impreso(Fact):
8        pass
```

Code3. Class example

In Code4, we show how the ES start, we used Forward Chaining to obtain the basic information about the problem of the user, in this way the ES goes in a different way as we can see in the Fig. 1. First of all, we want to know if the user has a problem or want to do another consult. For the negative answer we did have already an answer, but the idea is to send a help number or some information related to another travel bike issues, like workshop near of him.

```
1    class Travel(KnowledgeEngine):
2
3        @Rule(NOT(prob(ask1=W())))
4        def preg1(self):
5            self.declare(prob(ask1=input('¿Tienes un problema con tu bicicleta?(si/no)')))
6
7        @Rule(prob(ask1='si'))
8        def preg2(self):
9            self.declare(herra(ask2=input('¿LLevas herramienta basica?(si/no)')))
10
11       @Rule(herra(ask2='si'))
12       def preg3(self):
13           self.declare(herra_si(ask3=input('¿En que zona tienes problema? 1.-LLantas 2.-
14   Direccion  3.-Transmision')))
```

Code4. ES start.

As we can see in Code5, there are some Heuristic rules in a IF-THEN structure. Also, in this part of the code, we are making Forward Chaining to obtain information about the issue with the user's bike. In this part we start the Bayesian Engine to star collection the information about the user problem, and the global variable 'texto' start to appear.

So, we can see that for the Bayesian Engine start we need to have a problem to start it, and then we proceed to the Bayesian Rules.

The Knowledge Engine 'travel' is divided in the tree parts we aim to address to the repair module, that are: wheels, steering and transmission; each one contains his own forward chaining structure.

```
1    @Rule(herra_si(ask3='1'))
2        def preg4(self):
3            self.declare(llanta(ask4=input('¿Que sucede?    1.-Se poncho    2.- Se revento un rayo  3.-Se doblo
4    el rin')))
5                engine1 = Bayesiano_llantas()
6                engine1.reset()
7                engine1.run()
8    @Rule(llanta(ask4='1'))
9        def poncho(self):
10            global texto
11                texto +=  "  'El  procedimiento  para  parcharla  es  el  siguite:\n  Paso_1:\n  \n  Paso_2:\n   \n
12   Paso_3:\n  \n' "
13                print('El procedimiento para parcharla es el siguite:\n Paso_1:\n \n Paso_2:\n \n Paso_3:\n  \n')
14
15    @Rule(llanta(ask4='2'))
16        def preg_rayo(self):
17            self.declare(rayo(ask5=input('En caso de que sean menos de 2 rayos reventados retiralos para evitar
18   perforar la llanta, la bici puede seguir fincionando con precaucion hasta que cambies el rayo, ¿Quieres saber
19   como cambiar un rayo?(si/no)')))
20
21    @Rule(rayo(ask5='no'))
22        def rayo_no(self):
23            print('Muy bien te recomendamos acudir a tu mencanico lo antes posible para que realize las
24   reparaciones pertinentes')
```

Code5. Forward Channing Structure.

In Code6 the backward chaining structure is showed with some examples of application, in this case if at the beginning the user say 'no' for the question about a problem, we pass to this, to avoid all unnecessary information also giving another kind of help for the user, trying to help at the most.

```
1    @Rule(prob(ask1='no'))
2        def preg_bc1(self):
3            print('No podemos ayudarte con algo que no sea una reparacion.....   :(')
4    @Rule(herra(ask2='no'))
5        def preg_bc2(self):
6            self.declare(herra_no(ask17=input('Es necesario para la mayoria de reparaciones contar con algo de
7    herramienta basica, crees que tu problema se puede solucinar sin herramienta?(si/no)')))
8
9    @Rule(herra_no(ask17='si'))
10       def preg_bc3(self):
11            self.declare(val(ask18=input('¿En que zona tienes problema? 1.-LLantas  2.-Direccion    3.-
12   Transmision')))
13
14    @Rule(val(ask18=('14')))
15       def preg_bc4(self):
16            self.declare(llanta_bc(ask19=input('Lo maximo que puedes realizar es alinear el rin, ¿Te puedo
17   ayudar con eso?(si/no)')))
18
19    @Rule(llanta_bc(ask19='si'))
20       def preg_bc5(self):
21            print('Solo tienes que golpear de la parte donde esta mas pandeado y se alineara')
22
23    @Rule(val(ask18='2') | val(ask18='3'))
24       def fin_bc(self):
25            print('Este tipo de reparacion se ocupa forzosamente herramientas, lamentamos no poder ayudarte y
26   esperamos no tengas que caminar tanto.......')
```

Code6. Backward Chaining structure.

2.3 Bayesian Rules

To start working with the Bayesian rules, we need to create a database with the most common bike failures, we didn't find an exact database, so we create one based on research with some experts and some web information then we describe the evidence to formulate the hypothesis later and assign the LN and LS values to feed the equation in case of having a 'yes' answer we are going to use LS and LN in case of 'no' as showed in Code7. These values represent a probability purposed by us thinking in how many events can happened or happened in the database showed in Table 1, these database was made it by a research with the workshops in the area and asking for some cyclist about the most usual problems they get in his trips.

```
1    ˙    class Bayesiano_multiplicacion(KnowledgeEngine):
2             @DefFacts()
3             def inicializar(self):
4                 yield ruido()
5                 yield cambio()
6             @Rule(NOT(ruido(ask22=W())))
7             def preg2(self):
8                 self.juego = ruido(state=input('Esata bien apretada la multiplicacion?'))
9                 self.declare(self.juego)
10            @Rule(ruido(state='si')|ruido(state='no'))
11            def giro_info(self):
12                LS=0.7
13                LN=0.2
14                PH=0.6
15                PH = PH/(1-PH)
16                if self.juego['state']=='si':
17                    PH_T = PH*LS
18                    PH = PH_T/(1+PH_T)
19                if self.juego['state']=='no':
20                    PH_N = PH*LN
21                    PH = PH_N/(1+PH_N)
22                self.giro = cambio(OSr=PH)
23                self.declare(self.giro)
24                global texto
25                texto+= "Hay una posibilidad de '" + str(self.giro['OSr']) + "' de que el eje ya no funcione. Es
26    necesario revisarlo para un mejor analisis.' \n"
27                print('Hay una posibilidad de ' + str(self.giro['OSr']) + ' de que el eje ya no funcione. Es
         necesario revisarlo para un mejor analisis.')
```
Code7. Bayesian rules definition.

2.4 Fuzzy Inference System (FIS)

For this part, the effectiveness of a reparation it's calculated, based on the experience of the user and in the quantity of the tools that they own, this as the input universes or input values, using the UPAFuzzySystems library allow us to process this information in a easy way, were the inference system mechanism determine the information following the Fuzzy Logic Principles as we can see it in [22]. The experience universe goes since less than 2 years, between 2 or 6 years and more than 6 years, the tools are basic, moderated or professional, and as the output universe we have the facility for the user to do the repair as we can see in Code8. Then in Fig. 2, Fig. 3 and Fig. 4 we can see the graphics that these universes create.

Table 1. Problems data.

Failure	Description	Number of failures by 1000 Bicycles	Failure probability	Evidence (E)	Hypothesis (H)	P(H)	LS	LN
Flat tire	Tire puncture, causing air loss	250	25%	Flat tire	Tire puncture causing air leakage	0.7	0.7	0.3
Chain	Chain slipping out of proper position	100	10%	Loose pedals	Loose chain	0.5	0.7	0.5
Brakes	Brakes do not activate or do not brake properly	150	15%	The bike does not brake	Rubber less brake	0.7	0.3	0.6
Spoke wheel	Wheel spokes that break or come out of adjustment	50	5%	Break spoke	Impacted wheel	0.3	0.6	0.4
Derailleur	Derailleur not shifting correctly between the sprockets	100	10%	Desynchronized derailleurs	Loose Cable	0.5	0.2	0.1
Pedal	Pedals that loosen and become unstable	50	5%	Loose pedals	Worn thread	0.4	0.5	0.2
Tire	Tires with worn or worn-out treads	125	20%	Worn tire	Tire with more than 6 months of use	0.4	0.5	0.3
Rim	Wheel not rotating in proper plane, deflecting to one side	50	5%	Rhine bypassed	Received a hit	0.4	0.3	0.1
Bottom Bracket	Connecting rods loosening from the arms	50	5%	Loose Bottom Bracket	The frame or the notch is very worn out	0.6	0.2	0.7
Fork	steering misalignment affecting bicycle stability	75	10%	Loose Fork	Break Bearings	0.5	0.4	0.2

```
1     #Experience
2     Years_universe = np.arange(0,11,1)
3     Experience_Universe=UPAfs.fuzzy_universe('Experience',Years_universe,'continuous')
4     Experience_Universe.add_fuzzyset('Menos de 2 años','trapmf',[0,0,1,2]),
5     Experience_Universe.add_fuzzyset('Entre 2 y 6 años','trimf',[2,5,6])
6     Experience_Universe.add_fuzzyset('Mas de 6 años','trapmf',[5,6,10,10])
7
8     #Number of tools
9     Herram_universe = np.arange(0,4,1)
10    Num_herr_universe=UPAfs.fuzzy_universe('Number of tools',Herram_universe,'continuous')
11    Num_herr_universe.add_fuzzyset('Basica','trimf',[1,1,2])
12    Num_herr_universe.add_fuzzyset('Moderada','trimf',[1,2,3])
13    Num_herr_universe.add_fuzzyset('Profesional','trimf',[2,3,3])
14    Num_herr_universe.view_fuzzy()
15
16    #Repair Facility
17    Reparacion_universe = np.arange(0,101,1)
18    Facilidad_universe=UPAfs.fuzzy_universe('Repair
19 Facility',Reparacion_universe,'continuous')
20    Facilidad_universe.add_fuzzyset('Dificil','trapmf',[0,0,20,30])
21    Facilidad_universe.add_fuzzyset('Normal','trimf',[25,50,70])
22    Facilidad_universe.add_fuzzyset('Facil','trapmf',[60,70,100,100])
      Facilidad_universe.view_fuzzy()
```
Code8. Fuzzy universes defined.

In Fig. 2 for the experience we decide to use a trapezoidal function because the experience sometimes have a flat line if the necessities of the user didn't go further if they don't need it, for example if the user only use the bike for transportation, the repair are going to be only for maintenance, so it's not going to chance easy, so the trapezoid represent these flat lines the user could have.

Looking the Fig. 3 the quantity of tools the user had, we decide to use a triangle function because it the depends about the ability of the user, some specific tools could be replaced by ordinary tools, causing a less probability to get a good result in the repair.

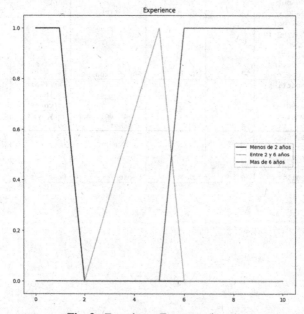

Fig. 2. Experience Fuzzy set (input).

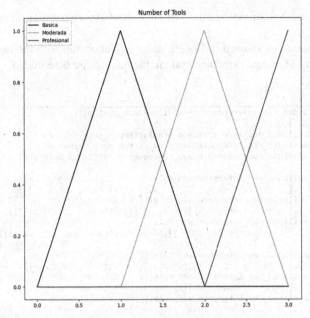

Fig. 3. Tools fuzzy set (input).

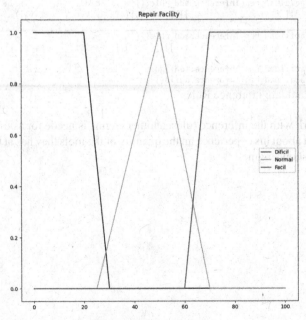

Fig. 4. Facility fuzzy set.

We can determine that theres is a big difference between less than 2 year and have 2 or 6 years for experience, but a small difference if the user have more than 6 years, so the FIS, have a less variation when the user have more experience than when it have less experience.

The inference rules showed in Code9, make a relationship with the two inputs and the output, using Mamdani inference system, the outputs are determined.

```
1   Facilidad_rep_Inference = UPAfs.inference_system('Probability of exit
2   repair')
3       Facilidad_rep_Inference.add_premise(Experience_Universe)
4       Facilidad_rep_Inference.add_premise(Num_herr_universe)
5       Facilidad_rep_Inference.add_consequence(Facilidad_universe)
6
7       Facilidad_rep_Inference.add_rule([['Habilidad','Menos      de      2
8   años'],['Numero','Basica']],['and'],[['Facilidad','Dificil']])
        Facilidad_rep_Inference.add_rule([['Habilidad','Menos      de      2
9   años'],['Numero','Moderada']],['and'],[['Facilidad','Dificil']])
        Facilidad_rep_Inference.add_rule([['Habilidad','Menos      de      2
10  años'],['Numero','Profesional']],['and'],[['Facilidad','Normal']])
11
        Facilidad_rep_Inference.add_rule([['Habilidad','Entre     2    y    6
12  años'],['Numero','Basica']],['and'],[['Facilidad','Dificil']])
        Facilidad_rep_Inference.add_rule([['Habilidad','Entre     2    y    6
13  años'],['Numero','Moderada']],['and'],[['Facilidad','Normal']])
        Facilidad_rep_Inference.add_rule([['Habilidad','Entre     2    y    6
14  años'],['Numero','Profesional']],['and'],[['Facilidad','Facil']])
15
        Facilidad_rep_Inference.add_rule([['Habilidad','Mas        de      6
16  años'],['Numero','Basica']],['and'],[['Facilidad','Normal']])
        Facilidad_rep_Inference.add_rule([['Habilidad','Mas        de      6
17  años'],['Numero','Moderada']],['and'],[['Facilidad','Facil']])

18      Facilidad_rep_Inference.add_rule([['Habilidad','Mas        de      6
19  años'],['Numero','Profesional']],['and'],[['Facilidad','Facil']])
20
        Facilidad_rep_Inference.configure('Mamdani')
        Facilidad_rep_Inference.build()
```

Code9. Mamdani inference Rules.

Next to work with the inference rules, another engine is neede for asking to the user the information about his experience an the quantity of the tools they has at that moment. In Code10 we show engine.

```
1    class EfectividadReparacion(KnowledgeEngine):
2        @DefFacts()
3        def inicializar_hechos(self):
4            yield Experiencia()
5            yield Cant_herramienta()
6            yield Efectividad()
7            yield Impreso()
8
9        @Rule(NOT(Impreso(ya_impreso=W())))
10       def ya_impreso(self):
11           self.impresoval = Impreso(ya_impreso='no')
12           self.declare(self.impresoval)
13
14       @Rule(NOT(Experiencia(state=W())))
15       def Preg_exp(self):
16           self.exp = float(input("Cuantos años tienes de experiencia en reparacion
     de Bicicletas"))
17           self.exp_value = Experiencia(state=self.exp)
18           self.declare(self.exp_value)
19
20       @Rule(NOT(Cant_herramienta(state=W())))
21       def Preg_cant(self):
22           self.tools = float(input("Cuanta herramienta tienes? 0.- Nada 1.-Basica
     2.-Moderada 3.-Profesional"))
23           self.tools_value = Cant_herramienta(state=self.tools)
24           self.declare(self.tools_value)
25
26       @Rule(Cant_herramienta(state=P(lambda x:x<=3) & P(lambda x:x>=0)),
27             Experiencia(state=P(lambda x:x<=10) & P(lambda x:x>=0)),
28             Impreso(ya_impreso='no'))
29       def Facilidad_rep(self):
30           global texto
31           Efectividad_y_facilidad           =
     Facilidad_rep_Inference.fuzzy_system_sim([self.exp,self.tools])
32           texto+=f"El  porcentaje  de  exito  de  la  reparacion  sera
     {Efectividad_y_facilidad[0][0]} %"
33           print(f"El  porcentaje  de  exito  de  la  reparacion  sera
     {Efectividad_y_facilidad[0][0]} %")
34
35           self.modify(self.impresoval,ya_impreso='si')
```
Code10. Engine for the FIS.

Finally, to use the Twilio library, we need to have an account to get the credentials that the code uses to make the API works. The structure of the code is showed in Code11.

```
1    #Twilio credencials
2    account_sid = 'ACcb47466fd62b7d316cdb04712b3680f0'
3    auth_token = '3d45c4f44325acae86a3239dd9d1e161'
4    twilio_phone_number = '+18595458757'
5    #twilio Client
6    client = Client(account_sid, auth_token)
7    #Send Function
8    def enviar_sms(recipeint_number, message_body):
9
10       message = client.messages.create(
11           body=message_body,
12           from_=twilio_phone_number,
13           to=recipeint_number
14       )
15
16       print('Mensaje enviado. SID:', message.sid)
17
18   #Send and Show message
19
20   def recibir_sms():
21       print("Esperando mensaje...")
22       messages = client.messages.list()
23       for message in messages:
24           print('Mensaje recibido:')
25           print('De:', message.from_)
26           print('Mensaje:' , message.body)
27           print('')
28       return messages
29
30   #Execute Code
31
32   messages = recibir_sms
33   enviar_sms("+524495720741" , texto)
```

Code11. Twilio structure.

Here is where the variable 'texto' is needed to show all the information that the ES get and to send the SMS.

Finally the Tkinter code that you can see it in Code12 with the structure is showed, we need to say that to work with Tkinter we need to work in python, —.py file, and not in.ipynb file, to work better and have a clean code we decide to use the Jupyter Notebook files, but in this case we migrate to python file. Also, we have to change the print command for "show_info" and the input command for "ask_info".

```
1   def sendInfo():
2       global information, root, info
3       info = information.get()
4       show_info(info)
5       root.destroy()
6
7   def showInfo():
8       global root
9       root.destroy()
10
11  def preg_fz(question):
12      global information,info,root
13      root =Tk()
14      root.title("Ask information")
15      root.geometry("300x200+20+20")
16
17
18      Ask_information = StringVar()
19      Ask_information.set(question)
20      Ask_information_label = ttk.Label(root, textvariable=Ask_information)
21      Ask_information_label.pack(pady=5)
22
23      information = StringVar()
24      information_entry = ttk.Entry(root, width=30, textvariable=information)
25      information_entry.pack(pady=5)
26
27      Send_information_button = ttk.Button(root,text="Accept",command=sendInfo)
28      Send_information_button.pack()
29      root.mainloop()
30      return info
31
32  def show_info(info):
33      global information,root
34      root = Tk()
35      root.title("Ask information")
36      root.geometry("300x200+20+20")
37
38      show_information = StringVar()
39      show_information.set(info)
40      show_information_label = ttk.Label(root,textvariable=show_information)
41      show_information_label.pack(pady=5)
42
43          Send_information_button    =    ttk.Button(root,    text="Accept",
    command=showInfo)
44      Send_information_button.pack()
45      root.mainloop()
```

Code12. TKINTER structure.

And finally, the way to start the full engine, the code to run is showed in Code13 the order of this code is important because the one that is on the top is the first one to run so we need to put the main engine at the top to start the others.

```
1   engine = Travel()
2   engine.reset()
3   engine.run()
4
5   engine = EfectividadReparacion()
6   engine.reset()
7   engine.run()
```

Code13. Engines starters.

3   Results

This is an example of a person with a flat tire, carries a professional tool, has 3 years of experience and has had this situation happen before with his tire (Figs. 5, 6, 7, 8, 9, 10, 11, 12 and 13).

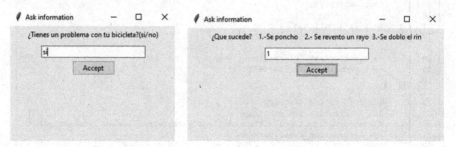

Fig. 5. Ask window for the ES.   **Fig. 6.** Ask window for the ES.

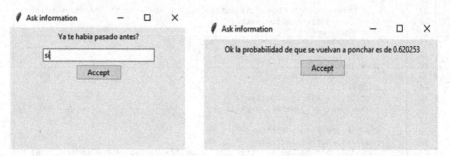

Fig. 7. Ask for the Bayesian rules.   **Fig. 8.** Showing the Bayesian Probability.

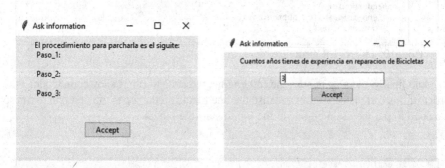

Fig. 9. Window showing the procedure.   **Fig. 10.** Asking the first input for the FIS.

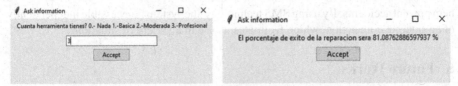

Fig. 11. Asking the second input for the FIS.

Fig. 12. Window showing facility of repair using FIS.

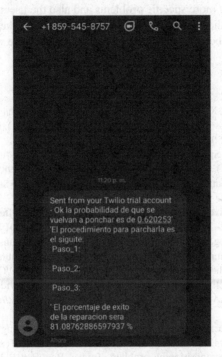

Fig. 13. SMS message sent with all the ES information.

4 Conclusion

In this work, an expert system has been developed to address mechanical problems related to bicycles. The results show user interaction with the system through pop-up windows, obtaining information and generating an SMS message with the probability of repair success and all the information needed to make the repair.

Cyclists, especially those who are not experts in bicycle mechanics, can often feel insecure when facing problems on their bicycle during a trip. This expert system gives cyclists the confidence that they can diagnose and solve problems on their own, which is especially beneficial in situations where cyclists rely on their bikes for daily transportation, as well as proper bike maintenance and timely resolution of mechanical problems help prevent accidents. By using SMS technology, the system is accessible independently of an Internet connection, which facilitates its use in a variety of situations.

5 Future Works

As a future works, thinking this as a mobile App, we want to try the use of Convolutional Neural Networks for image classification, a use the camera of a cellphone to take a picture an determine another kind of problems based on image classification like the rust in the chain, or the rubber level in the brakes. With this function we try to get a more complete ES that could help with more specific problems and also the users could help to build a better Data Base, with the images they get. Also try to improve the data base we already create to this problem, because we do it based on research in the workshop in the area.

Bibliography

1. Mao, G., Hou, T., Liu, X., et al.: How can bicycle-sharing have a sustainable future? A research based on life cycle assessment. J. Clean. Prod. **282**, 125081 (2021). https://doi.org/10.1016/J.JCLEPRO.2020.125081
2. Lee, K., Sener, I.N.: Strava Metro data for bicycle monitoring: a literature review. Transp. Rev. **41**, 27–47 (2021). https://doi.org/10.1080/01441647.2020.1798558
3. Bicycling Science, fourth edition - David Gordon Wilson, Theodor Schmidt - Google Libros. Accessed 25 Sep 2023
4. The potential for active commuting by bicycle and its possible effects on public health. J. Transp. Health **13**, 72–77 (2019). https://doi.org/10.1016/J.JTH.2019.03.012
5. Gatarin, G.R.: Beating the traffic: civil society participation in transport reforms and innovations in Metro Manila, Philippines. In: Advances in 21st Century Human Settlements, pp. 143–158 (2023). https://doi.org/10.1007/978-981-19-8726-7_9/COVER
6. Crash risk and subjective risk perception during urban cycling: evidence for congruent and incongruent sources. Accid. Anal. Prev. **142**, 105584 (2020). https://doi.org/10.1016/J.AAP.2020.105584
7. Shui, C.S., Szeto, W.Y.: A review of bicycle-sharing service planning problems. Transp. Res. Part C Emerg. Technol. **117**, 102648 (2020). https://doi.org/10.1016/J.TRC.2020.102648
8. A multiple type bike repositioning problem. Transp. Res. Part B Methodol. **90**, 263–278 (2016). https://doi.org/10.1016/J.TRB.2016.05.010
9. Aronson, J.E.: Expert systems. In: Encyclopedia of Information Systems, pp. 277–289 (2003). https://doi.org/10.1016/B0-12-227240-4/00067-8
10. Chhaya, K., Khanzode, A., Sarode, R.D.: Advantages and disadvantages of artificial intelligence and machine learning: a literature review. 9–10
11. View of Application of Expert Systems or Decision-Making Systems in the Field of Education. http://it-in-industry.org/index.php/itii/article/view/283/246. Accessed 26 Sep 2023
12. Mubarakali, A., Srinivasan, K., Mukhalid, R., et al.: Security challenges in internet of things: distributed denial of service attack detection using support vector machine-based expert systems. Comput. Intell. **36**, 1580–1592 (2020). https://doi.org/10.1111/COIN.12293
13. Ullman, T.D., Tenenbaum, J.B.: Bayesian models of conceptual development: learning as building models of the world. **2**, 533–558 (2020). https://doi.org/10.1146/ANNUREV-DEVPSYCH-121318-084833
14. Bayesian analysis | statistics | Britannica. https://www.britannica.com/science/Bayesian-analysis. Accessed 29 Jun 2023

15. Kruschke, J.K.: What is not on the BARG Bayesian Analysis Reporting Guidelines. https://doi.org/10.1038/s41562-021-01177-7

16. van Boekel, M.A.J.S.: On the pros and cons of Bayesian kinetic modeling in food science. Trends Food Sci. Technol. **99**, 181–193 (2020). https://doi.org/10.1016/J.TIFS.2020.02.027

17. Kalogirou, S.A.: Designing and modeling solar energy systems. Solar Energy Eng. 553–664 (2009). https://doi.org/10.1016/B978-0-12-374501-9.00011-X

18. Trillas, E., Eciolaza, L.: Fuzzy Logic An Introductory Course for Engineering Students (2015)

19. Python ORG. tkinter — Python interface to Tcl/Tk — Python 3.12.0 documentation. https://docs.python.org/3/library/tkinter.html. Accessed 18 Oct 2023

20. Amos D Python GUI Programming With Tkinter Working With Widgets Displaying Text and Images With Label Widgets Displaying Clickable Buttons With Button Widgets Getting User Input With Entry Widgets Getting Multiline User Input With Text Widgets Assigning Widgets to Frames With Frame Widgets Adjusting Frame Appearance With Reliefs Understanding Widget Naming Conventions Check Your Understanding Controlling Layout With Geometry Managers

21. Stringer, R.: Real-Time Twilio and Flybase. Real-Time Twilio and Flybase (2021). https://doi.org/10.1007/978-1-4842-7074-5

22. Rivera, M., Olvera-Gonzalez, M., Escalante-Garcia, E., et al.: UPAFuzzySystems: a python library for control and simulation with fuzzy inference systems. Machines **11**, 572 (2023). https://doi.org/10.3390/MACHINES11050572

Application of the Few-Shot Algorithm for the Estimation of Bird Population Size in Chihuahua and Its Ornithological Implications

Jose Luis Acosta Roman[1]([⊠]), Carlos Alberto Ochoa-Zezzatti[1,2], Martin Montes Rivera[3], and Delfino Cornejo Monroy[1]

[1] Autonomous University of Ciudad Juarez, Ciudad Juárez, Chihuahua, Mexico
al237866@alumnos.uacj.mx
[2] School of engineering of Anahuac University, Mexico, Mexico
[3] Polythecnic University of Aguascalientes, Aguascalientes, Mexico

Abstract. Biodiversity conservation is a major global concern, some of the problems it faces are simple only in definition and easy to observe as pollution, destruction and habitat fragmentation; and there are more complex ones such as climate change and global warming leading to the loss of biodiversity, but all these factors have something in common, are mainly anthropogenic consequences. For this reason, bioindicators of environmental health are used and this allows us to evaluate the causes and effects in the short or long term. In the state of Chihuahua, Mexico, birds are some of the organisms that fulfill the role of bioindicators, since they are found in large quantities and varieties in various parts of the state, this added to the anthropogenic causes, lead to decrease in population sizes or changes in birds distribution, makes necessary to determine or create a suitable way to estimate birds populations and changes that suffer over time. With all this context, the rapid advances of today's technologies make it possible for them to be used in various sciences, using them as tools that allow deeper analysis in a shorter time, the Few-Shot algorithm is one of these tools, since its application allows estimating population size using images and algorithms for counting and estimating a bird population, allowing decisions to be made with more and better information within ornithology and environmental conservation.

Keywords: Biodiversity · Chihuahua · Few-shot · Algorithm

1 Introduction

In recent years, biodiversity loss has increased significantly, and human activities are the main causes of this loss, including agriculture, livestock raising, deforestation and even technological development, focused on meeting the current needs of society, and which makes extensive use of natural resources [1]. One of the effects associated with the loss of biodiversity is the decrease in the population of species. The latter produces ethological changes, and the displacement of organisms to new areas that have all their habitat requirements in order to maintain their populations [2].

© The Author(s), under exclusive license to Springer Nature Switzerland AG 2024
H. Calvo et al. (Eds.): MICAI 2023 Workshops, LNAI 14502, pp. 152–158, 2024.
https://doi.org/10.1007/978-3-031-51940-6_12

Also, the changes are related to the movements of bird populations generated by the destruction of ecosystems; in these cases, when the population of species in an area is very large and the damage to the ecosystem is small, the population size will decrease slowly but steadily [3]. It should be noted that, if populations decrease, habitat welfare will also decrease, leading to effects that generate biodiversity loss at an accelerated rate and species being classified in a risk category [4].

An action associated with this loss is the release of chemicals into ecosystems, which lead to the generation of pathogens in the environment, which take advantage of the poor conditions of an ecosystem to reproduce and spread to most possible organisms [5]. This is of course potentially harmful to humans, as it leads to an imbalance in the environment and excessive pathogen production affecting organisms of different taxonomic levels and which may be transmitted by other organisms to humans, which has effects on society and its development [6].

Given the effect of anthropogenic actions on ecosystems, research has been carried out to establish the relationship between these actions and the decline in biodiversity and environmental health, to determine which activities, have the greatest impact on the environment. These studies established that the main cause of biodiversity loss is land use for various activities, such as agriculture and livestock farming, followed by direct exploitation of primary resources and mining, and finally general environmental pollution [7]. These anthropogenic effects are observed in the various ecosystems of the planet at different scales and have a direct impact on biodiversity loss [8].

Despite the negative effect that technological advances can have on ecosystems, technological applications have been reported that contribute to the sustainable development model. One application is the ability to estimate populations using algorithms that measure species variations and can even determine the causes of changes [9] and make inferences on the behavior of these by means of computational analyses that allow to monitor in efficient and novel ways this from the analysis of the data [10].

Some tools that have been shown to be useful for determining the behavior of a system based on data flow for habitat analysis and considering the influence of factors such as migration, are machine-learning and deep-learning systems, whose main characteristics allow the analysis of large volumes of data and with various variables that contribute to generate models that efficiently portray the state of natural systems, as well as providing an easy interpretation of the results obtained [11].

It should be noted that, the rapid development of deep-learning techniques allows to create algorithms such as Few-Shot, which is based on the use of images for the recognition and classification of data obtained from an image, labeling each data or body of interest and converting it into a value that allows its processing and interpretation by the computer to provide a statistic that can be compared and evaluated with other data [12].

2 Works Related

2.1 Population Ecology and Population Size Estimation

Population ecology involves many dynamics of interaction between all the organisms found in a specific area and the factors that compose it. These interactive systems are unpredictable because, although specific information is available on that population, biological interactions, such as competition for territory or food, it becomes very complicated to estimate within a pattern the dynamics that occur between the various populations that exist. But with the advance of technologies and algorithms applied to nature allows to create very complex programs or applications that allow to emulate those interactions and analyze the variables so that the analyses represent in a way more the behavior of a system organic [13].

2.2 The Few-Shot Algorithm: Concept and Application

The Few-Shot algorithm turns out to be a very useful tool to estimate the population sizes allowing the obtaining of a lot of information from a small set of data. This allows to build models in an effective way in a short time and with each repetition has a constant improvement when learning from the collected data allowing each photo to be representative for the data that can be relevant to the research reducing the variables which could be dissonant and generate an alteration on the result [14].

These algorithms have applications not only in field research but also within other areas such as health, these can be applied for the recognition of diseases in medical images that allow to provide a more concrete diagnosis and that allows to eliminate or reduce ambiguities in the images found [15].

In ornithology, these algorithms have a great field of exploitation and constant improvement, birds have very distinctive morphological characteristics among species, ranging from the shapes of their beaks to the patterns of colors they have, what is intended with these tools is to provide information that is relevant to the evaluation of bird populations and to generate an estimation model of the current population size to establish a direct relationship with the factors that alter the number of birds in that area and determine the actions to be carried out for the conservation of species and evaluation of all anthropogenic and natural effects that could cause these variations between populations [16].

The Few-Shot algorithm seeks to improve the identification of organisms that are in a certain area with little information obtained through photographs, which allows shortening the time of analysis and interpretation of data and taking advantage of deep-learning algorithm, significantly reducing misinterpretation of data. Unlike Many-Shot methods, which require the full interpretation of everything found in the images and analyze in depth if you capture what this tool needs, the few-shot algorithm allows to obtain the same or even greater amount of information delimiting and recognizing the subjects of interest quickly for their processing and labeling of the organisms shown, making the analysis more efficient, that the interpretation of data is simpler and allows to make a decision on which conservation methods can be applied to the organisms of interest [17] (Fig. 1).

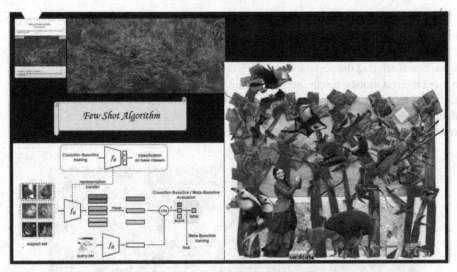

Fig. 1. Diagram showing the application of the Few-Shot model for the recognition of the birds of the state of Chihuahua.

2.3 Implications for Ornithology and Conservation

At present, it is undeniable that the use of technologies has quite high potential in many sciences and in the conservation of biodiversity is not the exception, you can select a device or several and connect them to a computer, establish which software will be used and how the data will be handled to then determine the algorithms and methods that will be applied to evaluate the state of the environment or the organisms that live in a certain area and which factors are those that have a greater impact by measuring them using sensors, images or other environmental sensitivity devices [18].

Birds have shown to be good ecological indicators, this is because most of them have migration periods during the year, which allows to evaluate in a relatively short period of time the conditions of the areas in which they are located. There are indices that are used to explain the structure, composition, function, abundance and richness of the flocks over time, which allows establishing a correlation between the groups of birds and the environment in which they are found, providing information about the conditions and changes that occurred during that time and the human activities that directly interfere transforming the environment, allows the evaluation of the activities that caused the decrease in population quantities or movement of birds to another area due to anthropogenic causes [19].

Thanks to technological advances it is possible to determine migratory routes and hot spots of flocks of various birds, this allows the establishment of a relationship between the presence of birds and the environmental conditions of the area to which they arrive, and migration, as an annual process, is a perfect tool to make constant estimates on the dependent variables and those that change with greater constancy in a relatively short period of time and with the application of algorithms a relation can be established on the decrease of the sizes population, movement of birds to new areas and dwindling natural resources make population observation and evaluation increasingly complicated [20].

The state of Chihuahua is of great importance for the investigation of the diversity of birds since it is a migration point in which a large number of these organisms arrive to the state thanks to the wide distribution of meadows and forests, however, in recent years it has been seen that the population of birds decreases over the years what has developed a great interest among researchers to establish a relationship between the conditions of the areas to which they arrive and if they are natural factors or anthropogenic factors that promote the decline of bird populations in the state to generate data that allow responsible development for the conservation of birds and their habitat [21] (Fig. 2).

Fig. 2. Population of Guacamayas of Chihuahua State

3 Conclusions, Challenges and Ethical Considerations Associated with This Topic

The technology plays a fundamental role nowadays, allowing the analysis of large amounts of data in a shorter time which is quite an important advance talking specifically about topics related to biology, the use of the Few-Shot algorithm for estimating the quantities and sizes of bird populations in the state of Chihuahua represents an important advance in ornithology, species conservation and the environment. This allows us to take advantage of the current power of computers and their machine learning for population estimates by removing the human bias, subjectivity and perception, which we have on that specific area but we understand that the human being is not only interpreting the data, is the one who determines the patterns and provides adequate and clear information for technology to be a support, which leads to the creation of more accurate and efficient methods for the actual estimation of population sizes, which has a very important impact when making decisions based on the actual description of the organisms in the area and how they behave over time. However, ethical and technical considerations must also be considered to ensure the appropriate and responsible use of these technologies with the aim of caring for and benefiting birds and their habitat.

Despite the extensive and novel applications that technology and algorithms have shown to have for the study of population estimates and identification of individuals with little information, there are several challenges and ethical considerations that must be taken into consideration, one of these considerations is that they are simple algorithms to use, but very complex in understanding what they are doing and why they do it in that way which means that the learning time has to be fast enough to possess the necessary knowledge to interpret and make decisions in a correct way, however it is difficult for an algorithm to have an adaptability when something that was not determined in its programming and this ends up generating a bias in part of the information and that is also limited by the power of the equipment with which research is being carried out [17].

Image detection programs and algorithms have proven to be very helpful but there is still much to be taken into consideration for their correct application, especially in areas of study such as biodiversity, as it must be sufficiently complex to provide relevant data with little information, which has shown that situations such as poor detection or labelling of organisms in images sometimes occur, if there are organisms in the images that are very similar can give misinterpretations of the data, and are subject to human error at the time of whether the image is clear enough to obtain the information that is needed, but understanding these points you can find areas in which improvement is needed and understand that there must be a balance between the application of technological tools in the study of living organisms but the factor in which the researcher determines the parameters of the research and is not only an interpreter of results [22] (Fig. 3).

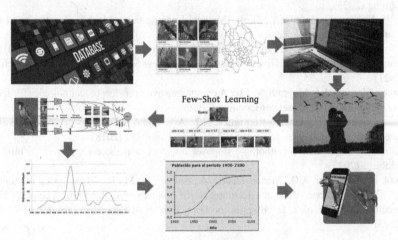

Fig. 3. Diagram showing the phases of the project.

References

1. Rocha, J.R., Martín M.P., Velasco, M.V.: La paradoja de la pérdida de biodiversidad y la aparición de nuevas formas de vida, ligadas a efectos antrópicos (2023)
2. Pecl, G.T., et al.: Biodiversity redistribution under climate change: impacts on ecosystems and human well-being (2023)
3. Chase, J.M., Blowes, S.A., Knight, T.M., Gerstner, K., May, F.: Ecosystem decay exacerbates biodiversity loss with habitat loss (2023)

4. Ali, J.R., Blonder, B.W., Pigot, A.L., Tobias, J.A.: Bird extinctions threaten to cause disproportionate reductions of functional diversity and uniqueness (2023)
5. Keesing, F., Ostfeld, R.S.: Impacts of biodiversity and biodiversity loss on zoonotic diseases (2020)
6. Meena, P., Jha, V.: Environmental Change, Changing Biodiversity, and Infections–Lessons for Kidney Health Community (2023)
7. Jaureguiberry, P., et al.: The direct drivers of recent global anthropogenic biodiversity loss (2022)
8. Sih, A., Ferrari, M., Harris, D.: Evolution and behavioural responses to human-induced rapid environmental change (2011)
9. Feng, J., Li, J.: An Adaptive Embedding Network with Spatial Constraints for the Use of Few-Shot Learning in Endangered-Animal Detection (2022)
10. Haipeng, W., Sizhe, C., Feng, X., Ya-Qiu, J.: Application of deep-learning algorithms to mstar data (2015)
11. Ullo, S.L., Sinha, G.R.: Advances in smart environment monitoring systems using IoT and sensors (2020)
12. Tang, B., Pan, Z., Yin, K., Khateeb, A.: Recent advances and Deep Learning in Bioinformatics and Computational Biology (2019)
13. Fisher, D.N., Pruit, J.N.: Insights from the study of complex systems for the ecology and evolution of animal populations (2019)
14. Tian, S., Lie, L., Lia, W., Ran, H., Ning, X., Tiwarif, P.: A survey on few-shot class-incremental learning (2023)
15. Latif, J., Xiao, C., Imran, A., Tu, S.: Medical Imaging using Machine Learning and Deep Learning Algorithms: A Review (2019)
16. Alayrac, J.-B., et al.: Flamingo: a Visual Language Model for Few-Shot Learning (2019)
17. Leng, J., et al.: A Comparative Review of Recent Few-Shot Object Detection Algorithms (2021)
18. Lahoz-Monfort, J.J., Magrath, M.J.L.: A comprehensive overview of technologies for species and habitat monitoring and conservation (2021)
19. Salas Correa, A.D., Mancera-Rodriguez, N.J.: Aves como indicadoras ecológicas de etapas sucesionales en un bosque secundario, Antioquia, Colombia (2018)
20. Robinson, S.K.: Radar ornithology, stopover hotspots, and the conservation of migratory landbirds (2023)
21. Pool, D.B., Panjabi, A.O., Macias-Duarte, A., Solhjem, D.M.: Rapid expansion of croplands in Chihuahua, Mexico threatens declining North American grassland bird species (2013)
22. Luccioni, A.S., Rolnick, D.: Bugs in the Data: How ImageNet Misrepresents Biodiversity (2023)

Searcher for Clothes on the Web Using Convolutional Neural Networks and Dissimilarity Rules for Color Classification Using Euclidean Distance to Color Centers in the HSL Color Space

Luciano Martinez[1]([✉]), Martín Montes[1], Alberto Ochoa Zezzatti[2,3], Julio Ponce[4], and Eder Guzmán[1]

[1] Research and Postgraduate Studies, Universidad Politécnica de Aguascalientes, Calle Paseo San Gerardo #201, Fracc. San Gerardo, 20342 Aguascalientes, Mexico
lucianomrtnz@gmail.com
[2] Universidad Autónoma de Ciudad Juárez, Ciudad Juárez, Mexico
[3] Centro de Alta Dirección en Ingeniería y Tecnologías, Universidad Anáhuac, Naucalpan, Mexico
[4] Universidad Autónoma de Aguascalientes, Aguascalientes, Mexico

Abstract. Searching and sorting objects, especially items of interest, on the web is an essential task in the digital age. Systems that facilitate this task are of great use to consumers and e-commerce companies. The use of convolutional neural networks (CNN) is a very useful tool when classifying many objects in different semantic fields according to their characteristics. If we add a module for obtaining the color of the base object to this classification and combine both results in the network, we will obtain as a result a system for searching for objects that is not only based on their characteristics but also on their exact color. In this study, we propose a literature-based methodology for obtaining clothing on the web. This system is based on the creation of a color base with different centers in the HSL (Hue, Saturation, Luminosity) space to obtain the color of an image based on dissimilarity rules. The base image previously passed through the convolutional neural network, plus the result of obtaining the color through the Euclidean distance will generate a character string, which will be placed in a search engine and will bring as a result similar garments to the base image. This work provides new perspectives and useful techniques for obtaining colors, which can significantly improve the accuracy and efficiency of object search on the web.

Keywords: Image searching · Convolutional neural networks · color classification · HSL space · Euclidean distance

1 Introduction

In today's information age, the vast universe of the Internet can make finding specific objects a challenge. The efficient sorting and searching of these objects are an essential task. Systems that facilitate this task are very useful as they allow for a more efficient and accurate search. In this paper, we present a search system that uses Convolutional Neural Networks (CNNs) to classify clothes based on images provided by the user. The CNN is trained to identify different types of clothing, such as trousers, shirts, dresses, etc., and the result of this classification returns a string of characters describing the type of clothing. In addition to classifying clothing, our sys-tem also detects the predominant color in an image. This is done by creating a color database in HSL (Hue, Saturation, Luminosity) space by classifying 12 colors with different color centers. Using dissimilarity rules based on Euclidean distance, the sys-tem can determine which color is more dominant in an image. The result is a string that combines the clothing classification and the predominant color. For example, if an image of blue trousers is entered, the system will generate the string "blue trousers". This string can then be entered into a pre-built web search engine to find similar garments available online. This work presents a new perspective on searching and classifying specific objects on the Internet. By combining object classification with color recognition, our system can significantly improve the accuracy and efficiency of online searches.

2 Related Word

Object classification using CNN has been an active area of study in recent years. This article [1] reviews CNN-based image classification algorithms. The authors cover the development of CNNs, from their predecessors to recent state-of-the-art network architectures.

Color detection in images using the HSL color space has been the subject of several studies. For example, this paper [2] introduces the HSL color space for better color recognition and processing instead of the CIELAB color space. In addition, they add flexible combinations of the weight coefficient for HSL to achieve different results.

In another study, new face detection methods based on HSL and HSI2 color spaces were presented. In this work [3], methods based on two main steps were proposed: first, skin-like regions are detected using the gradient values of the proposed color space. Then, the desired faces are determined from the recommended regions according to the main facial features, such as eyes, mouth, and nose.

On the Euclidean distance side, this other research focuses on face detection and recognition using texture features and Euclidean distance [4].

These related works demonstrate the potential and effectiveness of CNNs for object classification and color detection in images, in addition to the use of Euclidean distance as a dissimilarity rule. However, our work proposes a system that combines both techniques to improve the accuracy and efficiency of online clothing searches.

3 Materials and Methods

The methodology of our research is detailed below.

3.1 CNN

We use a sequential CNN model for garment image classification [5]. Our model consists of several layers, including Conv2D layers for feature extraction [6], MaxPooling2D for dimensionality reduction [7] and Dense layers for final classification [8]. As can be seen in Fig. 1, the sequential model of our CNN is presented.

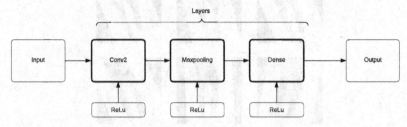

Fig. 1. Sequence model of the convolutional neural network

The hyperparameters and configurations used in our model include:

- Learning rate: We use the Adam optimizer, which adaptively adjusts the learning rate for each parameter.
- Batch size: 32
- Number of epochs: 15

In addition, our CNN model has the following configurations:

- Hidden layers: The model has three convolutional layers (Conv2D), three maximum pooling layers (MaxPooling2D), a flattening layer (Flatten), a dropout layer (Dropout), and two dense layers (Dense).
- Kernel size: 3×3
- Padding: We use the 'same' padding type in all convolutional layers. This means that zeros are added around the input image to allow the convolutional kernel to be applied at the edges and corners of the image.
- Strides: The default value for Conv2D and MaxPooling2D layers in Keras is (1, 1), which means that the kernel moves one pixel at a time.

To improve model performance and avoid over-fitting [9], we implement data augmentation techniques [10]. This involves creating new modified images from the original ones, e.g., by rotating, scaling, or cropping the original images. In Fig. 2, an image is shown that has been processed with various data augmentation techniques. These techniques are rotation, scaling, translation, and cropping. The images are randomly rotated in a range of -10 to $+10$ degrees, enlarged or reduced in a range of -10% to $+10\%$, and flipped horizontally with a 50% probability. The input images for the model are 180 $\times$ 180 pixels.

In addition to data augmentation, we also use segmentation in the HSV color space to improve the accuracy of the model [11]. This involves converting images from the RGB color space to the HSV color space and then segmenting the images based on hue, saturation, and value values.

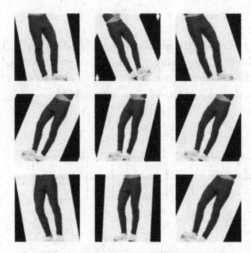

Fig. 2. Data augmentation techniques

The model was trained using a dataset consisting of images of different types of clothing. Table 1 shows the types of garments and the number of garments by color used as a database. The dataset is obtained from the Internet, more specifically from the Kaggle site [12].The data was divided into training, validation, and test sets. We evaluated the model performance by classification accuracy. We obtained a result of 90%.

Table 1. Clothing data set.

Cloth	Amount/Color
Dress	566 Yellow, 818 White, 800 Red, 450 Black, 520 Blue
Pants	246 Pink, 274 White, 308 Red, 870 Black, 798 Blue, 311 Brown, 227 Green
Shirt	230 Green, 332 Red,715 Black, 741 Blue
Shoes	455 Green, 766 Black, 464 Brown, 610 Red, 523 Blue, 600 White, 403 Silver
Shorts	328 Black, 299 Blue, 135 Green, 120 White, 195 Yellow
Hoodies	188 Brown, 347 Pink, 349 Red
Suit	243 Green, 320 Black, 354 White
Skirt	513 Pink, 361 Silver, 409 Yellow

3.2   Determining the Predominant Color of a Base Image

To determine the predominant color in an image, we use a Python program that calculates the percentage of colors in the image using the Euclidean distance in the HSL (Hue, Saturation, Lightness) color space [4]. The process starts by converting the image from RGB to HSL. Then, a database containing different colors in HSL format is read. This database represents a sample region for each color in HSL space and serves as color markers [13]. By comparing the colors in the image with the color markers in the database, we can determine the predominant color in the image.

This method allows us to obtain an accurate representation of the predominant color in an image, which is crucial for our task of searching for similar garments.

3.3 Color Database

This study uses 12 specific colors: purple, red, blue, cyan, green, yellow, orange, pink, white, brown, black, and gray. These colors have been selected to cover the most vibrant and saturated tones of the color spectrum. In addition, the inclusion of colors such as white, brown, black, and gray allows us to represent the lightest and darkest tones of the spectrum. Although this article mentions the use of 6 colors to represent different areas of the color spectrum in the HSL (Hue, Saturation, Lightness) model, we have expanded this approach to include 12 colors [13]. The HSL model is useful for representing colors in terms of their hue (Hue), saturation (Saturation), and lightness (Lightness). By using 12 colors instead of 6, we can cover a wider range of tones in the HSL spectrum and represent a wider variety of colors. Each of the 12 selected colors has several shades. Each of these shades is defined in the HSL format, which allows for an accurate and consistent representation of color.

In Table 2, we present an extract from our database showing a selection of 12 colors that are used in our database. Each color is presented with its name and its value in the HSL (Hue, Saturation, Lightness) format.

Table 2. Color centers in hsl space

Color	HSL Value
Purple	hsl(285, 37%, 34%)
Red	hsl(355, 96%, 63%)
Blue	hsl(239, 97%, 24%)
Cyan	hsl(196, 46%, 71%)
Green	hsl(89, 22%, 31%)
Yellow	hsl(50, 84%, 53%)
Orange	hsl(25, 75%, 65%)
Pink	hsl(355, 93%, 74%)
White	hsl(192, 22%, 95%)
Brown	hsl(17, 33%, 39%)
Black	hsl(131, 54%, 8%)
Gray	hsl(66, 13%, 69%)

3.4 Color Segmentation

In addition, a segmentation of the color regions in the image is carried out. This involves dividing the image into different areas, each of which contains pixels of a similar color. Figure 3 shows the base image we are working with.

Fig. 3. Base Image

Various filters are also applied to enhance the contrast of the image. These filters can include techniques such as brightness and contrast adjustment, histogram equalization, and edge enhancement. The application of these filters can improve the visual quality of the image and make details easier to see.

Figures 4a and 4b illustrate the results after segmentation and filter application. Figure 4a shows the different zones segmented by color tones, and Fig. 4b shows the image with filters applied to enhance contrast.

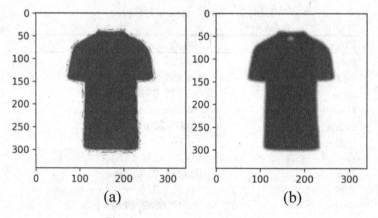

(a)   (b)

Fig. 4. a. Segmentation **b.** Filtering

3.5 Image Processing and Color Classification

Each pixel of the image is classified by calculating the Euclidean distance between that pixel and each color marker. The smallest distance indicates that the pixel is more like that color. In this way, each pixel is classified and the percentage of each color in the image is determined.

The resulting color will be represented as a string of characters. For example, if the predominant color in the image is blue, these bits will be represented as characters and will be concatenated with the result of the classification of the CNN. The combination of these two strings will form a keyword that will be entered into the search engine to obtain a corresponding image, in this case, several blue shirts.

3.6 Search Engineer

For practical research purposes, a keyword-based search engine was developed. This engine, implemented with SerpApi, analyzes the keywords entered by the user and compares them with the content of the indexed web pages [14].

An API request is made, and the results are returned in a Python dictionary. The code iterates through the returned images and adds each unique link to a list of results.

As shown in Fig. 5, the search engine first prompts the user to enter a keyword to search for. Once the keyword is entered, the search engine performs its function and provides a series of links as a result. These links correspond to images that match the keyword entered by the user. This process illustrates how the keyword-based search engine can facilitate the retrieval of relevant information in an efficient manner.

Fig. 5. Search engine

The purpose of the search engine is to aggregate the concatenation of the strings obtained from the garment classification results using the CNN and the base image color retrieval.

4 Results

4.1 Convolutional Neural Network Performance and Results in Image Classification

To evaluate the accuracy of our model, we tested the CNN using an image that was not part of the training database. The image, which represents a blue shirt (see Fig. 3), was processed by our model. The results indicated that our model correctly classified the garment with an accuracy of 97.69%. This high level of accuracy suggests that our model is capable of successfully identifying and classifying garments based on their type and color. In addition, you can see the result of the classification of the blue shirt in Fig. 6. This visual result provides a clear representation of how our CNN model was able to successfully classify the garment.

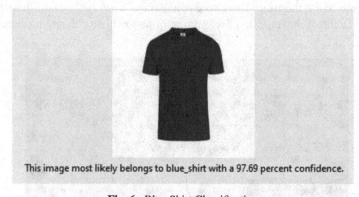

This image most likely belongs to blue_shirt with a 97.69 percent confidence.

Fig. 6. Blue Shirt Classification

4.2 Image Processing and Color Classification Using Euclidean Distance

As mentioned in the 'Image Processing and Color Classification' section, we take as an example obtaining the color of an image of a blue shirt. This section shows the result obtained by determining the color of an image using Euclidean distance and a set of color data with different spectrum centers in relation to their color.

A base image, in this case Fig. 3, is processed with code that determines the predominant color of the image. The code generates a data frame with 12 colors and the percentage of each color present in the image. According to the data from Table 2, the predominant color in the image is white, which represents approximately 67.42% of the image. The second most predominant color is blue, which represents approximately 30.01% of the image. Cyan and gray are also present, but in much smaller proportions.

The colors purple and pink are present in very small amounts, while the colors red, green, yellow, orange, brown and black are not present in the image.

The results obtained are shown in detail in Table 3. In this table, we can observe that the image is mainly composed of two large sections: the white background and the blue color, which is of interest to us. The latter will be the string that will be sent to the result of the image prediction and will be concatenated to search for images on the internet. The minority colors in the image can be appreciated in the image segmentation, which is visualized in Fig. 4a. Although these colors represent a small proportion of the total image, they can be important for understanding the complete composition of the image. The image segmentation can help highlight these details that might go unnoticed in a regular visual inspection of the image.

Table 3. Color Distribution in the Base Image

Color	Percentage
White	0.674221
Blue	0.301159
Cyan	0.013452
Gray	0.010138
Purple	0.000683
Pink	0.000346
Red	0
Green	0
Yellow	0
Orange	0
Brown	0
Black	0

4.3 Search for the Word Concatenated

To facilitate the linking of the above parts and for research purposes, we chose to create an application using Tkinter [15]. The interface consists of four buttons: "Select Image", "Predict Image", "Predict Color" and "Search Image", plus an image result field.

The "Select Image" button allows the user to select the image to predict. The "Predict Image" button inputs the selected image into the trained artificial neural network model to classify the garment into one of the CNN model labels. The "Predict Color" button passes the image to the Gaussian function to obtain the predominant color.

The "Search Image" button takes the string resulting from the garment classification and color retrieval, concatenates them and searches them using the SerpApi search engine on the web. The image result field displays the image selected from the "Select Image" button. At the end, there is a text box that will display the result of the image prediction.

As shown in Fig. 7, the final application integrates all the above processes. A base image is selected through the application, which then predicts that the image is a blouse and that the second predominant color is blue, since the main color is the white background. This coincides with the results obtained in the color elicitation process.

In addition, the application performs the concatenation of the two results, i.e., "blue skirt". Finally, the program displays a pop-up window with 100 links containing images of blue blouses. This demonstrates how the application can facilitate the search for relevant information based on the visual characteristics of an image.

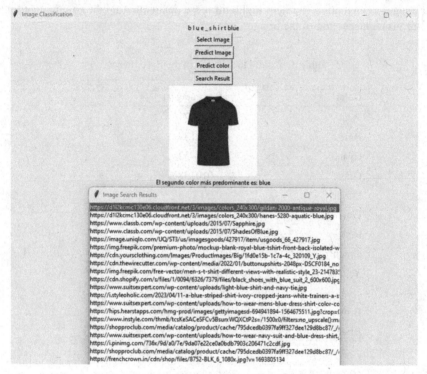

Fig. 7. CNN generated concatenated word search and color retrieval.

5 Conclusion

In this research, we present a methodology for obtaining objects on the web. This was achieved thanks to two important techniques: the creation of a CNN allows us to classify different types of garments and obtaining the dominant color of an image using Euclidean distance. Although this dissimilarity rule is widely used, the use of a database with 12 colors in HSL space gives a unique approach to our research.

This color basis is the core of this dissimilarity rule. Normally, only primary colors or RGB colors are used, but in this case, we opted for a wider range of twelve colors. These twelve colors have dozens of central color points.

The result was as expected. We made sure that the garment in the image was the correct one and standardized the colors to get the right ones. When entered the search engine, it is always those colors that are searched for. This demonstrates the efficiency and accuracy of our methodology in obtaining relevant objects on the web.

References

1. Chen, L., Li, S., Bai, Q., Yang, J., Jiang, S., Miao, Y.: Review of image classification algorithms based on convolutional neural networks. Remote Sens. **13**(22) (2021). https://doi.org/10.3390/rs13224712
2. Su, F., Xu, H., Chen, G., Wang, Z., Sun, L., Wang, Z.: Improved Simple Linear Iterative Clustering Algorithm Using HSL Color Space BT - Intelligent Robotics and Applications, pp. 413–425 (2019)
3. Elaw, S., Abd-Elhafiez, W.M., Heshmat, M.: Comparison of video face detection methods Using HSV, HSL and HSI color spaces. In: 2019 14th International Conference on Computer Engineering and Systems (ICCES), 2019, pp. 180–188 (2019). https://doi.org/10.1109/ICCES48960.2019.9068182
4. Yu, J., Li, C.: Face recognition based on Euclidean distance and texture features. In: 2013 International Conference on Computational and Information Sciences, pp. 211–213 (2013). https://doi.org/10.1109/ICCIS.2013.63
5. Sharma, N., Jain, V., Mishra, A.: An analysis of convolutional neural networks for image classification. Procedia Comput. Sci. **132**, 377–384 (2018). https://doi.org/10.1016/J.PROCS.2018.05.198
6. "Conv2D layer". https://keras.io/api/layers/convolution_layers/convolution2d/. Accessed 28 Sep 2022
7. "Capa MaxPooling2D". https://keras.io/api/layers/pooling_layers/max_pooling2d/. Accessed 28 Sep 2022
8. "Dense layer". https://keras.io/api/layers/core_layers/dense/. Accessed 28 Sep 2022
9. Bejani, M.M., Ghatee, M.: A systematic review on overfitting control in shallow and deep neural networks. Artif. Intell. Rev. **54**(8), 6391–6438 (2021). https://doi.org/10.1007/s10462-021-09975-1
10. Filipi Gonçalves, C., Santos, D.: Avoiding Overfitting: A Survey on Regularization Methods for Convolutional Neural Networks (2022). https://doi.org/10.1145/3510413
11. Pardede, J., Husada, M., Hermana, A., Rumapea, S.: Fruit Ripeness Based on RGB, HSV, HSL, L*a*b* Color Feature Using SVM (2019). https://doi.org/10.1109/ICoSNIKOM48755.2019.9111486
12. "Apparel Dataset | Kaggle". https://www.kaggle.com/datasets/kaiska/apparel-dataset/code. Accessed 28 Sep 2022
13. Lin, T., Liao, B.-H., Hsu, S.-L., Wang, J.: Experimental investigation of HSL color model in error diffusion. In: 2015 8th International Conference on Ubi-Media Computing (UMEDIA), pp. 268–272 (2015). https://doi.org/10.1109/UMEDIA.2015.7297467
14. "SerpApi: Google Search API." https://serpapi.com/. Accessed 29 Mar 2023
15. Roseman, M.: Modern Tkinter for busy python developers: quickly learn to create great looking user interfaces for Windows, Mac and Linux using Python's standard GUI toolkit, p. 257 (2020)

Identifying DC Motor Transfer Function with Few-Shots Learning and a Genetic Algorithm Using Proposed Signal-Signature

Martín Montes Rivera[1]([✉]) [iD], Marving Aguilar-Justo[2], and Misael Perez Hernández[1]

[1] Research and Postgraduate Studies Department, Universidad Politécnica de Aguascalientes (UPA), 20342 Aguascalientes, Mexico
martin.montes@upa.edu.mx
[2] Engineering Faculty, Universidad Veracruzana (UV), 94452 Ixtaczoquitlán, Veracruz, Mexico

Abstract. The most common actuators in precision or low-power applications are Direct Current (DC) motors. DC motors are in robots, commercial products, automobiles, and appliances for accurately controlling position or speed or in portable devices. Input tracking and system stabilization are the main goals of control theory. There are different approaches for classic control with their proprieties. However, classical controllers and sometimes others with probabilistic methods use mathematical models for designing or generating control laws. Alternatives that represent the model of a system include differential and difference equations, state space models, and transfer functions. Parametrizing those models implies supplying signals to the process and using deterministic or probabilistic algorithms. On the other hand, artificial intelligence and machine learning approaches like genetic algorithms have shown relevant results by tuning those models using the error with the system's frequency or time response. However, looking for reasonable solutions in search spaces with several dimensions, like in a five-dimensional problem optimizing the parameters of a DC motor, can be timely and computationally demanding. Alternatively, few-shots learning simplifies the data for optimization, resulting in less timely and computational training. In this work, we propose a novel method for transfer function DC motor identification using a few-shots learning approach based on a signature value unique for each motor. We obtained the motor signature by applying a signal that variates length and frequency for each engine and collecting the six numerical values with the exact time sampling. Since we reduce the components of the signals, we reduce the data size. Moreover, the search space becomes significantly simple, allowing us to find the DC motor transfer functions with R-square between 0.99 and 1.0 for at least 95% of the controllers with 6 component signatures.

Keywords: System identification · DC Motor · Few-Shots Learning · Genetic Algorithms

H. Calvo et al. (Eds.): MICAI 2023 Workshops, LNAI 14502, pp. 170–190, 2024.
https://doi.org/10.1007/978-3-031-51940-6_14

1 Introduction

The DC motor is a popular choice for position and speed control applications due to its precision and ease of use, especially in low power applications. However, controlling the DC motor requires a mathematical model, which is essentially a linear, second-order system [1–3].

To express the mathematical model of a DC motor, one can use different forms such as a differential equation, state space equation, or transfer function using Laplace transform. The transfer function comprises poles, zeros, and gain, which constitute the system parameters. The process of determining the parameter values is called identification or estimation, and one can perform it either online or offline [1, 4, 5].

According to Luengo in [6], more than 300 articles on parameter estimation research have been summarized using the Monte Carlo method. This method is considered effective for real-world applications involving statistical signal processing.

Timofeev in [7] compared three methods for identifying parameters in linear regression models using computational simulation: ordinary least squares, weighted least squares, and Williamson algorithm.

Odhano's in [8] deals with identifying parameters of oceanographic monitoring vehicles. This is crucial for intelligent control of the system. The process involves using empirical methods to estimate added mass and damping parameters, while rigid body parameters are estimated through computational modeling. To ensure accuracy, experimental tests are also conducted for parameter estimation.

In Gyuk et al. in [9], researchers utilized artificial intelligence to identify nine parameters of a mathematical model for the metabolism process of individuals who receive insulin due to type 2 diabetes. The approach combines Genetic Algorithms and brute force, resulting in better mean square error and Clarke's Error Grid Analysis compared to other studies on this topic.

Estimating motor parameters is a well-researched topic with numerous articles emphasizing the use of automatic control and artificial intelligence techniques.

The paper Obeidat in [10], presents a method for parameter estimation of a PMDC motor using quantized output observations from sensors. The method is validated through simulation and experimentation, achieving a TSE of less than 0.001 in some tests.

Lian et al. developed a method in [11] to determine the inertia of a synchronous permanent magnet motor. They found a sampling period of 25 ms and gains of kp = 0.008 and ki = 0.8 to be optimal through experimentation. The direct calculation method had a 0.7% percentage error, while the proportional integral regulation method had zero overshoot, a settle time of 13.1 ms, and a steady state error of 0.

Artificial intelligence (AI) tools have gained popularity for parameter identification of electrical motors. This technology has proven its effectiveness in optimizing motor performance and enhancing overall efficiency. With its ability to accurately determine key parameters, AI is a valuable asset in the field of motor engineering.

Rubaai and Kotaru in [12] proposed a method to identify a DC motor using a Multilayered Feedforward Artificial Neural Network with dynamic back propagation. The method accounted for nonlinearities, showed low tracking error, and had shorter convergence time than the literature-reported DBP learning algorithm.

Machine learning, an AI technique, has proven highly effective in various fields, including rotating electrical machines, due to its ability to analyze complex data and identify patterns.

In [13], Cosmin and Sorin conducted a study where they used reinforcement learning (RL) to control a DC motor with a two-quadrant converter. They trained the model offline to generate optimal control for speed regulation and achieved remarkable results. They recorded a rise time of 28 ms in 110 training episodes, with no long-term error. Furthermore, the tracking was admirable for frequencies below 20 Hz when a sine input was used.

Han et al. in [14] conducted a study on fault detection in a Brushless DC motor using XGBoost, neural network, and convolutional neural network to detect demagnetization failure by measuring mechanical radial vibration. The three machine learning methods showed satisfactory results in predicting the demagnetization values, with the convolutional neural network achieving the lowest error rates of 0.3 RMSE.

Integral Reinforcement Learning was used in [15] Bujgoi and Sendrescu to control DC motors in a method that can handle continuous domain systems. A critic neural network is used to evaluate performance. The study includes simulation and experimental tests using National Instruments hardware and software, showing that the controller can effectively regulate at varying levels.

Few-shot learning is an attractive subcategory of machine learning for identifying relevant parameters from a small amount of data. Object detection and image recognition are popular research fields that utilize this technique.

Li and Liu in [16] developed a few-shot learning method using the Topic Snowball Model to extract structured knowledge from text and create a knowledge graph. Their study found that the Topic Snowball model performed better than the Neural Snowball when few initial seeds were used, although precision was slightly lower.

In a study by Ma et al. in [17], Few-Shot Learning was explored as a fault diagnosis method. The study presented a classification of fault diagnosis methods, which included analytical model, qualitative knowledge, and data-driven approaches. The latter was further divided into statistical analysis, signal processing, and machine learning, where Few-Shot Learning was highlighted.

In this work, we propose a new technique for identifying the transfer function of a DC motor using a few-shot learning approach. Our method uses a unique signature value assigned to each motor. To test our approach, we applied a signal with varying length and frequency to 500 randomly generated motors and collected six numerical values using exact time sampling. By reducing the signal components and simplifying the search space, we were able to significantly reduce the data size. As a result, we successfully identified the DC motor transfer functions using a genetic algorithm with a mean absolute percentage error (MAPE) below 1e−5%, R-square in ranges 0.99–1.0 with maximum 5% of samples out of the range.

2 Materials and Methods

2.1 DC Motor Model and Transfer Function

A DC motor is a device that converts direct current electrical energy into mechanical energy in the form of rotation. The rotation occurs due to the interaction between the magnetic field of the stator and the current flowing in the rotor windings. In control systems, a mathematical representation that relates the input and output of a system, known as a transfer function, is often used to describe and analyze the behavior of dynamic systems. The DC motor model can be synthetized in the electrical (Eq. (1)), mechanical (Eq. (2)) and electrotechnical force (Eq. (3)) equations as shown in the Fig. 1 [4, 5].

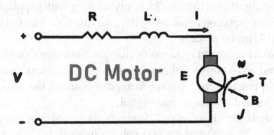

Fig. 1. Model of DC motor

$$V(t) = R \cdot i(t) + L\frac{di(t)}{dt} + e(t) \tag{1}$$

$$J\frac{d\omega(t)}{dt} = -b\omega(t) + k \cdot i(t) \tag{2}$$

$$e(t) = k \cdot \omega(t) \tag{3}$$

For obtaining the transfer function we start by combining electrical and electromechanical force equations as in Eq. (4).

$$V(t) = R \cdot i(t) + L\frac{di(t)}{dt} + k \cdot \omega(t) \tag{4}$$

Then we use mechanical and electrical equations fixed for current and relate them as in Eq. (5)

$$J\frac{d\omega(t)}{dt} + b\omega(t) = k \cdot \frac{V(t) - k \cdot \omega(t)}{R + L\frac{d}{dt}} \tag{5}$$

Applying Laplace transform and rearranging we get Eq. (6) with the transfer function of speed to voltage.

$$G(s) = \frac{\Omega(s)}{V(s)} = \frac{k}{\big((Js + b) * (Ls + R) + K^2\big)} \tag{6}$$

Similarly, we get Eq. (7) with the transfer function of position to voltage adding an extra integrator.

$$G(s) = \frac{\Theta(s)}{V(s)} = \frac{k}{\left((Js+b) * (Ls+R) + K^2\right)s} \tag{7}$$

2.2 Motor Signature

Identification with classical control methods uses test signals like the step response and measures parameters associated with the TFs, as described in Sect. 2.1. However, identifying those parameters demands working with signals with several samples.

Since supervised algorithms like the GA imply working with populations with several chromosomes across generations evaluating the cost function, training could result in a slow and inefficient learning process.

The motor signature proposed in this work changes those signals with several samples by six numerical components. Nevertheless, to still have representative information for identifying a transfer function from voltage to angular position, the six components must be unique for a motor and a voltage input signal.

Therefore, the signature s is a function that maps the signal containing the n_s samples of motor position $\theta(t)$ after receiving a $v(t)$ voltage input signal into six representative components of those changes, i.e., $s(v(t), \theta(t)) : R^{n_s} \mapsto R^6$.

In order to get signature components that are representative of the motor behavior, the $v(t)$ signal must produce changes in length and frequency, allowing an optimization algorithm to perceive the motor constants, as other identification methods in classical control described in Sect. 2.1.

Consequently, we use the $v(t)$ defined in Eq. (8) with three stages containing variations in length, faster, and slower frequencies, depending on the settling time with 2% of the variation ($ts_{2\%}$). The first one is a step signal maintained at 0 V during $\frac{ts_{2\%}}{2}$, then increases from 0 to 12 V in $\frac{ts_{2\%}}{10}$ ms and remains at 12 V until reaching $ts_{2\%}$. The second one variates length from 0 to 12 V progressively with $\frac{ts_{2\%}}{t}$ and frequency with the time-squared t^2 from 0 to $1{,}000 \cdot ts_{2\%}^2$ Hz. The third one variates in length from 0 to 12 V with $\frac{ts_{2\%}}{t}$ and frequency with t^2 from 0 to $0.5 \cdot ts_{2\%}^2$ Hz.

$$v(t) = \begin{cases} 0, & 0 \leq t \leq \frac{ts_{2\%}}{2} \\ 12 \cdot \min\left(\frac{t - \frac{ts_{2\%}}{2}}{\frac{ts_{2\%}}{10}}\right), & \frac{ts_{2\%}}{2} \leq t \leq ts_{2\%} \\ 12 \cdot \frac{ts_{2\%}}{t} \cdot \sin\left(1000t^2\right), & ts_{2\%} \leq t \leq 2ts_{2\%} \\ 12 \cdot \frac{ts_{2\%}}{t} \cdot \sin\left(0.5t^2\right), & 2ts_{2\%} \leq t \leq 3ts_{2\%} \end{cases} \tag{8}$$

Additionally, the v(t) signal can generate less data for simulation when reducing the number of samples, allowing it to require fewer training points in a supervised learning algorithm. Moreover, this approach reduces the number of training points used in few-shots learning.

The Fig. 2 shows the $v(t)$ signal with 5,000 samples generated for a motor with $J = 3.2284E{-}06$, $b = 3.5077E{-}06$, $K = 0.0274$, $R = 4.0000$, $L = 2.7500E{-}06$ and $ts_{2\%} = 0.08001600320064013$. Similarly the Fig. 3 shows $v(t)$ for the same motor but with six samples and Fig. 4 shows the six samples stem, showing directly the components of the signature.

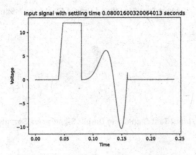

Fig. 2. Example of input signal v(t) for signature generation with 5,000 samples.

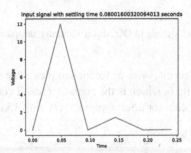

Fig. 3. Example of input signal v(t) for signature generation with six samples.

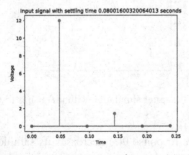

Fig. 4. Stem of example input signal v(t) for signature generation with six samples

$v(t)$ produces different results with different samples, but even with fewer points, it is still representative, reducing the required data for training as occurs in the few-shots approach in Fig. 5.

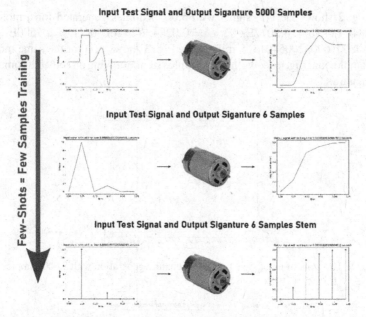

Fig. 5. Signature with test signals in DC Motor reducing samples for few-shots learning

$\theta(t)$ also has different results with different samples after simulation. The result is the obtained response in Fig. 6, which is the signature result containing 15,000 samples with the angular position behavior after applying $v(t)$ with 15,000 samples.

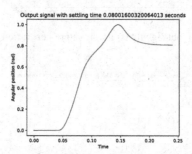

Fig. 6. Output $\theta(t)$ after simulation with $v(t)$ using 15,000 samples.

Similarly, the signature response $\theta(t)$ decreases its samples or components if $v(t)$ decreases too. Figure 7 shows the output signature obtained with six samples in $v(t)$, and Fig. 8 shows the stem of the same output with precise signature components used as training points in our proposal. However, as fewer components in $\theta(t)$, less representative is the signal. Therefore, the user must find a good value of samples for training and converge to a suitable candidate solution.

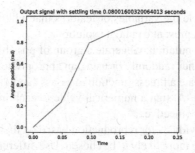

Fig. 7. Output $\theta(t)$ after simulation with $v(t)$ using 6 samples.

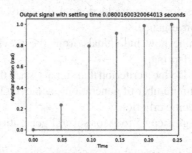

Fig. 8. Stem of $\theta(t)$ after simulation with $v(t)$ using 6 samples.

2.3 Genetic Algorithm

The genetic algorithm is a search technique based on Charles Darwin's theory of natural selection. It selects the fittest candidates for reproduction to generate the next generations [18].

In the natural selection process, the fittest candidates from a population are selected to produce better offspring. The offspring inherit the characteristics of their parents and, if the parents have better fitness, their offspring will have a better chance of surviving [18]. This process is repeated to create a generation with the fittest candidates. This concept can be applied to a hunt problem by selecting the best results from a set. The genetic algorithm involves five phases [19, 20].

1. Original population
2. Fitness function
3. Selection
4. Crossover
5. Mutation original Population

In the application to solve a problem with genetic algorithms it may vary depending on the problem but in general the following steps are followed [19, 21]:

1. Problem Definition: To begin with a genetic algorithm, you must first identify the problem you want to solve, set a clear objective, and specify the variables to optimize or discover.

2. Gene Encoding: Define the variables or features that will make up each individual in the population to represent candidate solutions.
3. Generate an Initial Population: Generate a group of possible solutions to start your genetic algorithm, either randomly or using an appropriate initialization method.
4. Fitness Function: Create a fitness function to assess candidate solutions' quality for problem optimization. Assign a numerical value to each individual that reflects its quality relative to the objective.
5. Selection: Select individuals for reproduction based on their fitness, with higher fitness individuals being more likely to be chosen. Use different methods for selection, such as roulette wheel or tournament selection.
6. Crossover: Create new offspring by combining selected candidate solutions in pairs.
7. Mutation: This allows for the introduction of diversity into the population and prevents premature convergence.
8. Replacement: Determine how individuals from the previous generation will be replaced by the new offspring.
9. Termination Criterion: Define a criterion that signals when the algorithm should halt. This can be a maximum number of generations, achieving a target fitness value, or any other problem-specific criterion.
10. Iteration: Iterate through steps 5 to 9 for several generations until the termination criterion is met.
11. Result: Upon meeting the termination criterion, you will obtain a solution that approximates the optimal solution according to your fitness function. This solution represents the best solution found by the genetic algorithm [19, 21].

Key operators in a Genetic Algorithm include selection, crossover, and mutation. These operators are responsible for generating new diversity and driving the evolution of the population towards optimal solutions [22].

Selection: Selection methods, such as roulette wheel, tournament, or rank-based selection, determine which individuals are chosen for reproduction based on their fitness [23].

Crossover: Crossover combines genetic information from two or more parents to create offspring. Common types of crossovers include single-point, multi-point, and uniform crossover [23].

Mutation: Mutation introduces random changes into a chromosome's genes to explore new solutions in the search space [23].

In this work we use a genetic algorithm with tournament selection, single-point crossover and unform mutations all those mechanisms are described below:

Selection Method

Tournament selection (Algorithm 1) involves randomly choosing a small number of individuals from the population (P) with limit in the population size (P_s), referred to as the tournament size (t_s), and then selecting the best individual from that group for reproduction. This process is repeated multiple times to form the new generation as shown in the pseudocode below [19, 21, 22].

```
1     function Tournament (P, t_s)
2     for 0 to P_s
3         Set best to 0
4         for 0 to t_s
5             Get current random element from population
6             If current element is fitness > best fitness
7                 Current = best
8             End if
9         End for
10        Add best to mating pool.
11    End for
```

Algorithm 1 Tournament Selection

- Advantages: Tournament selection maintains good genetic diversity, allowing even less fit individuals a chance of being selected. It can adapt well to multimodal problems with multiple optimal solutions [19, 21, 22].
- Disadvantages: The choice of tournament size can significantly impact algorithm performance, and determining the optimal size can be challenging. It does not guarantee that the fittest individuals are selected on every occasion [19, 21, 22].

Crossover

The primary goal of crossover is to generate offspring that inherit favorable traits from both parents, potentially leading to the convergence towards optimal or improved solutions within the search space [24]. The crossover mechanism with one single crossover point work with the below elements:

- Selection of Crossover Points: In the case of single-point or multi-point crossovers, random crossover points are selected on the parents' chromosomes.
- Offspring Creation: Segments of the parents' chromosomes before and after the crossover points are exchanged to form the chromosomes of the offspring.
- Number of Offspring: Depending on the genetic algorithm's design, crossover can generate one or more offspring for each pair of parents.

In this method, one single random crossover point is chosen on the chromosomes of the parents, and the segments before and after that point are exchanged to create offspring. The Algorithm 2 shows how it works [25].

```
1     function SinglePointCrossover(parent1, parent2):
2         length = length(parent1)
3         crossover_point = random_integer(1, length - 1)
4         child1 = []
5         child2 = []
6         for i in range(length):
7             if i < crossover_point:
8                 child1.append(parent1[i])
9                 child2.append(parent2[i])
10            else:
11                child1.append(parent2[i])
12                child2.append(parent1[i])
13        return child1, child2
14    end SinglePointCrossover
```

Algorithm 2 Single-Point Crossover

Mutation
Mutation is a fundamental operator in genetic algorithms that introduces randomness into the population by individually altering genes within an individual. Mutation plays a critical role in maintaining genetic diversity within the population and helps prevent premature convergence towards a suboptimal solution [26]. The mutation process implies working with the below elements:

- *Selection of the Individual.* First, one or more individuals from the population are selected to undergo mutation. Selection can be random or follow specific criteria.
- *Selection of Genes.* Next, one or more genes are chosen from the selected individual to be mutated. Gene selection is often random, with a mutation rate determining the probability of mutating each gene [27].
- *Gene Modification.* The selected genes are then modified according to a mutation process. The nature of this modification depends on the problem and can vary widely. Some common examples of mutation include [28]:
- Bit Mutation. - In. binary problems, such as combinatorial optimization problems, a bit is flipped, changing 0 to 1 or vice versa.
- Value Mutation. In numerical problems, gene values are altered by adding or subtracting a small random value.
- Permutation Mutation: In problems involving permutations (e.g., the traveling salesman problem), two elements in a sequence may be swapped.
- Insertion or Deletion Mutation: In problems with sequences (e.g., genetic sequence design), an element can be inserted or removed from the sequence [28].

Creation of the New Individual
The mutated individual is created by taking the original genes of the individual, with the mutated genes replacing the previously selected genes. This results in a new individual that is now part of the population [28].

Mutation Rate
The mutation rate (P_m) is a critical parameter that controls the probability of an individual gene being mutated. A low mutation rate can maintain genetic stability within the population, while a high rate can increase exploration of the search space for solutions. The choice of the mutation rate depends on the problem and should be experimentally tuned to strike an appropriate balance between exploration and exploitation [28]. The Algorithm 3 describe the process of mutation.

```
1  function Mutation (individual, p_m):
2      mutated_individual = copy(individual)
3      for i in range(length(mutated_individual)):
4          random_value = random_uniform (0, 1)
5          if random_value < p_m:
6              mutated_gene = mutate_gene(mutated_individual[i])
7              mutated_individual[i] = mutated_gene
8      return mutated_individual
9  end Mutation
```

Algorithm 3 Mutation

3   Results and Discussion

The first stage in the proposed method starts by generating a 500 DC motor dataset for validating the transfer function identification achieved with the few-shots signature approach described in Sect. 2.2. The first ten randomly generated transfer functions and their corresponding six signature components (c_1, c_2, c_3, c_4, c_5, and c_6) are in Table 1.

After that, we use the GA to optimize the 500 randomly generated TFs, obtaining the TFs and signature components in Table 2. These results use the same input parameters experimentally selected for the GA ($P_s = 1000$, $t_s = 2$, $n_g = 5$, $n_a = 16$, $p_m = 0.08$, $dec = 1,000$, $g = 10,000$, $c_d = 1.00 + \text{E}-05$) and the cost function is the MAPE between the coefficients of the desired signature (c_1, c_2, c_3, c_4, c_5, and c_6) and those obtained (c_{1o}, c_{2o}, c_{3o}, c_{4o}, c_{5o}, and c_{6o}).

Table 3 shows the cost value for the best and worst chromosomes in the population, the number of generations required to reach the desired cost $c_d = 1.00 + \text{E}-05$, the average time per generation, and the MSE, MAE, MAPE, and R-square metrics for the obtained six-component signature. Additionally, we show in Fig. 9 the convergence diagram comparison in best fitness for the first ten motors.

Similarly, to identify the effect using the few-shots approach for training, we re-evaluate the candidate solutions using a signature with 15,000 components, i.e., without reducing the data for measuring. The MSE, MAE, MAPE, and R-square in this evaluation are in Table 4, showing the error increments changing from six to 15,000 components. However, the trained TFs per DC motor remain below 1% MAPE in the 15,000 samples evaluation.

Furthermore, we convert the transfer functions acquired by the GA that express the relationship between position and voltage to those that relate speed and voltage by eliminating an s integrator in the transfer functions. This conversion enables us to model and contrast the step response, which is a widely used criterion for system identification in classical control. The outcomes of this conversion are provided in Table 5.

Table 1. First ten randomly generated transfer functions with the components of their signature.

ID	Randomly Generated Transfer Function	c_1	c_2	c_3	c_4	c_5	c_6
1	$\dfrac{0.3917961944446569}{9.823440813714 \cdot 10^{-10}s^2 + 0.00126344210671416s + 0.15515071791453}$	0.00E+00	1.35E-01	4.50E-01	6.34E-01	9.51E-01	1.00E+00
2	$\dfrac{7.43436312741763 \cdot 10^{-5}}{2.58571841214713 \cdot 10^{-10}s^2 + 0.00053268563922 91194s + 0.0009201157 0628111}$	0.00E+00	1.13E-01	4.05E-01	5.81E-01	8.76E-01	1.00E+00
3	$\dfrac{0.357280610621161}{7.74350474154462 \cdot 10^{-10}s^2 + 0.000801898049318912s + 0.127697119213499}$	0.00E+00	1.49E-01	5.01E-01	6.76E-01	9.55E-01	1.00E+00
4	$\dfrac{0.189088080293414}{2.07799573291231 \cdot 10^{-9}s^2 + 0.001188760389752 19x + 0.0372826597621439}$	0.00E+00	2.83E-01	1.00E+00	8.84E-01	2.30E-01	9.62E-02
5	$\dfrac{0.632244833977175}{2.850711405111024 \cdot 10^{-9}s^2 + 0.0029205260832 7723s + 0.401386224878578}$	0.00E+00	1.38E-01	4.86E-01	6.65E-01	9.44E-01	1.00E+00
6	$\dfrac{0.134367450970627}{4.58091182819256 \cdot 10^{-10}s^2 + 0.000861586393993046s + 0.0214340381014662}$	0.00E+00	1.66E-01	5.89E-01	7.55E-01	9.56E-01	1.00E+00
7	$\dfrac{0.533798929987591}{4.06116969101815 \cdot 10^{-10}s^2 + 0.000157293755739376x + 0.284999781955137}$	0.00E+00	3.79E-01	9.02E-01	9.47E-01	9.99E-01	1.00E+00
8	$\dfrac{0.284966000436581}{3.15306285748907 \cdot 10^{-10}s^2 + 0.000233250220978434s + 0.0835895526754455}$	0.00E+00	2.31E-01	6.82E-01	8.09E-01	9.85E-01	1.00E+00
9	$\dfrac{0.564976451139048}{8.58823205088309 \cdot 10^{-10}s^2 + 0.0019588544900002s + 0.321370696582876}$	0.00E+00	1.52E-01	5.04E-01	6.78E-01	9.57E-01	1.00E+00
10	$\dfrac{0.322252640537195}{6.22884258808504 \cdot 10^{-12}s^2 + 5.8661566754782 \cdot 10^{-6}s + 0.10413053916331}$	0.00E+00	4.46E-01	9.09E-01	9.54E-01	1.00E+00	1.00E+00

Table 2. Results of the first ten TFs and signature coefficients obtained with GA.

ID	Transfer Function Obtained with GA	c_{1o}	c_{2o}	c_{3o}	c_{4o}	c_{5o}	c_{6o}
1	$\dfrac{0.39465}{2.26899992 \cdot 10^{-10} s^3 + 0.00127727766984489 8s^2 + 0.156278426208s}$	0.00E+00	1.35E-01	4.50E-01	6.34E-01	9.51E-01	1.00E+00
2	$\dfrac{0.00034}{1.63502955 \cdot 10^{-9} s^3 + 0.002436012971434437 s^2 + 0.00420807071077s}$	0.00E+00	1.13E-01	4.05E-01	5.81E-01	8.76E-01	1.00E+00
3	$\dfrac{0.35287}{1.06542313 \cdot 10^{-9} s^3 + 0.0007920327020404111 s^2 + 0.126119553609s}$	0.00E+00	1.49E-01	5.01E-01	6.76E-01	9.55E-01	1.00E+00
4	$\dfrac{0.19668}{3.8234262 \cdot 10^{-10} s^3 + 0.00123651982205871 s^2 + 0.0387780234136s}$	0.00E+00	2.83E-01	1.00E+00	8.84E-01	2.30E-01	9.62E-02
5	$\dfrac{0.63093}{1.486766944 \cdot 10^{-9} s^3 + 0.00291444768139098 8s^2 + 0.400553157376s}$	0.00E+00	1.38E-01	4.86E-01	6.65E-01	9.44E-01	1.00E+00
6	$\dfrac{0.15551}{1.90768675 \cdot 10^{-10} s^3 + 0.0008997181219258281 s^2 + 0.024806743912s}$	0.00E+00	1.66E-01	5.89E-01	7.55E-01	9.56E-01	1.00E+00
7	$\dfrac{0.53201}{4.0680684 \cdot 10^{-11} s^3 + 0.000156837315841788 s^2 + 0.284044123978s}$	0.00E+00	3.79E-01	9.02E-01	9.47E-01	9.99E-01	1.00E+00
8	$\dfrac{0.29155}{3.61943768 \cdot 10^{-10} s^3 + 0.000238615417425186 s^2 + 0.085521838031s}$	0.00E+00	2.31E-01	6.82E-01	8.09E-01	9.85E-01	1.00E+00
9	$\dfrac{0.56325}{1.14732653 \cdot 10^{-10} s^3 + 0.00195307263427662 s^2 + 0.32038746885s}$	0.00E+00	1.52E-01	5.04E-01	6.78E-01	9.57E-01	1.00E+00
10	$\dfrac{0.32313}{1.367769 \cdot 10^{-11} s^3 + 5.866448428706 \cdot 10^{-6} s^2 + 0.104414467493s}$	0.00E+00	4.46E-01	9.09E-01	9.54E-01	1.00E+00	1.00E+00

Table 3. Metrics of the first ten candidate solutions with six component signatures.

id	best	worst	MSE	MAE	MAPE	R2	Generations	Average time per generation (seconds)
1	8.84E−06	2.31E−03	3.74E−11	4.33E−06	8.84E−06	1.00E+00	3.62E+03	2.64E−02
2	8.23E−06	6.72E−04	1.57E−11	2.98E−06	8.23E−06	1.00E+00	7.49E+03	2.45E−02
3	9.93E−06	5.33E−04	2.71E−11	3.93E−06	9.93E−06	1.00E+00	5.23E+03	2.52E−02
4	9.62E−06	6.68E−04	1.70E−10	7.79E−06	9.62E−06	1.00E+00	5.52E+03	2.49E−02
5	8.16E−06	1.36E−03	1.50E−11	3.17E−06	8.16E−06	1.00E+00	3.37E+03	2.62E−02
6	6.14E−06	4.18E−04	1.65E−11	3.54E−06	6.14E−06	1.00E+00	8.24E+03	2.58E−02
7	8.90E−06	4.50E−04	2.39E−10	7.61E−06	8.90E−06	1.00E+00	6.53E+03	2.50E−02
8	8.86E−06	1.44E−04	4.22E−11	5.36E−06	8.86E−06	1.00E+00	5.15E+03	2.31E−02
9	3.12E−06	1.31E−03	5.43E−12	1.77E−06	3.12E−06	1.00E+00	3.39E+03	2.50E−02
10	8.91E−06	4.52E−04	6.47E−11	4.99E−06	8.91E−06	1.00E+00	4.67E+03	2.55E−02

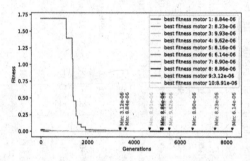

Fig. 9. Convergence diagram comparison in best fitness for the first 10 motors

Table 4. Metrics of the first ten candidate solutions with 15,000 component signatures.

ID	MSE	MAE	MAPE	R2
1	1.50E−06	1.07E−03	1.26E−05	1.00E+00
2	2.49E−10	1.32E−05	1.10E−05	1.00E+00
3	1.19E−06	9.57E−04	7.62E−06	1.00E+00
4	9.72E−07	8.63E−04	1.48E−05	1.00E+00
5	2.50E−07	4.38E−04	3.56E−06	1.00E+00
6	9.41E−09	8.53E−05	3.20E−06	1.00E+00
7	1.18E−06	9.50E−04	1.60E−06	1.00E+00
8	7.09E−06	2.33E−03	9.57E−06	1.00E+00
9	1.96E−07	3.87E−04	3.09E−06	1.00E+00
10	5.58E−06	2.07E−03	3.37E−06	1.00E+00

Table 5. Metrics of the first ten candidate solutions among the step responses of the speed in the DC motors.

ID	MSE	MAE	MAPE	R2
1	2.30E−09	4.03E−05	5.98E−05	1.00E+00
2	8.13E−13	8.16E−07	1.45E−05	1.00E+00
3	1.50E−09	2.94E−05	5.21E−05	1.00E+00
4	9.71E−09	9.55E−05	5.30E−05	1.00E+00
5	3.75E−10	1.36E−05	5.86E−05	1.00E+00
6	1.71E−09	3.79E−05	1.28E−05	1.00E+00
7	4.32E−07	1.74E−04	7.95E−04	1.00E+00
8	2.79E−09	4.44E−05	3.65E−05	1.00E+00
9	3.66E−10	1.23E−05	4.43E−05	1.00E+00
10	3.44E−06	1.72E−04	2.05E−04	1.00E+00

Figure 10 presents a boxplot that helps us to understand the variability of the metrics obtained from the 500 motors. The analysis reveals that the R-square metrics, namely R2r, R2so, and R2s, show an average of 1.0 for the system response, regression with a signature of 6, and regression with 15,000 points.

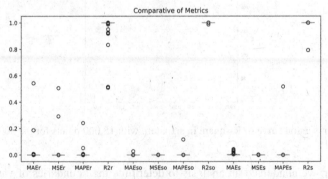

Fig. 10. Boxplot comparison of the different metrics with the system response, the signature with 6 and with 15,000 points, for the 500 motors.

Figure 11, Fig. 12 and Fig. 13 show the dispersion curves for the system response (R2r), regression with signature of six points (R2so), and with 15,000 points (R2s), respectively. These figures provide insights into the averages of the R-square metrics for the 500 motors. The curves' means are around 1.0, with upper and lower boundaries of 0.99 and 1.02.

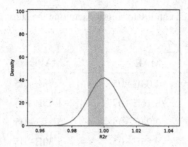

Fig. 11. Dispersion curve of R-square in the system response for the 500 motors.

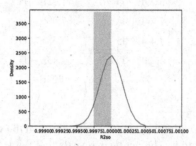

Fig. 12. Dispersion curve of R-square in signature with 6 points for the 500 motors.

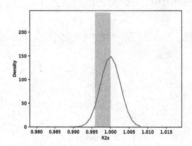

Fig. 13. Dispersion curve of R-square in signature with 15,000 points for the 500 motors.

Moreover, we conducted an analysis to determine the boundaries of variation and the quality of regression obtained from GA transfer function using the six components signature. To evaluate the R-square metrics, we used a binomial test with an acceptable range of variation between 0.99 and 1.0 for the 500 randomly generated motors. We accepted up to a 5% probability of having a metric outside this range. The results for the system response (R2s) are in Table 6, regression with 6 components signature in Table 7, and 15,000 components signature in Table 8.

Our analysis showed that in all the regressions, we successfully maintained outliers in the range of 0.99–1.0 below 5%.

Table 6. Results of binomial test for R2 in the response of the system

Data	Value
Range	0.99–1.0
Values within the range	492
Values outside the range	8
Null hypothesis p	0.05
Expected value	25
Observed p value	8
p-value result	0.00012196053082001523
F-statistic	0.016

Table 7. Results of binomial test for R2 in the signature regression with 6 points

Data	Value
Range	0.99–1.0
Values within the range	492
Values outside the range	8
Null hypothesis p	0.05
Expected value	25
Observed p value	8
p-value result	4.132695012586317e−10
F-statistic	0.002

Table 8. Results of binomial test for R2 in the signature regression with 15,000 points

Data	Value
Range	0.99–1.0
Values within the range	492
Values outside the range	1
Null hypothesis p	0.05
Expected value	25
Observed p value	1
p-value result	4.132695012586317e−10
F-statistic	0.002

4   Conclusions

In this work, we developed a novel approach for determining the transfer function of a DC motor using a few-shot learning technique. Our method involves assigning a unique signature value to each motor. We generate this signature value by subjecting the motor to a signal of varying length and frequency and collecting six numerical values through precise time sampling. This approach simplifies and reduces the search space by decreasing the signal components to six parameters.

To validate our approach, we tested 500 randomly generated motor transfer functions and optimized the mean absolute error between the transfer function generated and the one optimized with the GA. We collected metrics for the signature and the step response of the transfer functions with speed and voltage, using 6 and 15,000 components. We evaluated the results and found that our approach successfully identified the DC motor transfer functions using with the genetic algorithm, with a mean absolute percentage error (MAPE) below 1e−5%, and R-square values between 0.99–1.0 with a probability of 5% of obtaining metrics out of the range.

We found that the results optimized the motor transfer functions using the signature method maintain quality when increasing from six to 15,000 components. The optimization processing time was an average of 2.51E−02 s per generation, with an identification time of 183.6891794 s per run. Our approach offers a significant improvement over existing methods, as it simplifies the process and reduces the time required for identification.

Based on these results, we are confident that this approach holds immense potential for future applications in the field of motor control and automation. The findings suggest that this methodology can be effectively employed to achieve enhanced precision, efficiency, and safety in motor control systems. These promising outcomes underscore the need for further research and development in this area to fully realize its potential and explore its broader implications. Overall, this study sheds new light on the opportunities and challenges of integrating advanced technologies into existing motor control frameworks, and highlights the importance of interdisciplinary collaboration and innovation in this field.

4.1   Future Work

Within the scope of this study, we have exclusively identified transfer functions of DC motors. Nevertheless, we believe that the process signature method, in conjunction with few-shot approaches, has the potential to be extended to other models or transfer functions of any kind. We recommend repeating the process and assessing the results with different systems to evaluate its effectiveness.

References

1. Sra, I.K., Sra, J.S.: Position control of DC motor by using PID, FLC, ANN controller techniques and the comparison of performances. Int. J. Eng. Res. Technol. **4** (2018). https://doi.org/10.17577/IJERTCONV4IS15018

2. Rahmatullah, R., Ak, A., Serteller, N.F.O.: SMC controller design for DC motor speed control applications and performance comparison with FLC, PID and PI controllers. In: Nagar, A.K., Singh Jat, D., Mishra, D.K., Joshi, A. (eds.) Intelligent Sustainable Systems, vol. 579, pp. 607–617. Springer, Singapore (2023). https://doi.org/10.1007/978-981-19-7663-6_57

3. Suman, S.K., Giri, V.K.: Speed control of DC motor using optimization techniques based PID controller. In: Proceedings of 2nd IEEE International Conference on Engineering and Technology, ICETECH 2016, pp. 581–587 (2016). https://doi.org/10.1109/ICETECH.2016.7569318

4. Semenov, A.S., Khubieva, V.M., Kharitonov, Y.S.: Mathematical modeling of static and dynamic modes DC motors in software package MATLAB. In: 2018 International Russian Automation Conference, RusAutoCon 2018 (2018). https://doi.org/10.1109/RUSAUTOCON.2018.8501666

5. Poovizhi, M., Senthil Kumaran, M., Ragul, P., et al.: Investigation of mathematical modelling of brushless dc motor (BLDC) drives by using MATLAB-SIMULINK. In: International Conference on Power and Embedded Drive Control, ICPEDC 2017, pp. 178–183 (2017). https://doi.org/10.1109/ICPEDC.2017.8081083

6. Luengo, D., Martino, L., Bugallo, M., et al.: A survey of Monte Carlo methods for parameter estimation. EURASIP J. Adv. Sig. Process. **2020**, 2020:1–2020:62 (2020). https://doi.org/10.1186/S13634-020-00675-6

7. Timofeev, V.S., Shchekoldin, V.Y., Timofeeva, A.Y.: Identification methods for linear regression based on data of household budget surveys. In: 2018 14th International Scientific-Technical Conference on Actual Problems of Electronic Instrument Engineering, APEIE 2018 – Proceedings, pp. 311–314 (2018). https://doi.org/10.1109/APEIE.2018.8545726

8. Rojas, J., Eichhorn, M., Baatar, G., et al.: Parameter identification and optimization of an oceanographic monitoring remotely operated vehicle. In: 2018 OCEANS - MTS/IEEE Kobe Techno-Oceans, OCEANS - Kobe 2018 (2018). https://doi.org/10.1109/OCEANSKOBE.2018.8559202

9. Gyuk, P., Vassanyi, I., Kosa, I.: Blood glucose level prediction with improved parameter identification methods. In: IEEE 30th Jubilee Neumann Colloquium, NC 2017, pp. 85–88 (2018). https://doi.org/10.1109/NC.2017.8263257

10. Obeidat, M.A., Wang, L.Y., Lin, F.: Real-time parameter estimation of PMDC motors using quantized sensors. IEEE Trans. Veh. Technol. **62**, 2977–2986 (2013). https://doi.org/10.1109/TVT.2013.2251431

11. Lian, C., Xiao, F., Gao, S., Liu, J.: Load torque and moment of inertia identification for permanent magnet synchronous motor drives based on sliding mode observer. IEEE Trans. Power Electron. **34**, 5675–5683 (2019). https://doi.org/10.1109/TPEL.2018.2870078

12. Rubaai, A., Kotaru, R.: Online identification and control of a dc motor using learning adaptation of neural networks. IEEE Trans. Ind. Appl. **36**, 935–942 (2000). https://doi.org/10.1109/28.845075

13. Cosmin, B., Sorin, T.: Reinforcement learning for a continuous DC motor controller. In: 15th International Conference on Electronics, Computers and Artificial Intelligence, ECAI 2023 - Proceedings (2023). https://doi.org/10.1109/ECAI58194.2023.10193912

14. Han, L., Fu, X., Tian, S., Shen, Z.: Demagnetization fault detection research of brushless DC motor based on machine learning. In: 2022 5th World Conference on Mechanical Engineering and Intelligent Manufacturing, WCMEIM 2022, pp. 634–638 (2022). https://doi.org/10.1109/WCMEIM56910.2022.10021516

15. Bujgoi, G., Sendrescu, D.: DC motor control based on integral reinforcement learning. In: 2022 23rd International Carpathian Control Conference, ICCC 2022, pp. 282–286 (2022). https://doi.org/10.1109/ICCC54292.2022.9805935

16. Li, H.G., Liu, B.: Knowledge extraction: a few-shot relation learning approach. In: Proceedings - 2022 International Conference on Machine Learning and Knowledge Engineering, MLKE 2022, pp. 261–265 (2022). https://doi.org/10.1109/MLKE55170.2022.00057

17. Ma, L., Shi, F., Wu, Z., Peng, K.: A survey of few-shot learning-based compound fault diagnosis methods for industrial processes. In: Proceedings - 2023 IEEE 6th International Conference on Industrial Cyber-Physical Systems, ICPS 2023 (2023). https://doi.org/10.1109/ICPS58381.2023.10128105

18. Khatri, K.C.A., Shah, K.B., Logeshwaran, J., Shrestha, A.: Genetic algorithm based techno-economic optimization of an isolated hybrid energy system. ICTACT J. Microelectron. **4** (2023). https://doi.org/10.21917/ijme.2023.0249

19. Dey, N.: Applied Genetic Algorithm and Its Variants: Case Studies and New Developments. Springer, Singapore (2023). https://doi.org/10.1007/978-981-99-3428-7

20. Aziz, R.M., Mahto, R., Goel, K., et al.: Modified genetic algorithm with deep learning for fraud transactions of ethereum smart contract. Appl. Sci. **13**, 697 (2023). https://doi.org/10.3390/APP13020697

21. Jebari, K., Madiafi, M.: Selecztion methods for genetic algorithms. Int. J. Emerg. Sci. **3**, 333–344 (2013)

22. Alzanin, S.M., Gumaei, A., Haque, M.A., Muaad, A.Y.: An optimized Arabic multilabel text classification approach using genetic algorithm and ensemble learning. Appl. Sci. **13**, 10264 (2023). https://doi.org/10.3390/APP131810264

23. Altarabichi, M.G., Pashami, S., Nowaczyk, S., Mashhadi, P.S.: Fast genetic algorithm for feature selection - a qualitative approximation approach. In: GECCO 2023 Companion - Proceedings of the 2023 Genetic and Evolutionary Computation Conference Companion, pp. 11–12 (2023). https://doi.org/10.1145/3583133.3595823

24. Kieszek, R., Kachel, S., Kozakiewicz, A.: Modification of genetic algorithm based on extinction events and migration. Appl. Sci. **13**, 5584 (2023). https://doi.org/10.3390/APP13095584

25. Ali, W., Saeed, F.: Hybrid filter and genetic algorithm-based feature selection for improving cancer classification in high-dimensional microarray data. Processes **11**, 562 (2023). https://doi.org/10.3390/PR11020562

26. Zeng, D., Yan, T., Zeng, Z., et al.: A hyperparameter adaptive genetic algorithm based on DQN. J. Circ. Syst. Comput. **32** (2022). https://doi.org/10.1142/S0218126623500627

27. Greenstein, B.L., Elsey, D.C., Hutchison, G.R.: Best practices for using genetic algorithms in molecular discovery (2023). https://doi.org/10.26434/CHEMRXIV-2023-67HFC

28. Gen, M., Lin, L.: Genetic algorithms and their applications. In: Pham, H. (ed.) Springer Handbook of Engineering Statistics, pp. 635–374. Springer, London (2023). https://doi.org/10.1007/978-1-4471-7503-2_33

Real-Time Emotion Recognition Using Convolutional Neural Network: A Raspberry Pi Architecture Approach

Antonio Romero[1] and Ángel Armenta[2]($\boxtimes$)

[1] Instituto Tecnológico Superior de Misantla, Veracruz, Mexico
[2] División de Estudios de Posgrado e Investigación, Tecnológico Nacional de Mexico/Instituto Tecnológico Superior de Misantla, Misantla 93821, Veracruz, Mexico
ramelendeza@itsm.edu.mx

Abstract. Facial identification of emotions is an area of study that has several cutting-edge solutions, aimed at applications in areas such as security, marketing and robotics. In the literature, there are numerous articles that present algorithms from various perspectives to perform this task to improve the understanding of human behavior and improve efficiency in the performance of tasks in the future.

Emotions can manifest themselves in a variety of ways, such as textual, vocal, verbal, and facial expressions. It is particularly crucial to pay attention to facial expressions as they provide useful information for social interaction and proper communication by observing a person's actions.

This paper presents an architecture that could identify and classify seven facial emotions in children. To achieve this, the Raspberry Pi 3b+ development board was used, and real-time feature extraction is performed using a Convolutional Neural Network (CNN). The process consists of three main stages: face detection, facial feature extraction and facial emotion classification using the FER-2013 dataset. The results obtained for facial emotion recognition show a satisfactory identification rate of 95%, which demonstrates that the use of this architecture is feasible to implement in future robotics projects.

Keywords: Face Recognition · Real-Time Systems · Feature extraction

1 Introduction

Currently, technological advances have progressed rapidly, enabling the creation of applications and systems, one of the fields that has experienced significant growth is Artificial Intelligence (AI), a branch of computer science that seeks to provide machines with the ability to learn and make decisions like those made by a human being [1]. Within AI, one of the most interesting areas is computer vision, which focuses on allowing machines to "see" and understand the world visually, processing images and videos to extract relevant information [2]. Computer vision has revolutionized a wide range of areas, from medicine and medicine to the human body [3] to the automotive industry [4], and one of its most exciting applications is the identification of emotions in real time.

H. Calvo et al. (Eds.): MICAI 2023 Workshops, LNAI 14502, pp. 191–200, 2024.
https://doi.org/10.1007/978-3-031-51940-6_15

Emotion identification is an innate human ability and essential to our social interactions. AI and computer vision have made significant progress in this field, the process of emotion identification using computer vision generally involves the use of machine learning algorithms, such as convolutional neural networks (CNNs), to analyze facial images or gestures in real time [5]. These algorithms are previously trained with large, labeled datasets containing images of faces expressing various emotions, these can be defined as intense feelings that have a great impact on our environment and play a fundamental role in numerous daily activities, such as decision making, learning, perception, reasoning, among many others. Consequently, facial expression is manifested through gestures and facial muscle movements, which act as a form of non-verbal communication and allow us to express emotions, feelings, intentions, goals and opinions to others. Real-time detection of facial expressions using Artificial Intelligence has the potential to transform the way we interact with technology and with each other, providing new opportunities to improve the quality of life and enrich our human experiences [6].

Therefore, the aim of this research is to show the architecture used to perform real-time success predictions of 7 facial expressions (anger, disgust, fear, happiness, sadness, surprise and neutral) by using Raspberry Pi 3b+ development board, 2 IMX219 cameras, facial emotion recognition dataset (FER-2013) and a CNN real-time convolution neural network with OpenCV library, viz: TensorFlow and Keras.

This paper is organized as follows: Section II describes some of the most relevant works in computer vision and face detection, Section III discusses the proposed method, the results and analysis are presented in Section V. The conclusion and future work are given in the last section.

2 Related Work

Computer vision and object detection have been constantly evolving areas of research in recent years. Due to advances in Artificial Intelligence and deep learning, great strides have been made in the accuracy and performance of algorithms used in these disciplines. The following are some of the most outstanding achievements and recent trends in both areas:

Convolutional Neural Networks (CNNs): they are the driving architecture behind many of the advances in computer vision [7]. These networks have proven effective in classifying and localizing objects in images, including human faces. CNNs have been widely adopted for tasks such as face recognition [8], emotion detection, face tracking [9] and identity verification [10].

On the other hand, real-time face detection: has reached impressive levels of efficiency and accuracy. Algorithms such as Single Shot Multibox [11], Detector (SSD) [12] and You Only Look Once (YOLO) [13], have enabled real-time detections even in devices with limited resources, such as surveillance cameras and cell phones. Among the works found, the classification of emotional expressions of people with physical disabilities (deaf, mute and bedridden) was discovered. As well as children with autism by providing a solution to be implemented to understand the emotional states of a group of people with special needs with therapies and treatments [14]. In addition, devices that analyze and read the facial expressions of children when they enter day care centers through cameras in no more than five seconds [15]. Proposed methods include face detection using Deep Learning [14], Viola Jones, Haar cascade, Active Shape Model (ASM) for feature extraction, and AdaBoost for real-time classification [16], and implement robots such as NAO, a combination of face and emotion recognition for real-time application on humanoid robots which use convolutional neural network architectures [14] and compare the performance of these with other algorithms.

In summary, computer vision and object detection have experienced rapid progress due to deep learning and convolutional neural networks. These advances have led to the creation of more efficient and accurate systems for tasks such as face recognition, emotion detection and real-time face tracking.

3 Proposed Method

3.1 Dataset

The FER2013 dataset, was introduced by Pierre-Luc Carrier and Aaron Courvill and provided by Kaggle, consists of grayscale images of faces categorized into different emotions [14]. Each image has a resolution of 48 pixels by 48 pixels. The dataset contains 35,887 images, with 4,953 "Anger", 547 "Disgust", 5,121 "Fear", 8,989 "Happiness", 6,077 "Sadness", 4,002 "Surprise", and 6,198 "Neutral" images. These represent 7 different types of emotional micro expressions. Each image is labeled with a numerical rating from 0 to 6, corresponding to the 7 different emotional categories (see Fig. 1). We used this dataset to train the model with the help of a convolutional neural network.[1]

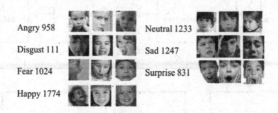

Fig. 1. FER-2013 data set.

The faces have been automatically registered so that the face is centered and occupies approximately the same amount of space in each image. Likewise, the task consists of categorizing each face according to the emotion shown in the facial expression into one of seven categories: anger, disgust, fear, happy, sad, surprise, and neutral, is described in Table 1. The training set consists of 28,709 examples and the public test set consists of 3,589 examples.

The objective of this task is to assign an emotional category to each face based on facial expression, using a set of seven previously predefined categories.

---

[1] https://www.kaggle.com/datasets/msambare/fer2013.

Table 1. Number of data in the FER-2013 dataset.

Classification of micro expressions	Validation Data		Training Data	Dataset Total
	Private	Public		
Angry	491	467	3395	3953
Disgust	55	56	436	547
Fear	528	496	4097	5121
Happy	879	895	7215	8989
Sad	594	653	4830	6077
Surprise	416	515	3171	4002
Contempt	626	607	4965	6198
	3589	3589	28709	35887

3.2 Algorithm

A Convolutional Neural Network (CNN) implemented in TensorFlow and Keras follows a common structure. In general, the CNN works in the following way using some libraries (see Fig. 2).

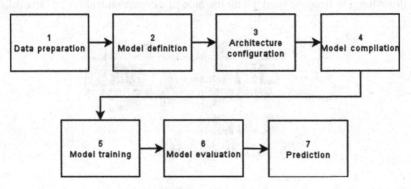

Fig. 2. General structure of CNN.

In data preparation, prior to training a CNN, the data must be properly prepared. This involves loading the training and test images, performing preprocessing (such as resizing or normalizing), and dividing the data into training and validation sets, to give way to model definition, where a CNN model is created using the layers provided by TensorFlow and Keras [16]. A typical CNN consists of convolutional layers, pooling layers, activation layers and fully connected (dense) layers.

Thus, the architecture configuration, where the architecture of the convolutional neural network is defined by establishing the number and size of the convolutional filters, the activation function, the grouping size and the structure of fully connected layers. Then in the model compilation, the model training parameters are configured, such as the optimizer, the loss function and the metrics to evaluate the model performance. Likewise, the model training, in which the training set is fed to the CNN and the weights of the layers are adjusted by backpropagating the error. During training, the parameters are optimized to minimize the loss function and improve the model performance. For the evaluation of the previously trained model, the model performance is evaluated using the validation or test set. This involves calculating performance metrics such as accuracy, accuracy per class or confusion matrix. Consequently, prediction, this is performed once the model is trained and evaluated, can be used to make predictions on new unlabeled data. This involves feeding the input data to the network and obtaining the corresponding predictions.

It is important to note that TensorFlow provides the underlying infrastructure for defining and training neural network models, while Keras provides a high-level interface to simplify the construction and training of deep learning models, including CNNs. The combination of the two libraries enables an efficient and flexible workflow for developing CNN models in TensorFlow.

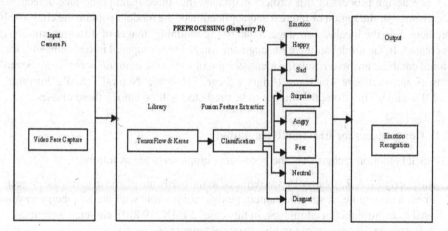

Fig. 3. Real Time Emotion Recognition Architecture.

In this real-time emotion recognition system, the photo captured by the digital camera is used as input and processed by computer code. This code is fed into the Raspberry Pi 3b+ to perform emotion recognition. As a result, a specific emotion classification is obtained as output. Finally, the identified emotion is displayed for visualization, the above process (see Fig. 3).

3.3 Steps to Perform Facial Expression Recognition on Raspberry P

To implement Expression Recognition on Raspberry Pi, the following three steps must be followed.

Step 1: detect the faces in the input video stream.
Step 2: find the region of interest (ROI) of the faces.

Step 3: apply the facial expression recognition model to predict the infant's expression (see Fig. 4).

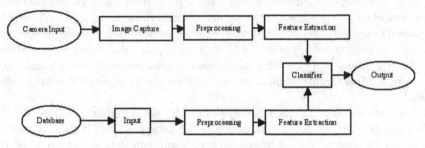

Fig. 4. Architectural overview.

The design process of this project is divided into three main parts: face detection, localization of the region of interest and application of the model to obtain the emotional prediction. In the localization stage, a database containing images of different emotions is created. In the model application stage, the images are compared in real time with the stored database and as a result the identification of a specific emotion is obtained. Seven classes are used here which are 'Angry', 'Fear', 'Happy', 'Neutral', 'Sad', 'Surprise' and 'Disgust'. Therefore, the images to be predicted will be among these classes.

3.4   Components for Emotion Recognition

For facial emotion recognition, the necessary components are as follows:

1. Raspberry Pi: A Raspberry Pi was required as the hardware platform to run the project.
2. Pi camera module: A specific camera module compatible with the Raspberry Pi was used to capture the facial images, in this case, 2 IMX219 8MP cameras were used.
3. A Raspberry Pi compatible multi-camera adapter (See Fig. 5).

Fig. 5. Hardware and elements used for facial emotion recognition.

In addition to these components, you need to install and configure the appropriate software on the Raspberry Pi:

4. OpenCV: It is necessary to install the open source OpenCV library on the Raspberry Pi. OpenCV is widely used for digital image processing and provides functions and algorithms for tasks such as object detection, face recognition and image analysis.
5. Python: Installation of the Python programming language is required to develop the code and operate the board with the components.
6. TensorFlow y Keras: It is also necessary to install these open-source libraries on the Raspberry Pi.
7. Dataset: The FER-2013 data set was used.

With these elements and configurations in place, facial images can be captured using the Pi camera module, and then use OpenCV's functions and algorithms to process and analyze those images, with the goal of recognizing and classifying facial expressions.

4 Workflow

The Keras library was imported using the TensorFlow API. The pre-trained FER-2013 model was loaded from the Keras library. In addition, a dictionary was created that assigns labels to the 7 available emotional classes the process is described below (see Fig. 6).

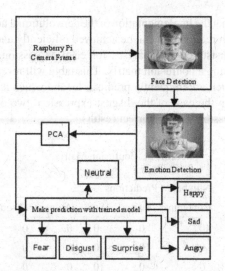

Fig. 6. Architecture operation (Color figure online)

Additionally, the Haarcascade classifier path was provided using a function from the OpenCV library. In the "detect image" function, the "color" function was used to convert the input image to grayscale. The region of interest (ROI) of the image faces is then extracted. This function returns three important elements: the ROI of the faces, the coordinates of the faces and the original image, in which a green rectangle was drawn around the detected face (see Fig. 7).

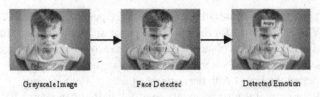

Grayscale Image   Face Detected   Detected Emotion

Fig. 7. Gray scale image. (Color figure online)

The model is applied using the ROI values as input. Two functions are used to obtain the input image and pass it to the face detection function. Then, the output result is displayed with the detected faces. The model output is obtained from the label dictionary that was created earlier, which contains the 7 possible emotions. A function is used to draw labels on the detected faces using a green box. The "show" function is used to display the window and visualize the detected emotion.

A frame is taken from the input video stream and entered the "face in video" function. The "predict" function of the classifier is then used to predict the expression of the detected faces. Labels are assigned to each prediction on the faces and the window with the recognized expression on each face is displayed.

5   Results and Analysis

The findings revealed that the implementation of the convolutional neural network (CNN) model in facial emotion recognition can be achieved efficiently and in real time.

The evaluation consists of ten analyses for each expression, and the results are presented in Table 2 using a confusion matrix. This table will reveal which expressions are easy to predict, which are frequently predictable, and which are difficult to predict. Looking at the table, in the case of the disgust expression, two out of ten attempts to predict the facial expression yield incorrect results.

Table 2. Confusion Matrix.

		Number of Predictions						
		Contempt	Happy	Sad	Angry	Disgust	Surprise	Fear
Expression	Contempt	10	0	0	0	0	0	0
	Happy	0	10	0	0	0	0	0
	Sad	0	0	10	0	0	0	0
	Angry	0	0	0	9	0	0	0
	Disgust	0	0	0	2	8	0	0
	Surprise	0	0	0	0	0	10	0
	Fear	0	0	2	0	0	0	8

The FER-2013 dataset contains a total of 547 data, with a special focus on the disgust expression. However, there are other factors that contribute to the fact that disgust presents a higher number of failures, since this facial expression is quite like the angry expression (See Table 2). The success rate of the architecture in recognizing the user's face is 95%, so it can be concluded that the system shows good results.

The architecture successfully recognized the tests performed. While the results for the anger emotion had an error once, the disgust expression had an error twice, as did the fear expression. In addition, in the tests based on the conditions or the position of the face tilted to the left or to the right, the face looking up or looking down, and the face turning to the left or to the right considering the viewing angle of the cameras. Also, there are 7 emotion ratings and 7 possible positions so that the total test is 287 attempts. Based on the point of view captured by the camera, the correct prediction result was 251 times, while it failed 36 times. So, the overall emotion recognition success rate is 87.45%. The effect of the number of failures is because the image contained in the FER-2013 dataset has very little variation in the viewpoint of the dataset.

6 Conclusion and Future Work

An approach has been presented in this paper that allows detecting 7 emotions in real time using a Raspberry Pi 3b+ with Convolutional Neuronal (CNN) using the FER-2013 dataset. An overall accuracy of 95% has been obtained. The Raspberry Pi 3b+ is a compact and lightweight device that can be embedded in mobile robots. The application of this proposed system is wide and beneficial to society, as emotion recognition plays an important role in various applications.

In future research, it is possible to implement different algorithms to improve the detection accuracy. Also, integration of this approach can be carried out in robots, allowing them to recognize emotions and combine other functions such as audio recognition and image fusion to increase the accuracy of the system.

During the process of facial expression recognition, the results are displayed to visualize the expressions detected by the system. However, it is suggested to use new datasets or add more training data to improve the accuracy of the system as the used dataset presents an imbalance in the amount of both test and training images, especially in tests involving distances greater than 3 m, viewing angles and image rotations, to improve the recognition in real time.

It is advisable to select hardware, such as a digital camera with a higher resolution and functions such as autofocus, to obtain sharp images even when the object is moving, which will improve the performance of the network in both detection and recognition of facial expressions, here is why to have used both cameras, its main objective was to improve the viewing angle and somehow improve the final aspect of future work, as it is planned to implement in a social robot working with children, where the cameras serve as eyes of a bio-inspired robot.

References

1. Langner, S., Beller, E., Streckenbach, F.: Artificial intelligence and big data. Klin. Monbl. Augenheilkd. **237**(12), 1438–1441 (2020). https://doi.org/10.1055/a-1303-6482

2. Matsuzaka, Y., Yashiro, R.: AI-based computer vision techniques and expert systems. AI **4**(1), 289–302 (2023). https://doi.org/10.3390/ai4010013

3. Mintz, Y., Brodie, R.: Introduction to artificial intelligence in medicine. Minim. Invasive Ther. Allied Technol. **28**(2), 73–81 (2019). https://doi.org/10.1080/13645706.2019.1575882

4. Ishikura, H., Khare, A.: Does the Japanese work ethic conflict with the needs of retail in the digital era? In: Khare, A., Ishikura, H., Baber, W.W. (eds.) Transforming Japanese Business. FBF, pp. 71–87. Springer, Singapore (2020). https://doi.org/10.1007/978-981-15-0327-6_6

5. Khan, S., Javed, M.H., Ahmed, E., Shah, S.A.A., Ali, S.U.: Facial recognition using convolutional neural networks and implementation on smart glasses. In: 2019 International Conference on Information Science and Communication Technology, ICISCT 2019, pp. 1–6 (2019). https://doi.org/10.1109/CISCT.2019.8777442

6. Zhao, X.M., Wei, C.B.: A real-time face recognition system based on the improved LBPH algorithm. In: 2017 IEEE 2nd International Conference on Signal and Image Processing, ICSIP 2017, pp. 72–76 (2017). https://doi.org/10.1109/SIPROCESS.2017.8124508

7. Bhatt, D., et al.: CNN variants for computer vision: history, architecture, application, challenges and future scope. Electronics **10**(20), 1–28 (2021). https://doi.org/10.3390/electronics10202470

8. Gaddam, D.K.R., Ansari, M.D., Vuppala, S., Gunjan, V.K., Sati, M.M.: Human facial emotion detection using deep learning. In: Kumar, A., Senatore, S., Gunjan, V.K. (eds.) ICDSMLA 2020. LNEE, vol. 783, pp. 1417–1427. Springer, Singapore (2022). https://doi.org/10.1007/978-981-16-3690-5_136

9. Wu, J., Liu, C., Long, Q., Hou, W.: Research on personal identity verification based on convolutional neural network. In: 2019 IEEE 2nd International Conference on Information and Computer Technologies, ICICT 2019, pp. 57–64 (2019). https://doi.org/10.1109/INFOCT.2019.8711104

10. Mehmood, R., Selwal, A.: Fingerprint biometric template security schemes: attacks and countermeasures. In: Singh, P.K., Kar, A.K., Singh, Y., Kolekar, M.H., Tanwar, S. (eds.) Proceedings of ICRIC 2019. LNEE, vol. 597, pp. 455–467. Springer, Cham (2020). https://doi.org/10.1007/978-3-030-29407-6_33

11. Jiang, Z., Wang, R.: Underwater object detection based on improved single shot multibox detector. In: ACM International Conference Proceeding Series (2020).https://doi.org/10.1145/3446132.3446170

12. Zhai, S., Shang, D., Wang, S., Dong, S.: DF-SSD: an improved SSD object detection algorithm based on DenseNet and feature fusion. IEEE Access **8**, 24344–24357 (2020). https://doi.org/10.1109/ACCESS.2020.2971026

13. Sarda, A., Dixit, S., Bhan, A.: Object detection for autonomous driving using YOLO algorithm. In: Proceedings of 2021 2nd International Conference on Intelligent Engineering and Management, ICIEM 2021, No. Icicv, pp. 447–451 (2021). https://doi.org/10.1109/ICIEM51511.2021.9445365

14. Goodfellow, I.J., et al.: Challenges in representation learning: a report on three machine learning contests. Neural Netw. **64**, 59–63 (2015). https://doi.org/10.1016/j.neunet.2014.09.005

15. Gong, Q., Zhang, J., Zeng, Z.: CNN model design of gesture recognition based on tensorflow framework. In: Information Technology, Networking, Electronic and Automation Control Conference (2019)

16. Zeng, Z., Gong, Q., Zhang, J.: CNN model design of gesture recognition based on tensorflow framework. In: Proceedings of 2019 IEEE 3rd Information Technology, Networking, Electronic and Automation Control Conference, ITNEC 2019, No. Itnec, pp. 1062–1067 (2019). https://doi.org/10.1109/ITNEC.2019.8729185

Optimization of CO_2 Capture Efficiency Through Analysis of Temperature Variables in a Packed Absorption Column

Rafael Terrero Mariano[1]([⊠]) [iD], José Ismael Ojeda Campaña[1] [iD],
Carlos Alberto Ochoa Ortiz[2] [iD], Miriam Navarrete Procopio[3] [iD],
and Víctor Manuel Zezatti Flores[3] [iD]

[1] Tecnológico Nacional de México/ITES de Los Cabos, Los Cabos, México
20381191@loscabos.tecnm.mx
[2] Autonomous University of Ciudad Juarez, Ciudad Juárez, Mexico
[3] Autonomous University of the State of Morelos (UAEM), Cuernavaca, Mexico

Abstract. With the advent of the industrial revolution, climate change has accelerated alarmingly, leaving significant traces on the ozone layer and increasing levels of carbon dioxide (CO_2) and other greenhouse gases. It becomes crucial to reduce and optimize CO_2 capture efficiency in order to mitigate the impacts of climate change and decrease emissions of greenhouse gases responsible for this global phenomenon. The development of effective strategies for CO_2 capture becomes a fundamental priority. This analysis focuses on the effect of temperature variations in a packed absorption column on CO_2 capture efficiency. Through the calibration of temperature sensors using Arduino, data acquisition and analysis tools are employed to gather and analyze temperature data in the absorption process. The obtained data is then compared between steady-state absorption conditions and saturation conditions to evaluate CO_2 capture. The findings of this analysis contribute to the development of more effective CO_2 capture strategies in post-combustion processes, supporting the goal of reducing greenhouse gas emissions and mitigating climate change. By focusing on the analysis of temperature variations in the absorption column, this study provides valuable information to inform CO_2 capture processes and promote more sustainable practices in the industry.

Keywords: CO_2 capture · temperature analysis · absorption column · greenhouse gas emissions

1 Introduction

The growing concern about global warming, driven by the increasing concentrations of greenhouse gases, particularly carbon dioxide (CO_2), in the atmosphere, has spurred the search for strategies to mitigate climate change [1]. The latest report from the Intergovernmental Panel on Climate Change (IPCC) highlights numerous options to reduce greenhouse gas emissions and adapt to climate change caused by human activities [2,

© The Author(s), under exclusive license to Springer Nature Switzerland AG 2024
H. Calvo et al. (Eds.): MICAI 2023 Workshops, LNAI 14502, pp. 201–217, 2024.
https://doi.org/10.1007/978-3-031-51940-6_16

3]. The impact of climate change is particularly pronounced in regions where almost half of the world's population resides, experiencing greater vulnerability to climate-related disasters such as floods, droughts, and storms [4–6]. Reducing the effects of CO_2 emissions from fossil fuel burning is the primary goal of carbon capture and storage (CCS) technologies [7–9]. The current form of energy generation relies on fossil fuel combustion [10, 11], which is one of the main sources of CO_2 emissions according to government institutions reports [12, 13]. In this context, the development of CO_2 capture technologies in post-combustion processes has become a key strategy for both climate change mitigation and economic development [14]. This initiative is backed by various national and international organizations. It is widely recognized that CO_2 capture is crucial for industries such as steel, iron, cement, natural gas processing, and refineries, as reported by the Elcano Royal Institute [15]. There are various technological path-ways for converting CO_2 into commercial products, including catalytic, electrochemi-cal, mineralization, biological (using microorganisms and enzymes), photocatalytic, and photosynthetic processes [16–18]. Post-combustion CO_2 capture processes encompass a variety of technologies, including absorption, adsorption, membrane separation, and cryogenics [19, 20]. Among these options, chemical absorption processes are the most widely used methods in the industry for separating CO_2 from gas streams, standing out as the most extensively applied technique [21]. This study focuses on optimizing CO_2 capture efficiency by analyzing the impact of temperature variations within a packed absorption column that uses MEA as an absorbent at low concentrations due to its high corrosion at high concentrations. The most relevant is to optimizing data acquisition and analysis processes in steady-state absorption and saturation experiments. In this context, an Arduino circuit has been developed to measure temperatures, representing a significant advance in improving efficient CO_2 capture.

2 Methodology

The aim of this work is to provide the user with a tool that allows them to interpret and make decisions based on the obtained results. This section outlines the methodology developed to achieve the objective.

2.1 Concept Diagram

Figure 1 describes the methodology for optimizing CO_2 capture through the analysis of temperatures obtained in the absorption column. The methodology is divided into six stages, starting with the materials and measurement instruments for temperature acqui-sition during the absorption process. Subsequently, a program was developed to acquire temperatures from a user-friendly graphical interface for the computer. Temperature sensors were calibrated and integrated into the absorption system. Experimental tests in the absorption system were conducted using a design of experiments approach, and the collected data were analyzed to optimize the efficiency of CO_2 capture.

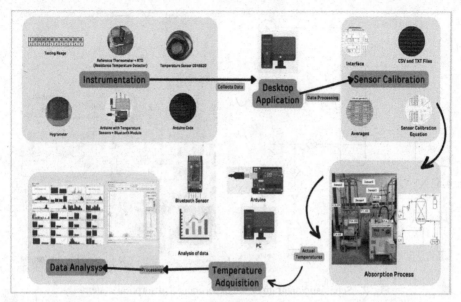

Fig. 1. Conceptual diagram of the methodology for the development of data acquisition optimization.

2.2   System Description

Figure 2 presents the detailed design of the absorption system described by Navarrete Procopio et al. (2023) [22]:

> "The system is a column packed with ½-inch ceramic Berl saddles (PT-101). Tank 1 (TK-101) contains the CO$_2$ that mixes with the airflow supplied by the piston compressor (C-101). Tank 2 (TK-102) contains the MEA solution that feeds the absorption column from the top (stream 1) through a peristaltic pump (P-101). The concentration of CO$_2$ in the gas stream is controlled with the rotameters (R-101, R-102). Tank 3 (TK-103) contains the MEA solution saturated with CO$_2$ that exits from the bottom of the absorption column (stream 6). The gas mixture is fed at the bottom of the column (stream 4). The purified gas is vented to the atmosphere from the top of the absorption column through a gas analyzer to determine the CO$_2$ concentration (stream 5)"

The main difference between original study by Navarrete Procopio et al. and this is the replacement of DS18B20 sensors instead thermocouples, located in current flow 1, 4, 5 and 6.

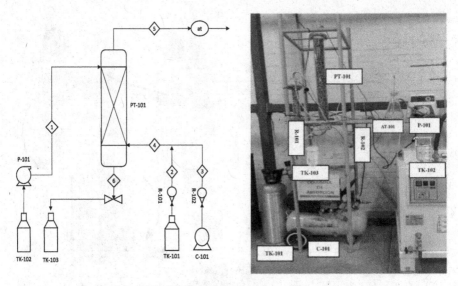

Fig. 2. Schematic diagram of the experimental absorption system.

2.3 Design of Experiments

The absorption process, using MEA (Monoethanolamine) as the absorbent, is an essential technique in the industry for separating and purifying gases through the capture of specific components. Figure 3 below shows an outline of the experimental design performed and the variables to be considered.

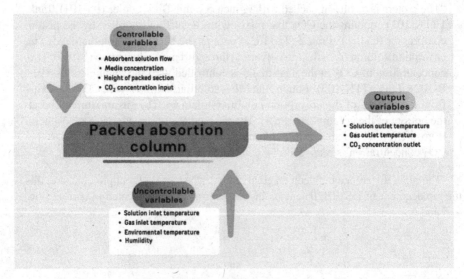

Fig. 3. Design of experiments diagram

In this section, we present the details of the experiments performed to determine the saturation time and to reach the steady state. Table 1 shows the operating parameters of the experiments.

Table 1. Experimental operating parameters.

Parameter Data	
Packing Type	Ceramic Berl saddles, 1/2"
Absorber Diameter (m)	0.08
Packed Absorber Height (m)	0.7
Gas Flow Rate (L/min)	16
MEA Concentration (w/w. %)	10 y 15
Pressure (kPa)	80

2.3.1 Saturation Tests

When the MEA solution reaches its saturation point, it loses the ability to absorb more gas molecules, indicating that it has reached its maximum retention capacity. Knowing the saturation time is essential to optimize the adsorption process and to ensure a flow rate for maximum MEA adsorption capacity. During the experimental tests, two MEA concentrations, one at 10% and the other at 15%, were evaluated to analyze their influence on the adsorption capacity and the saturation time. The gas flow is continuous at the CO$_2$ concentration set under the operating conditions, while the liquid to be saturated is a quantity of sorbent solution loaded in the sorption column.

2.3.2 Continuous Testing

The continuous tests performed with 10% and 15% w/w MEA solutions were designed to operate the sorption column continuously to achieve a steady state condition. Consequently, achieving and maintaining a controlled steady state guarantees the maximum sorption capacity of the sorbent solution.

2.4 Software Development

A tool was developed to assist the user in interpreting and making decisions based on temperatures acquired through an Arduino circuit by automating the calculations and generating the graphs in real time, the waiting time was reduced, allowing early detection of potential errors in the experimental test. Table 2 describes the applications and programming language used in the development of this software. The choice of these tools is essential to ensure the efficiency and functionality of the data processed by the software.

Table 2. Description of the tools used and programming language for software development.

Tools	Description	Version
Programming Language	Python	3.8.10
Pandas Library	Library for handling and analyzing tabular data	1.3.0
Matplotlib Library	Library for generating charts and visualizations	3.4.2
User Interface	QT5 Designer	5
Arduino IDE	Integrated Development Environment for programming Arduino	1.8.13
Excel	Microsoft spreadsheet application for data export	2019, 365, etc.

2.4.1 Electrical Diagram

With the aim of increasing the efficiency of manual processes related to data acquisition and processing, the design of the electrical circuit includes an Arduino Uno connected to four DS18B20 temperature sensors, a 4.7 kΩ providing the correct amount of pull-up current for 1-wire system, and HC-05 Bluetooth module. This setup allows communication and data transfer from the sensors to the Arduino. The component layout of the circuit is shown in Fig. 4. With this configuration, a more efficient and accurate temperature acquisition is achieved.

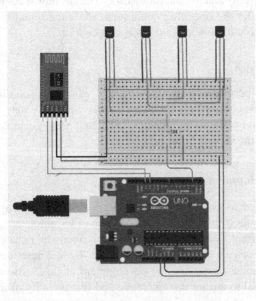

Fig. 4. Diagram of the electrical circuit for precise temperature acquisition.

2.4.2 Acquisition Code

The Arduino code was developed using the "DallasTemperature" library with DS18B20 sensors to acquire temperature data and send temperature readings to a mobile device via a Bluetooth module, allowing their visualization through a serial application that receives data from the Bluetooth connection.

2.4.3 Acquisition Interface of the Absorption System

Figure 5 shows the main interface used for the absorption process experiments, with data acquired from each sensor in the absorption column. The interface is divided into specific sections: in section (a) is where the Arduino port is configured to acquire data, the start button to begin temperature acquisition and allows to set the configuration and duration of the experiment is (b), on (c) has a tooltip with information about process experiments. Section (d) is responsible for calculating the saturation time based on the parameters entered in the form. The time of each test is displayed in section (e), and the acquired temperatures during the specified time in seconds are visualized in section (f).

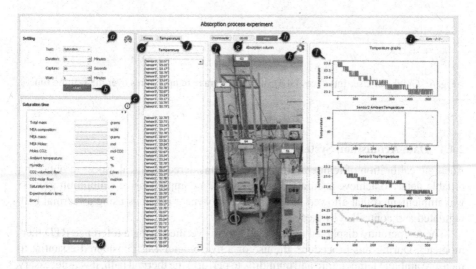

Fig. 5. Main GUI Interface for Absorption Process Experiments with Data Acquired from Each Sensor Located in the Absorption Column.

The interface also includes a timer in section (g), which continues to run even after data acquisition has been completed. (h) it's a button to stop the timer and thus the ongoing experiment. The actual computer date is displayed in section (i), providing a time stamp of the tests performed. Section (j) includes an illustrative picture of the absorption column, with colors representing the location of each sensor. The button (k) generates a graph representing the indicated capture time, calculating the average of the acquired temperatures. Finally, section (l) graphically displays the acquired temperatures, with each sensor represented by a different color for easy identification.

Figure 6 shows a graphical user interface (GUI) displaying the average temperatures over the waiting time, based on data obtained from the four sensors.

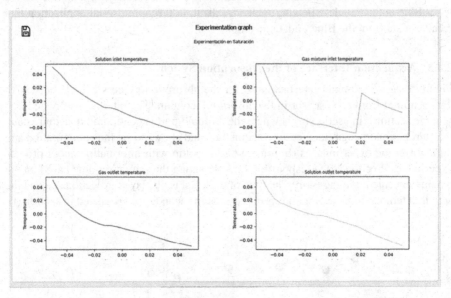

Fig. 6. Graphical User Interface (GUI) showing average temperatures over the waiting time.

2.4.4   Data Acquisition System Calibration

Accurate calibration of temperature sensors is considered a critical aspect in obtaining reliable data. To achieve accurate measurements, a comprehensive calibration process was implemented that allowed the adjustment of sensor readings using a thermal bath regulated and controlled by a cryostat.

The temperature displayed on the cryostat was verified using a reference RTD thermometer. During this procedure, the use of a hygrometer was considered essential to measure ambient temperature and humidity, to perform the thermal bath, it was necessary to define a temperature range in which the sensors would operate within the absorption column.

In this case, the range defined was from 0 °C to 65 °C, this range determines the intervals at which temperature measurements are taken for subsequent averaging and curve fitting. Figure 7 shows the sensors placed in the cryostat to perform the thermal bath and a reference RTD thermometer is used to verify the temperature and ensure that the tests are performed within the specified temperatures.

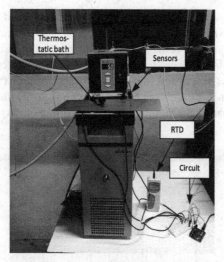

Fig. 7. Configuration of the thermal bath and verification of the cryostat temperature using a reference sensor before starting the temperature sensor tests.

A comparative calibration approach was implemented to establish a direct correspondence between the sensor measurements and the reference values. This approach allowed the application of an individual fitting curve for each sensor, as shown in Fig. 8.

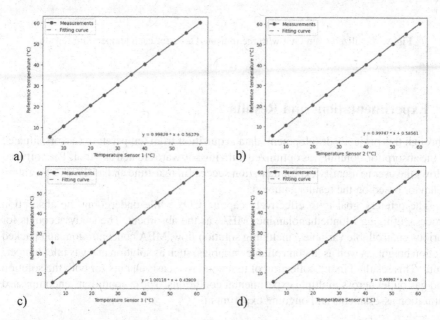

Fig. 8. Fitting curve for each sensor: a) Sensor 1, b) Sensor 2, c) Sensor 3, and d) Sensor 4.

Figure 9 shows the visual interface of the software developed to acquire and adjust the temperatures of the sensors in the Arduino circuit. The interface includes the configuration of data acquisition (a), specific tests for each sensor with visualization of averages and errors (b), (c) display of overall averages (d), the option to configure the Arduino port (e), count readings (f), start and stop temperature acquisition (g), (h), and individual temperature acquisition for each sensor (i), (j), (k), (l). It also includes functions for redirecting (m) and clearing the configuration of performed tests (n).

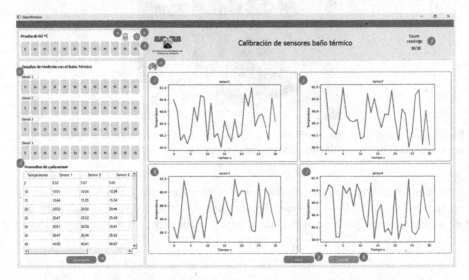

Fig. 9. Finalized main GUI with the indicated tests for each temperature sensor.

3   Experimentations and Results

Through this project's development, data acquisition from temperature sensors situated in the absorption column was optimized utilizing software created as a tool. The software allows the user to identify experimentation success in real-time and analyze temperature behavior based on the results gathered.

The primary goal is to effectively capture CO_2 while undergoing the absorption process, utilizing Monoethanolamine (MEA) as the absorbent. The study accounts for various controllable variables, including solution flow, MEA concentration, and packed section height, as well as uncontrollable variables, such as solution and gas inlet temperature. This resulted in the collection of temperature measurements for both the solution and gas outlet across multiple experimental conditions. These conditions encompassed saturation assessments and ongoing experiments.

3.1 Temperature

The paper presents absorption test results under steady-state and saturation conditions, examining the temperature behavior of gas outlet and solution under different experimental settings.

The software-collected data shows the results of experimental tests conducted at 10% MEA in both continuous and saturation stages. The saturation test indicates when the system has reached its maximum gas absorption capacity, while the continuous test involves a temperature increase to reach a stable equilibrium point. However, equilibrium cannot be fully observed due to the short duration of the experimental test. Refer to Fig. 10 for an illustration of this trend.

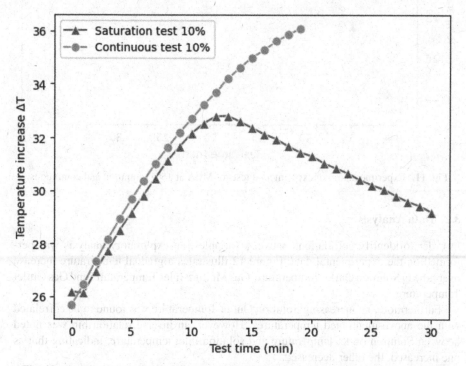

Fig. 10. Experimentation of experimental tests of MEA at 10% saturation and continuous.

An additional experiment was conducted using a 15% MEA solution. The obtained temperatures during the experiment are displayed in Fig. 11, illustrating the establishment of a steady state once equilibrium was reached. Subsequently, the continuous state indicated temperature stabilization, due to ongoing adsorption of MEA molecules. This explains the temperature maintenance. In saturation experiments, the temperature rises initially, reaches a peak, and then declines.

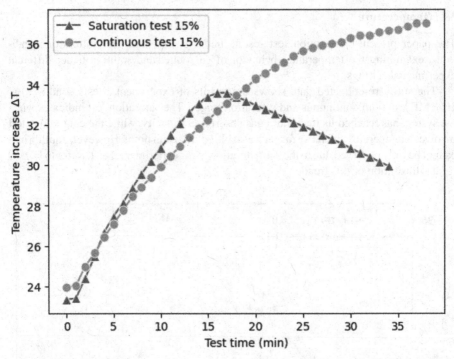

Fig. 11. Experimentation of experimental tests of MEA at 15% saturation and continuous.

3.2 Data Analysis

In order to identify correlations between variables, an exploratory analysis was performed on the experimental data. Figure 12 illustrates a gradual temperature increase over time in Solution Outlet Temperature, Gas Mixture Inlet Temperature, and Gas Outlet Temperature.

Furthermore, an increasing Solution Outlet Temperature was found to be correlated with the above-mentioned temperatures. However, an inverse relationship was noted between Solution outlet temperature and solution inlet temperature, indicating that as one increased, the other decreased.

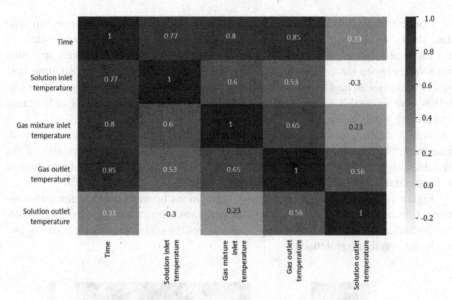

Fig. 12. Correlation of Variables in 10% MEA Saturation Test

In Fig. 13, the relationships are more pronounced, with CO$_2$ gas temperature increasing over time, pointing to a faster absorption rate. Furthermore, a stronger negative correlation was observed between the concentrated solution temperature and the solution temperature, signifying higher sensitivity at greater concentrations.

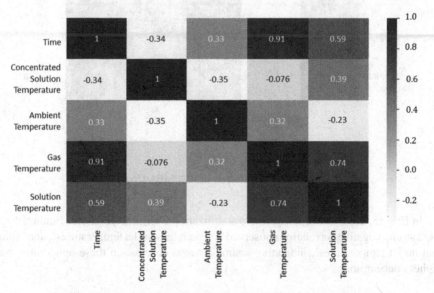

Fig. 13. Correlation of Variables in 15% MEA Saturation Test

Concentrating the MEA solution was found to be a more effective approach. The data shows a positive relationship between time and CO_2 gas temperature, indicating faster absorption at higher concentrations. Furthermore, the negative correlation observed between the temperature of the concentrated solution and the solution suggests more efficient heat transfer from the gas to the solution. Additionally, the text will follow conventional academic structure and format, with clear and objective language and precise word choice.

In both continuous experiments utilizing 10% and 15% MEA solutions, a significant negative correlation was discovered between time and the temperatures of the variables being analyzed. This finding suggests a trend toward stabilization or reduction in temperatures as the experiment progressed.

Figure 14 displays important negative correlations between the concentrated solution's temperature, the ambient and gas temperatures, illustrating the substance's noteworthy influence on the surrounding temperatures. Notably, we observed strong and moderate negative correlations.

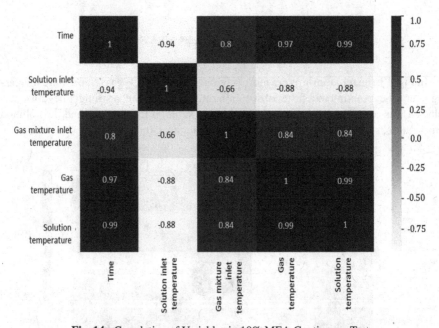

Fig. 14. Correlation of Variables in 10% MEA Continuous Test

In Fig. 15, the trend of temperature stabilization over time persisted. Finally, there were strong negative correlations observed between 'gas outlet temperatures', and 'solution outlet temperatures', indicating a simultaneous decrease in these temperatures at higher concentrations.

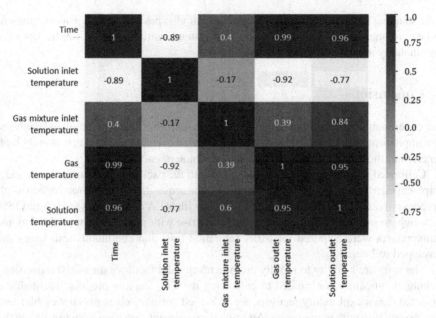

Fig. 15. "Correlation of Variables in 15% MEA Continuous Test"

Both experiments emphasize the importance of maintaining a controlled steady state to maximize efficiency in the continuous absorption of CO$_2$ with MEA solutions. The higher concentration (15%) exhibited more pronounced correlations, suggesting greater thermal sensitivity and more efficient heat transfer compared to the 10%. These findings are crucial for industrial and environmental applications, ensuring an optimal CO$_2$ capture process and efficient temperature management within the absorption system.

4 Future Work

Integration of Artificial Intelligence for CO$_2$ Absorption Process Optimization, based on the results obtained from the exploratory analysis of saturation experiments with 10% and 15% MEA solutions, alongside the continuous state, significant opportunities to enhance the efficiency of the CO$_2$ absorption process have been identified. Despite identifying crucial correlations between variables, a broader dataset is required to pinpoint additional key points. So far, the analysis has focused on linear relationships and observable patterns from existing data. However, the complexity of the CO$_2$ absorption process and the dynamic interaction of various variables necessitate more sophisticated approaches, such as constructing Machine Learning models, neural networks, deep learning algorithms, or support vector machine algorithms capable of capturing complex patterns and nonlinear behaviors in the data. These models could predict system behavior under various conditions, enabling real-time adaptation to maximize CO$_2$ absorption efficiency.

Optimization algorithms can explore a vast and complex parameter space far beyond the capabilities of traditional exploratory analysis. These tools will not only enhance the understanding of the CO_2 absorption process but also pave the way for more efficient and environmentally friendly carbon capture systems, with significant implications for sustainability and environmental management.

5 Conclusions

The experimental process was optimized through real-time data collection and analysis. The implementation of specialized software enabled accurate data collection and graph generation, showing the error margins and precision of each sensor.

Calibrated temperature sensors integrated into the packed absorption column and a graphical interface for data acquisition enabled the acquisition of temperatures across all 4 process streams. Experiments were conducted with MEA contractions at 10% and 15% by weight to mitigate the corrosion issues that arise with more concentrated solutions. Temperatures were acquired for MEA saturation tests and continuous tests using the developed software.

The software's ability to rapidly provide results and facilitate data-driven decision-making significantly contributed to enhancing the CO_2 capture process. Through the collected data's exploratory analysis, we identified correlations between variables and their association with temperature. Additional experiments are necessary to validate the observations; nonetheless, the results provided will facilitate future implementations of Artificial Intelligence techniques. These techniques offer a more advanced and flexible approach to optimizing the CO_2 capture process.

References

1. Organización Mundial del Comercio, El Comercio y el Cambio Climático, Secretaría de la OMC, Suiza (2009)
2. Intergovernmental Panel on Climate Change (IPCC): Urgent climate action can secure a liveable future for all, Switzerland (2023)
3. Intergovernmental Panel on Climate Change (IPCC): AR6 Synthesis Report Climate Change 2023, IPCC (2023)
4. Intergovernmental Panel on Climate Change (IPCC): Climate Change 2023: Synthesis Report. Contribution of Working Groups I, II and III to the Sixth, H. Lee and J. Romero, Geneva, Switzerland (2023)
5. Iberdrola, Países más afectados por el cambio climático (2020). https://www.iberdrola.com/sostenibilidad/paises-mas-afectados-cambio-climatico. Accessed 8 July 2023
6. United Nations, Causes and Effects of Climate Change (2023). https://www.un.org/en/climatechange/science/causes-effects-climate-change. Accessed 17 July 2023
7. Salaet Fernández, S., Roca Jusmet, J.: Agotamiento de los combustibles fósiles y emisiones de CO_2: algunos posibles escenarios futuros de emisiones. Revista Galega de Economía **19**(1), 1–19 (2019)
8. Rosca Jusmet, J.: La política climática y los combustibles fósiles: una perspectiva desde la oferta. Revista de Economía Crítica (34), 9–25 (2022)
9. Díaz Cordero, G.: El Cambio Climático. Ciencia y sociedad **XXXVII**(2), 227–240 (2012)
10. Laguna Monroy, I.: La generación de energía eléctrica y el ambiente. Gaceta Ecológica (65), 53–62 (2002)

11. Arango, M.A.: Model risk assessment projects in thermal power generation. Revistas Espacios **37**(9), 26 (2016)
12. Samaniego, J., Schneider, H.: Cuarto informe sobre financiamiento para el cambio climático en América Latina y el Caribe, 2013–2016. Comisión Económica para América Latina y el Caribe (CEPAL) (2019)
13. Olcina Cantos, J.: Cambio climático y riesgos climáticos en España. Investigaciones Geográficas (49), 197–220 (2010)
14. Romeo, L.M., Bolea, I.: Overview post-combustion CO$_2$ capture. Boletín del Grupo Español del Carbón (35), 8–11 (2015)
15. Álvarez Pelegry, E.: La captura y almacenamiento de CO$_2$: una solución eficiente para luchar contra el cambio climático. Real Instituto Elcano, pp. 1–21 (2010)
16. Zhu, Q.: Desarrollos en tecnologías de aprovechamiento de CO$_2$. Energía limpia **3**(2), 85–100 (2019)
17. Alper, E., Orhan, O.Y.: Utilización de CO$_2$: desarrollos en procesos de conversión. Petróleo **3**(1), 109–126 (2017)
18. Chauvy, R., Meunier, N., Thomas, D., De Weireld, G.: Selección de productos emergentes de utilización de CO$_2$ para implementación a corto y mediano plazo. Energía aplicada (236), 662–680 (2019)
19. Ayala Blanco, E., Martínez Ortega, F.: Tecnologías de captura de CO en procesos de postcombustión de gas natural. MET&FLU (14), 22–33 (2019)
20. Chao, C., Deng, Y., Dewil, R., Baeyens, J., Fan, X.: Post-combustion carbon capture. Renew. Sustain. Energy Rev. **138**(C), 110–490 (2021)
21. Environment LIFE Programme: LIFE-CO$_2$-INT-BIO - CO$_2$ emissions reduction by industrial integration and value chains creation, Ministerio para la Transición Ecológica y el Reto Demográfico, Madrid, España
22. Navarrete Procopio, M., Urquiza, G., Castro, L., Zezatti, V.: Saturation of the MEA solution with CO$_2$: absorption prototype and experimental technique. Results Eng. **19**, 101286 (2023). https://doi.org/10.1016/j.rineng.2023.101286

Fuzzy-Bayesian Expert System for Suggesting Personalized Training Plans with Exercises and Routines

Rosa Lizeth Estrada Ortega[✉]

Polytechnic University of Aguascalientes, Aguascalientes, Mexico
mc220010@alumnos.upa.edu.mx

Abstract. In the dynamic world of fitness, the quest for tailor-made exercise plans has never been more critical. Expert systems are stepping up to this challenge, offering a promise of personalized routines for both seasoned athletes and newcomers. In this article, we introduce an ingenious fuzzy-Bayesian expert system, armed with fuzzy and Bayesian logic, that tackles the uncertainties lurking within fitness data. It's a system with a dual focus: for fitness veterans, it conducts a meticulous analysis of their training history, past achievements, and current ambitions. Guided by the nuanced power of fuzzy and Bayesian logic, it makes decisions rooted in probabilities. As for beginners, the system builds a strong foundation, considering your present fitness status and your future aspirations. This forward-thinking approach speaks directly to the increasing demand for adaptable fitness solutions. For seasoned athletes, it paves the way to peak performance and injury prevention. Meanwhile, beginners embark on their fitness journey with gradual and secure steps. The Fuzzy-Bayesian Expert System offers a comprehensive answer that embraces data diversity and individual fitness aspirations in both scenarios. Furthermore, the seamless integration of the UPAFuzzySystems library and Twilio elevates the system's functionality and its connection with users, turning it into a personal fitness enhancement tool. Dive into a world where fitness meets technology for a custom-made future of health and well-being.

Keywords: Exercise Routines · Healthy Recommendation System · Expert Systems · Fuzzy Logic · Bayesian Logic

1 Introduction

An expert system is a computer system capable of reasoning and acting at the level of an expert in a specific field or activity. Expert systems are designed to address specific problems by applying Artificial Intelligence (AI) techniques that simulate the human reasoning process. These systems earn their name, "expert systems", due to their ability to emulate the decision making performed by experts in a particular field [1, 2]. Approaching the expert system with the training plans, we realize that there is a close relationship, then, some concepts that will be discussed later:

Physical exercise is an essential activity for human beings, as it strengthens muscles, improves blood circulation, oxygenates organs and increases skin nutrition. In addition, it contributes to the development of muscle mass, increases endurance and has positive effects such as weight loss, mood improvement and increased body energy [1]. With the growth of the fitness community, there have been technological advances that facilitate activities related to nutrition and exercise. However, there is still an opportunity to improve care for these needs [2]. The following is a summary of the findings.

Exercising, participating in sports activities and maintaining constant physical activity and even vibration of hole body routines offer a valuable opportunity to preserve our health and improve both our physical condition and mental well-being. There is strong evidence to support the positive impact of physical activity on our health, helping to prevent and mitigate various diseases and health problems. Regardless of our age, we all have the ability to integrate this healthy habit into our daily routine, whether we are just starting or resuming physical activity after a period of inactivity [2, 3].

Among modern technologies, the Internet of Things (IoT) stands out, which enables the connection to the network and the incorporation of computing capabilities to virtually any object of daily use. This makes it possible to record all interactions with these objects and generate a valuable historical database [4], within the development of the system we will take up some topics such as fuzzy logic and Bayesian logic and how they were applied in the process until obtaining a result.

Fuzzy Logic offers an inference mechanism that mimics human reasoning in knowledge-based systems. Its theory provides a mathematical framework for modeling uncertainty in human cognitive processes [5].

Bayesian inference, founded on Bayes' theorem, is a statistical inference strategy with properties that make it especially relevant in scientific research [6, 7].

We decided to create the system of training plans with the objective of providing training focused on beginners and athletes taking into consideration that many of those who start or not in the gym tend to perform the exercises incorrectly or without taking into consideration very important aspects such as rest, health aspects, objectives and levels.

Multivariate analysis plays a crucial role in our research by allowing us to simultaneously examine multiple variables and their interactions. As we explore the data collected and evaluate the results of our study, it is essential to consider the joint influence of several variables on the phenomena studied. This technique gives us a deeper understanding of the complex relationships that can exist between different factors and allows us to identify patterns, trends and effects that might otherwise go unnoticed. By applying multivariate analysis in our work, we are equipped to address complex questions and answer our research objectives with greater precision and robustness. This translates into greater reliability and validity in our conclusions, which strengthens the robustness of our research and the quality of the recommendations and training plans we provide in our expert system.

Experimental design is going to play a fundamental role in this research, as it focuses on developing an expert system that suggests personalized training plans. Determining the appropriate design will allow the collection of accurate and representative data to support fitness decision making. To achieve this, careful consideration must be given to variables such as the athletes' level of experience, their individual goals, proposed exercise routines, and other factors that influence the effectiveness of training plans.

The choice of a sound experimental design will allow meaningful analysis of how decisions based on fuzzy and Bayesian logic influence athletes' progress, providing valuable information for continuous improvement of the expert system.

2 Theoretical Framework

2.1 Forward Chaining and Backward Chaining Inference Techniques

Forward Chaining
Is an inference method used in expert systems to make decisions or reach conclusions based on a set of initial rules and facts, this approach starts with known facts or input data and combines them with inference rules to generate new conclusions or actions [7].

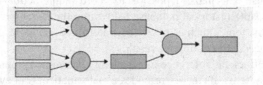

Fig. 1. Forward Chaining

Backward Chaining
A method of inference used in expert systems to reach a conclusion or decide by working backward from the desired goal to the available facts or data. Instead of starting with the initial facts and applying rules to reach a conclusion, backward chaining starts from a goal and looks for rules and facts that support or satisfy that goal [8] (Fig. 2).

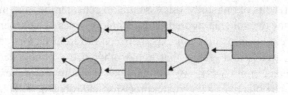

Fig. 2. Backward chaining

2.2 Bayes Theorem

Bayes' theorem is used to calculate the probability of an event, having information beforehand about that event. We can calculate the probability of an event A, knowing also that A complies with a certain characteristic that conditions its probability. Bayes' theorem understands probability inversely to the total probability theorem. The total probability theorem makes inference about an event B, from the results of the events A. On the other hand, Bayes calculates the probability of A conditioned to B, as detailed in Eq. (1) [9]

$$P(A|B) = \frac{P(B|A)P(A)}{P(B)} \tag{1}$$

where $P(A|B)$ is the conditional probability of A given B. $P(B|A)$ is the conditional probability of B given A, $P(A)$ and $P(B)$ are the prior probabilities of A and B, respectively.

2.3 Fuzzy Logic

The use of fuzzy logic in expert systems has proven to be especially effective in areas where uncertainty and subjectivity are common, such as medical diagnosis, engineering decision making, and route planning in navigation systems. These systems have enabled greater accuracy and adaptability in a variety of practical applications [9].

2.4 Tkinter

Tkinter is a Python library designed for creating and developing desktop applications with a graphical user interface. This library simplifies the creation of graphical user interfaces in Python and is used to interface with the Tcl/Tk (Tk) toolkit. It is the standard Python package that facilitates the construction of graphical user interfaces in desktop applications [10].

2.5 Twilio

Twilio provides an easy entry point into the world of telephony [11], it is a development platform that enables programmers to create cloud communication applications and web systems. Its application programming interfaces (APIs) make it possible to deliver personalized communication experiences for users through web and mobile applications. [12] Twilio's APIs are also available in the cloud. (Fig. 1 Dashboard Twilio) (Fig. 3).

Fig. 3. Dashboard Twilio

3   Methodology

In addition, the key steps that make up the implementation process will be discussed, providing a complete overview of the system.

Facts are the basic unit of information of experta. They are used by the system to reason about the problem. We start by adding the required classes to define the facts that the expert system requires to identify rules and conclusions in the proposed expert system. Those classes include level of exercise, pressure, training, future health, and age, impression, satisfaction, quality, time, target, routine the Code 1 shows the definition of the classes in Python using experta library.

Code 1 Rule-based expert system

```
1    class Level(Fact):
2
3        pass
4    class Pressure(Fact):
5        pass
6    class Training (Fact):
7
8        pass
9    class HealthFuture (Fact):
10       .pass
11   class Age(Fact):
12
13       pass
14   class Printed Fact):
15       pass
16
17   class Satisfaction (Fact):
18       pass
19   class Quality (Fact):
20       pass
21
23   class Time (Fact):
24       pass
25   class Objective (Fact):
26
27       pass
28   class Routine (Fact):
29       pass
30
31
```

Most of the time expert systems needs a set of facts to be present for the system to work. This is the purpose of the DefFacts decorator.

You can subclass Fact to express different kinds of data or extend it with your custom functionality. Those sub-facts could help to work with the functions defined in each rule of the expert system. In Code 2 there is an example defining the sub-class facts required when the expert system is initialized.

Code 2 Definition of sub-facts for the initialization of the expert system.

```
1    @DefFacts()
2
3    def initialize_facts (self):
4        yield Level ()
5        yield Pression()
6
7        yield Quality ()
8        yield Time()
9        yield Satisfaction ()
```

A more efficient and effective approach to inferring causes and taking actions based on accumulated knowledge, backward chaining is a crucial technique for decision making, problem solving and development of intelligent systems. Within question_objective1, the ask_info function is used to ask the user about his or her goal ("What is your goal (to lose weight/gain muscle mass/fitness)?"). Here, the goal "lose weight" is set as a possibility. The following Code 3 shows one of the forward chaining rules in the Exercise class:

Code 3 The rules that are of type backward chaining

```
1    @Rule(Training(training ="yes"))
2
3        def question_objective1 (self):
4            self.declare(Objetive(objetive=ask_info("¿What is your
5    goal (to lose weight/gain muscle mass/fitness)?)")))
6        @Rule(Objetive(objetive=" lose weight "),
7
8    NOT(Level(level=W())))
9        def level_question11(self):
10           self.declare(Level(level=ask_info("What is your expe-
11   rience level? (1/2/3)")))
```

Backward chaining is used to make knowledge-based decisions. Its capability is to work backward from a target.

First, rules are established containing conditions (premises) and actions (consequences). The rules are activated if the condition of having a "lose weight" goal and a specific level ("1" or "2") is met. When one of these conditions is met, the action corresponding to the rule is executed. For example, if the objective is "lose weight" and the level is "1", the rule recommendation_route11 is executed.

Code 4 The rules that are of type forward chaining

```
1   @Rule(Objetive(objetive=" lose weight "), Level(level="1"))
2
3       def recommendation_route11 (self):
4           show_info("We recommend low intensity cardio and
5   strength training routines.")
6           self.declare(Rutine(rutine="1"))
7
8       @Rule(Objetive(objetive=" lose weight "), Lev-
9   el(level="2"))
10      def recommendation_routine21 (self):
11          show_info("We recommend doing cardio and strength
12  training routines with moderate intensity.")
13          self.declare(Rutina(rutina="2"))
```

A rule based on Bayes' theorem is mentioned that involves a person's age and the probability of experiencing difficulties in his body; {LS = 1.17, LN = 0.37}: These numerical values are the probability factors associated with the above condition. In this case, two values are provided: LS and LN. The rule is shown below in Code 5.

Examples and possible results:

"What is your goal (lose weight /gain muscle mass /fitness)"

Objective	Nivel	Result
lose weight	"1"	We recommend low intensity cardio and strength training routines
lose weight	"2"	We recommend doing cardio routines and strength exercises with moderate intensity
lose weight	"3"	We recommend functional training routines and high intensity exercises
gain muscle mass	"1"	We recommend strength routines with moderate weight and high number of repetitions
gain muscle mass"	"2"	We recommend strength routines with high load and low repetitions
gain muscle mass	"3"	We recommend strength routines with high load and low repetitions
fitness	"1"	We recommend basic exercise routines to improve your physical condition
fitness	"2"	We recommend doing combined exercise routines to improve your fitness
fitness	"3"	We recommend advanced exercise routines to maintain a high level of physical fitness

Code 5 Bayesian rules IF Age is Old {LS = 1.17 LN= 0.37} THEN difficulties
in their body {0.38}

```
1   @Rule(Age(state="old") | Age(state="young"))
2
3       def EdadFit(self):
4           LS=1.17
5           LN=0.37
6           OS=0.38
7
8           OS=OS/(1-OS)
9           if self.today_edad['state'] == 'old':
10              OS_T = OS*LS
11
12              OS = OS_T/(1+OS_T)
13          if self.today_edad['state'] == 'young':
14              OS_N = OS*LN
15
16              OS = OS_N/(1+OS_N)
17          self.future = SaludFuturo(OSr=OS)
18          self.declare(self.future)
19          print("Old, probability problems in body %f"
20  %self.future['OSr'])
21
```

This rule is based on the condition of a person's age and uses probability factors based on Bayes' theorem to calculate the probability of needing training. If the age is classified as "young", the posterior probability of needing training is 0.61, according to the probability factors provided. Show Code 6.

Code 6 IF Age is young {LS = 2.67 LN= 0.85} THEN training {0.61}

| 1<br>2<br>3<br>4<br>5<br>6<br>7<br>8<br>9<br>10<br>11<br>12<br>13<br>14<br>15<br>16<br>17<br>18<br>19<br>20<br>21 | ```python
@Rule(Age(state="young") | Age(state="old"))
    def EdadFit1(self):
        LS=2.67
        LN=0.85
        OS=0.61
        OS=OS/(1-OS)
        if self.today_edad['state'] == 'young':
            OS_T = OS*LS
            OS = OS_T/(1+OS_T)
        if self.today_edad['state'] == 'old':
            OS_N = OS*LN
            OS = OS_N/(1+OS_N)
        self.future = SaludFuturo(OSr=OS)
        self.declare(self.future)
        print("Young, probability able to training %f"
%self.future['OSr'])
``` |

Rule-Based Expert System with Fuzzy Rules

The proposed expert system has 2 rules, which use at least two fuzzy universes in the input to calculate the acceptance of each of the rules. For knowledge genera-tion, rules supported on a dataset were used.

Rules.
1. Classify user satisfaction on the basis of two variables: quality and time.

Fuzzy sets

Quality_Universe: Represents the fuzzy universe for training quality, with the fuzzy sets "bad", "fair" and "good".
Universe_Time: Represents the fuzzy universe for training outcome time, with the fuzzy sets "low", "medium" and "high".
Satisfaction_Universe: Represents the fuzzy universe for customer satisfaction status, with the fuzzy sets "bad", "fair" and "good" (Fig. 4).

quality_universe = np.arange(1, 101, 1)
tiempoRes_universe = np.arange(1, 7, 1) [Days]
satisfaction_universe = np.arange(0,100.1,0.1)
```

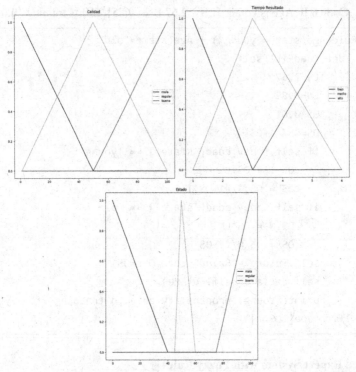

Fig. 4. Universe's quality, time, satisfaction

The following Code 7defines a rule-based inference system using the UPAFuzzySystems library. The inference system is called "Result Quality_Time Inference" and is used to evaluate satisfaction as a function of result quality and time.

Assumptions are added to the inference system, which include three universes: Quality_Universe, TimeRes_Universe (Result Time) and Satisfac-tion_Universe (Satisfaction).

Rules are added to the inference system that establish relationships between quality, time-result and satisfaction status. For example, if the quality is "bad" and the result time is "low", the satisfaction state is "bad".

Multiple rules are defined covering different combinations of quality and outcome time, and how they affect the satisfaction state. The inference system is set up using the "Mamdani" method.

Code 7 Definition of the Rule 1

```
1   Satisfaction_Inference    =UPAfs.inference_system('Result   Quali-
2   ty_Time Inference')
3
4   Satisfaction_Inference.add_premise(Quality_Universe)
5   Satisfaction_Inference.add_premise(TimeRes_Universe)
6   Satisfaction_Inference.add_premise(Satisfaction_Universe)
7
8   Satisfaction_Inference.add_rule([['Quality','bad'],['Time   Re-
9   sult','low']],['and'],[['State','bad']])
10  Satisfaction_Inference.add_rule([['Quality','bad'],['Time   Re-
11  sult','medium']],['and'],[['State','regular']])
12
13  Satisfaction_Inference.add_rule([['Quality','bad'],['Time   Re-
14  sult','high']],['and'],[['State','good']])
15  Satisfaction_Inference.add_rule([['Quality','regular'],['Time
16  Result','low']],['and'],[['State','bad']])
17
18  Satisfaction_Inference.add_rule([['Quality','regular'],['Time
19  Result','medium']],['and'],[['State','regular']])
20  Satisfaction_Inference.add_rule([['Quality','regular'],['Time
21  Result','high']],['and'],[['State','regular']])
22
23  Satisfaction_Inference.add_rule([['Quality','good'],['Time   Re-
24  sult','low']],['and'],[['State','good']])
25  Satisfaction_Inference.add_rule([['Quality','good'],['Time   Re-
26  sult','medium']],['and'],[['State','good']])
27
28  Satisfaction_Inference.add_rule([['Quality','good'],['Time   Re-
29  sult','high']],['and'],[['State','good']])
30
31  Satisfaction_Inference.configure('Mamdani')
32  Satisfaction_Inference.build()
```

2. Calculate blood pressure status and level of training.

Fuzzy sets

Universe_Level: Represents the fuzzy universe for the training level, containing three fuzzy sets: "low", "medium" and "high".
Arterial_pressure_Universe: Represents the fuzzy universe for blood pressure, containing three fuzzy sets: "Low", "Normal" and "High" (Fig. 5).

nivel_universe = np.arange(1,4,1)
presion_arterial_universe = np.arange(60, 161,0 1)

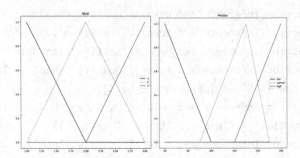

Fig. 5. Level and Pression Universes

We add premises to the inference system, Code 8 which include two universes: Level_Universe and Pressure_Universe.

A consequence is established in the inference system, which is the Estate_Universe.

Rules are added to the inference system that establish relationships between level and pressure, and how they affect the state. For example, if the level is "1" and the pressure is "low", then the state is "Low".

Multiple rules are defined that cover different combinations of level and pressure, and how they affect the state.

Code 8 Definition of the Rule 2

```
1
2   Estate_Inference =UPAfs.inference_system('Results pres-
3   sion_level Inference')
4   Estate_Inference.add_premise(Level_Universe)
5   Estate_Inference.add_premise(Pression_Universe)
6   Estate_Inference.add_consequence(Estate_Universe)
7
8   Estate_Inference.add_rule([['Level','1'],['Pre-
9   sion','low']],['and'],[['Estate','Low']])
10  Estate_Inference.add_rule([['Level','1'],['Presion','nor-
11  mal']],['and'],[['Estate','Low']])
12
13  Estate_Inference.add_rule([['Level','1'],['Pre-
14  sion','high']],['and'],[['Estate','Low']])
15  Estate_Inference.add_rule([['Level','2'],['Pre-
16  sion','low']],['and'],[['Estate','Low']])
17
18  Estate_Inference.add_rule([['Level','2'],['Presion','nor-
19  mal']],['and'],[['Estate','Middle']])
20  Estate_Inference.add_rule([['Level','2'],['Pre-
21  sion','high']],['and'],[['Estate','High']])
22
23  Estate_Inference.add_rule([['Level','3'],['Pre-
24  sion','low']],['and'],[['Estate','High']])
25  Estate_Inference.add_rule([['Level','3'],['Presion','nor-
26  mal']],['and'],[['Estate','High']])
27
2-  Estate_Inference.add_rule([['Level','3'],['Pre-
29  sion','high']],['and'],[['Estate','High']])
30
31  Estate_Inference.configure('Mamdani')
32    Estate_Inference.build()
```

Python Expert Systems.
Two modules related to the Python graphical interface are imported using the tkinter library:

From tkinter import *: This line imports all the elements and functions available in the tkinter module. This includes classes and functions to create windows, botons, labels and other GUI elements.

From tkinter import ttk: In addition to tkinter, this line imports the ttk submodule of tkinter, which provides widgets.Code 9.

Code 9 Tkinter in python

```
1   from tkinter import *
2   from tkinter import ttk
```

These modules allow you to create graphical user interfaces (GUI) in a simple way Code 10,

sendInfo(): This function is used to collect the information entered by the user through a text input window and then print that information to the console. It then destroys the main window.

showInfo(): This function is used to close the main window without collecting additional in-formation.

ask_info(question): This function creates a new dialog box to ask the user a question. The question is passed as an argument in the variable question.

Code 10 Interfaces

```
1    def sendInfo():
2
3        global information, root, info
4        info = information.get()
5        print(info)
6        root.destroy()
7
8    def showInfo():
9        global root
10       root.destroy()
11
12   def ask_info(question):
13       global information,info,root
14       root =Tk()
15       root.title("Ask information")
16
17       root.geometry("700x300+300+300")
18       Ask_information = StringVar()
19       Ask_information.set(question)
20
21       Ask_information_label = ttk.Label(root, textvaria-
22   ble=Ask_information)
23       Ask_information_label.pack(pady=5)
24       information = StringVar()
25
26       information_entry = ttk.Entry(root, width=30, textvaria-
27   ble=information)
28       information_entry.pack(pady=5)
29       Send_information_button =
30   ttk.Button(root,text="Accept",command=sendInfo)
31       Send_information_button.pack()
32
33       root.mainloop()
34
35       return info
```

These functions are designed to obtain information entered by the user, store it in a variable called info, print that information on the console and then close the main window where the application or graphical interface is located.

The GUI function developed in Python using the tkinter library. The function is called show_info(info) and is used to display information in a dialog box. Code 11.

This function creates a pop-up window to request information from the user, obtains the information entered by the user and returns it as a result.

Code 11 Windows

```
1   def show_info(info):
2
3       global information,root
4       root = Tk()
5       root.title("Ask information")
6
7       root.geometry("700x300+300+300")
8       show_information = StringVar()
9       show_information.set(info)
10      show_information_label =
11  ttk.Label(root,textvariable=show_information)
12
13      show_information_label.pack(pady=5)
14      Send_information_button = ttk.Button(root, text="Accept",
15  command=showInfo)
16
17      Send_information_button.pack()
        root.mainloop()
```

Twilio

This code Code 12imports and uses the twilio, flask, and twilio.twiml libraries to integrate the Twilio platform into a Python application from twilio.rest import Client: Imports the Client class from the twilio.rest library, which allows communication with the Twilio API.

From flask import Flask, request, jsonify: Imports the Flask, request, and jsoni-fy classes from the flask library, which are used to create a web application and handle HTTP requests.

From twilio.twiml.messaging_response import MessagingResponse:Importsthe

MessagingResponse class from the twilio.twiml library, which is used to generate responses to incoming SMS messages.

account_sid and auth_token: These variables store the authentication credentials of your Twilio account.

client = Client(account_sid, auth_token): Creates a Client object using the provided credentials. This object is used to interact with the Twilio API and send SMS messages.

Code 12 Twilio SMS

```
1  from twilio.rest import Client
2
3  from flask import Flask, request, jsonify
4  from twilio.twiml.messaging_response import MessagingResponse
5  import time
6
7  account_sid = 'AC1ea0fd99cb15e7a138a1035e14dd4904'
8  auth_token = 'df9e074b1e9d0dcd86f7e15919aa93a0'
9  client = Client(account_sid, auth_token)
```

Sends an SMS message with the content provided in the parameter ask from the phone number + 18148139551 to the phone number + 524492182792, and then prints the SID of the sent message. Code 13.

message = client.messages.create(…): This line uses the client object (which was previously initialized with your Twilio credentials) to create a new SMS message. It specifies the sender's number (from_), the message content (body), and the recipient's number (to).

print(message.sid): Once the message is sent, the identifier (SID) of the message is printed.

Code 13 Parameters

```
1  def send_sms(ask):
2
3      print('test)
4      message = client.messages.create(
5          from_ = '+18148139551',
6          body=ask,
7
8          to=  '+524492182792'
9      )
10     print(message.sid)
```

The function returns the contents of the last message received as a string. If no message has been received, the function will return None.

The while loop will run as long as test is equal to 0, which means that it will keep looking for messages until one is received. Code 14.

If there are no messages in the list, this code block will be executed.

Previous variables is used in a while loop to control the reception of messages.

while test = = 0::: This loop will be executed as long as test is equal to 0, that is, until a message is received.

messages_list = list(messages): Converts the list of SMS messages to a Python list for easy manipulation.

if messages_list::: Checks if there are messages in the list.

last_message = messages_list[0]: If there are messages in the list, gets the last message, which is the first item in the list.

respose = last_message.body: Gets the content of the message and stores it in the respose variable.

Code 14 Content

```
1   def recive_sms():
2       print('prueba salida')
3       time.sleep(5)
4       messages = client.messages.list(from_= '+524492182792' ,
5   limit=5)
6
7       test = 0
8       while test == 0 :
9           messages_list = list(messages)
10          if messages_list:
11              ultimo_mensaje = messages_list[0
12              respose = ultimo_mensaje.body
13              test = 1
14          #return str(respose)
15          else:
16              #respose = "No se encontraron mensajes."
17              test = 0
18              time.sleep(5)
19              messages = client.messages.list(from_=
20   '+524492182792' , limit=5)
21          #print=(respose)
22      return str(respose)
```

4   Results

Possible answers that the user can enter.

Example when a user enters this information:

Information is being collected on whether the user wishes to train Fig. 6.

Fig. 6. Ask the user if he/she is ready to train.

The user's goal (e.g., to lose weight, gain muscle mass or improve fitness) is being collected and incorporated as part of the expert system's knowledge for further processing and recommendations Fig. 7.

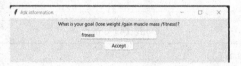

Fig. 7. User's objective: Lose weight, Gain muscle mass Fitness

Determine a user's level of expertise based on his or her training objective Fig. 8.

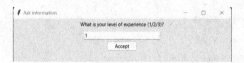

Fig. 8. Experience level 1 beginner, 2 Intermediate 3 Expert

Differentiating between beginners, intermediates and experts allows training programs to be tailored effectively. Each group has specific needs and goals, and this classification serves as a starting point for designing customized fitness plans.

By categorizing individuals into these three levels, it is easier to measure and communicate progress. Athletes can clearly see their progress as they move from one level to the next, which can be a source of motivation.

Beginners often have different needs than experts in terms of exercise, intensity and volume. This classification helps ensure that newcomers are not faced with routines that are too intense and potentially dangerous.

This rule Fig. 9 provides a specific exercise routine recommendation for a user who has the goal of improving their fitness and is a beginner (level 1). The recommendation is to follow basic exercise routines to achieve that goal.

Fig. 9. Results for user

This rule is used to ask the user about the weekly training time in days (in a range from 1 to 6) if the system has no previous information about this data Fig. 10.

Fig. 10. Real time for training

To ask the user about the quality of the training on a scale from 1 to 100. Figure 11 The user's answer is converted into a decimal value and stored in an instance of the "Quality" entity. This allows the expert system to gain knowledge about the quality of the training as perceived by the user.

Fig. 11. Percent for training quality

Final customer satisfaction is determined as the minimum value between the quality and time memberships. This means that satisfaction will be limited by whichever aspect has the lowest membership Fig. 12.

Fig. 12. Result for user

It is used to collect information about the user's age if the system has no previous information about it. The user's response is stored for further processing in the expert system Fig. 13.

Fig. 13. Range for user old or young

Assess the likelihood of problems in the body of a young person with high or low blood pressure.a message is displayed to the user informing about the probability of problems in the body of a young person with high blood pressure. The probability value is obtained from the "FutureHealth" instance Fig. 14 (Figs. 15, 16 and 17).

Fig. 14. Results for user

Fig. 15 Results with presion low problem

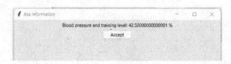

Fig. 16 Results in your blood pressure and training

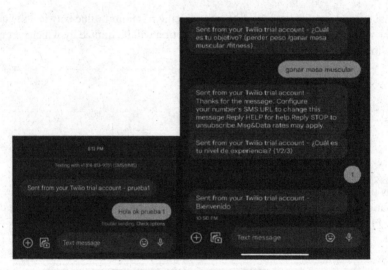

Fig. 17. Results with Twilio SMS

Note: The situation in which the system fails to recognize the text entered by the user becomes a critical point that can lead to the deterioration of the user experience and negatively affect the effectiveness of the system as a whole. When the system is unable to correctly interpret user requests, questions or commands, a number of unintended consequences are generated. This can include irrelevant or incorrect responses, redundant requests for clarification, and even the inability to complete tasks or provide crucial information. This lack of proper acknowledgment can also result in wasted user time and increased frustration, ultimately destroying the reliability and usefulness of the system. Therefore, effectively addressing the improvement of text recognition capability is essential to ensure a smooth and successful interaction between the user and the artificial intelligence application or system.

The conceptual diagram, presented below, is a visual map that identifies the key components of our research, the interactions between them and how they contribute to the achievement of our objectives.

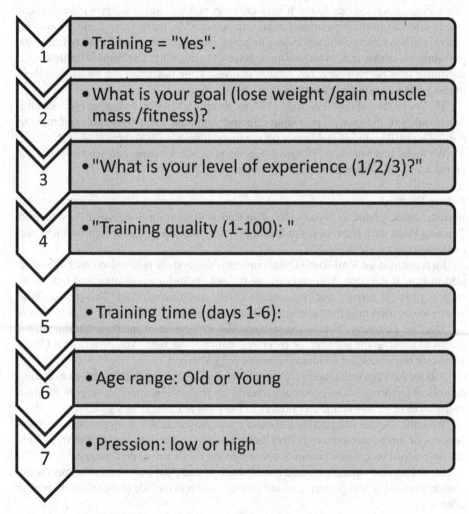

As we move forward in exploring our research, the conceptual diagram will help us illustrate the relationship between variables, processes, influencing factors, and expected results. It will also serve as a guide for readers, allowing them to immediately understand the general structure of our study before delving into the details.

5 Conclusions

In this work we defining an expert system that uses fuzzy and Bayesian rules to help expert athletes or beginners to find the right routines for their level of goal they are looking for. The proposed system has successfully achieved the goal of providing personalized recommendations for both experienced athletes and beginners.

The use of fuzzy and Bayesian logic has effectively addressed the inherent uncertainty in the data and optimized fitness decision making. Since we provide rules that take in consideration common and imprecise language by using fuzzy logic. Moreover, in terms of the conclusions associated to health there are rules that can only lead with probabilities, in this type of rules we use Bayesian reasoning. Experienced athletes can maximize their performance and prevent injuries, while beginners can embark on their fitness journey gradually and safely.

The integration of the UPAFuzzySystems library and Twilio has further enhanced the functionality of the system, providing effective communication with users and making it more accessible and valuable as a personalized fitness improvement tool.

We found that our expert system offers the possibility to save and send information on the user's cell phone thanks to twilio library.

Future Work

In future work related to "Fuzzy-Bayesian Expert System for Suggesting Personalized Training Plans with Exercises and Routines", the following areas of improvement and expansion can be considered:

Incorporation of Real-Time Data: currently, the system is based on user-provided information at the time. You can explore the possibility of integrating real-time data, such as physical activity tracking, energy levels, or biometric data. This would allow you to further fine-tune the recommendations.

Machine Learning: You can investigate the integration of machine learning algorithms to improve the accuracy of recommendations over time. The system could learn from user responses and results to refine its suggestions.

Advanced Personalization: Enhance personalization capabilities by considering additional factors, such as personal preferences, physical limitations, or even location preferences (e.g., whether a user prefers to work out at home or at a gym).

Wearable Device Integration: Consider integrating wearable devices, such as smartwatches or fitness trackers, to collect real-time physical activity and health data. This can help adjust recommendations according to the user's actual performance.

More Exercise Options: expand the database of exercises and routines to encompass a wider range of fitness preferences and goals. This could include niche or sport-specific exercises.

Long-Term Progress Monitoring: Develop a feature that allows users to track their progress over the long term and set future goals. This could motivate users to stay committed to their training plans.

Nutritional Tips: Add a section for personalized dietary recommendations to complement training plans. Nutrition plays an important role in fitness, and this addition could improve the effectiveness of the system.

Data Quality Analysis: Perform an in-depth analysis of the quality of the input data. Ensure that the system can handle incomplete or inconsistent data effectively.

References

1. Wahyuni, R., Irawan, Y.: Web-based heart disease diagnosis system with forward chaining method (case study of ibnu sina islamic hospital). J. Appl. Eng. Technol. Sci. (JAETS) **1**, 43–50 (2019). https://doi.org/10.37385/JAETS.V1I1.19
2. Unsworth, H., Dillon, B., Collinson, L., et al.: The NICE evidence standards framework for digital health and care technologies – developing and maintaining an innovative evidence framework with global impact. Digit Health 7 (2021) https://doi.org/10.1177/205520762110 18617/ASSET/IMAGES/LARGE/10.1177_20552076211018617-FIG1.JPEG
3. Praja, H.N., Yudha, R.P.: Sports education learning program evaluation in senior high school. JUARA : Jurnal Olahraga **6**, 222–238 (2021). https://doi.org/10.33222/juara.v6i2.1215
4. Zhang, W.: Investigation of intelligent service mode of digital stadiums and gymnasiums in the context of smart cities. Int. J. Data Warehous. Mining **19** (2023). https://doi.org/10.4018/IJDWM.322393
5. View of Heart rate variability series analyzing by fuzzy logic approach. https://periodicals. karazin.ua/medicine/article/view/18013/16561. Accessed 29 Sept 2023
6. Grunau, G., Linn, S.: Detection and diagnostic overall accuracy measures of medical tests. Rambam Maimonides Med. J. **9**, e0027 (2018). https://doi.org/10.5041/RMMJ.10351
7. Perbawawati, A.A., Sugiharti, E., Muslim, M.A.: Bayes theorem and forward chaining method on expert system for determine hypercholesterolemia drugs. Sci. J. Inf. **6**, 116–124 (2019). https://doi.org/10.15294/SJI.V6I1.14149
8. Vista de Desarrollo y evaluación de simuladores virtuales para la enseñanza de competencias en el campo de la salud. https://revistas.unicordoba.edu.co/index.php/assensus/article/view/1284/1611. Accessed 29 Sept 2023
9. Morales-Luna, G.: Introducción a la lógica difusa (2002)
10. Moore, A.D.: Python GUI Programming with Tkinter Second Edition Design and build functional and user-friendly GUI applications (2021)
11. Beccaria, M.: How to provide live library information via SMS using Twilio. Code4Lib J. (2011)
12. Romeiser, J.L., Cavalcante, J., Richman, D.C., et al.: Comparing email, SMS, and concurrent mixed modes approaches to capture quality of recovery in the perioperative period: retrospective longitudinal cohort study. JMIR Form Res **5**(11), e25209 (2021). https://formative.jmir.org/2021/11/e252095:e25209. https://doi.org/10.2196/25209

Proposal of a Storage Methodology for Asset Management, Using Artificial Intelligence Techniques, to Make Unit Load Work Processes More Efficient

Ismael Cardona[1]([✉]) and Edgar Gonzalo Cossio Franco[2]

[1] Centro de Tecnología Avanzada A.C., CIATEQ, Zapopan, México
ismaelcardona66@yahoo.com.mx
[2] Instituto de Información Estadística y Geográfica de Jalisco, Zapopan, México
edgar.cossio@iieg.gob.mx

Abstract. This paper proposes an intelligent system that gives solution to a problem of management and work performance, administrative for the technology area of the Urban Electric Train System (SITEUR), addressing the problem, it is proposed to generate a tool to capture online assets, through an app, integrating predictive skills with the help of Machine Learning specifically in the area of electric route, the analysis of this research is oriented to the proposed solution for the area described below and generate an impact with scientific contribution to society.

Keywords: Artificial Intelligence · Assets · Mobile Applications · Machine Learning

1 Introduction

This proposal seeks to generate and integrate a digital tool to address the problems identified in the area of labor management and administrative procedures within the work area of a governmental public transportation institution (SITEUR) in the state of Jalisco, Mexico.

In the control and technology area, where the daily tasks of each of the employees become routine activities at different points in the metropolitan area of Guadalajara, the staff attends 3 transportation systems currently known as: SITREN System, MI MACRO CALZADA, MI MACRO PERIFERICO and LINEA ELÉCTRICA.

Integral urban electric train system (SITREN); LINE 1, 2, 3 and 4: this system consists of urban truck routes, in which the area in question monitors the video surveillance of the unit, radio frequency communication and control of the routes, in the event of a road accident, vandalism, crime or internal control, the area is requested to intervene to retrieve recording information to clarify the facts, the area goes to the site if it is urgent and the evidence is presented to the competent authorities requesting it.

H. Calvo et al. (Eds.): MICAI 2023 Workshops, LNAI 14502, pp. 242–258, 2024.
https://doi.org/10.1007/978-3-031-51940-6_18

Mi Macro Calzada and Mi Macro Peripheric systems: This is a bus rapid transit system or BRT (Bus Rapid Trans-port), in this area employees perform specific activities; control of the video surveillance system, maintenance and support for station breakdowns, lighting system, state Wi-Fi networks and supervision of contractors such as: private security for station users, cleaning service and collection of valuables.

Electric Route System: In this area the collaborators are in charge of charging the electric buses, monitoring and maintenance of electrical equipment such as bus chargers, monitoring the video surveillance system, support for radio communication breakdowns, road accidents, vandalism, crime or internal control of operators and passengers, control of route control which shows the route passage, in messages of lateral projection led to the upper front of the unit, and electrical system of the buses (headlights, internal bells, bodywork such as handrails, seat belts and ramps for the disabled).

2 Background

For field activities, which are those described above for each system, they are recorded on physical paper forms, where the breakdown is highlighted, to whom it is assigned, total time of the employee's shift to address the problem, setbacks detected, reason why the breakdown is not solved in turn, signature of station managers and area manager for validation and closure of the incident or if necessary reassignment of the activity to another shift to be addressed and if necessary conclude the problem.

In the case of administrative procedures in the technology area, the most important are those related to vacations, economic days, justifications, registration of off-shift overtime, unpaid leave, working on holidays, birthday incentives and work breaks. These procedures are handled directly in the area's offices for validation and response of acceptance or rejection of the request.

In both areas of activities and administrative procedures, time is lost in assigning, validating, and accepting activities or requests due to the time of each person who presents the process filters required for each format, since the signatures of the heads and managers are essential to validate each process that is carried out. If any of the area managers is not present when required to authorize or deny different activities or procedures, the operation is delayed, generating labor wastage by personnel who do not work without authorization or generating problems of delays in administrative areas for the assignment of requests.

The digital tool will take the form of a digital application for use in Android operating systems and in the future it will be developed for iOS, in which all the activities carried out on a daily basis will be compiled, generating a database with which the heads and managers of the area will be able to see the progress, time spent per activity and pending generated by each collaborator.

It will also be available to generate administrative incidences, such as vacation requests, permits, registration of vehicle use, overtime, among others that are handled within the institution.

Mexico has a registered part-time work margin of 17.2% only in (2017), 0.8% above the average recorded in the surveyed countries and a percentage much higher than the average of 3.3% of the most developed countries. According to what is specified in [1] where it is denoted that Mexico is not so rooted to staff turnover, in 2022 in Mexico there is a margin of 17%, 0.9% above the average of the surveyed countries.

Given the above context, it is clear that fairness in Mexican labor is present; economic logic would lead one to think that by law the use of overtime in the country would be limited. But as can be consulted in [2] specifically in the section on working hours, it is mentioned "The working day may also be extended for extraordinary circumstances, without ever exceeding 3 h a day or 3 times a week" or "The extension of the overtime exceeding nine hours a week, However, it is also specified that "Workers are not obligated to render their services for a longer period of time than that allowed".

According to [3] a census conducted by the Economic Organization for Cooperation and Development (OECD) says that in the period 2018–2022, Mexico is the second place in hours worked with a total of 2,226, one place below Colombia, which leads the survey with 2,405 h, being Mexico by 474 h above the OECD average, showing that no matter how much laws may protect the worker, different factors may intervene in the excess of work performance in the country, giving Mexico by 474 h above the OECD average, 405 h, with Mexico being 474 h above the OECD average, demonstrating that even though the laws may protect the worker, different factors may intervene in the excess of labor performance in the country, giving in some indirect way a normal practice in the country to work overtime from the normal shift established weekly.

2.1 Definition of the Problem

SITEUR (SISTEMA DE TREN ELECTRICO URBANO), is recognized for being a public organism of the state, dedicated to public transportation in the state of Jalisco, at the service of the citizens. The department of the multimodal transportation technologies department faces a problem in managing the information of the action logs that workers fill out when performing their daily tasks and carrying out the administrative procedures of this department, decreasing the efficiency in work performance due to factors of authorization of activities or the administrative branch, causing low productivity of the department and a rise in their economic reports that negatively impact the system, such as: rework tables, delay of reported activities and complaints of internal administrative inconsistency.

Computer applications are present unnoticed in an everyday environment, directly impacting the lifestyle of those who work in an area where cell phone access is not restricted during working hours as mentioned in [1]. This is the case of the control and technology department, which has a dead time impact on work since, as mentioned in the following research, it shows that interruptions or distractions in work centers can occur due to the use of cell phones, a notable example of this would be the use of cell phones such deficiencies would be: phone calls, text messages, non-productive conversations, (social networks) causing the loss of more than a quarter of the working time [2].

The rise of digital applications has forced organizers, managers, and managers to incorporate digital environments that facilitate their interconnection with globalized environments, "digital transformation is nothing more than seeking the adaptation of companies to the technological changes of the environment and lead them towards competitiveness" [3].

Hence the need to involve all the human talent around technologies, towards new processes and labor management models as a way to grow the organization of the environment where the area of transport SITEUR is dedicated to is circumscribed.

Focused on looking for possible gaps that generate low productivity in addition to the processes identified above, this research focuses on the personnel, analyzing the employees in the technology area, which is made up of 32 employees, 21.9% of whom are over 45 years old, 31.3% are under 30 years old and only 46.9% are under 26 years old, respectively, as shown in Fig. 1.

Please select the age range you are in

32 answers

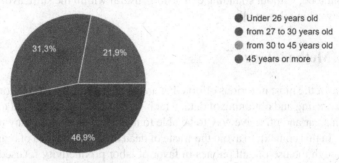

Fig. 1. Age Ranges. Self-Creation.

As explained in [4] "it is documented that, as people age, certain skills related to the ability to perform physical work, reading and numerical comprehension or the use of new technologies deteriorate. For this reason, it is planned to take advantage of the experience acquired by the personnel of the area in their work career to reform a better planning, planning and reaction against time, to give shape to the structure of the digital tool that is planned to be integrated" (Figs. 2 and 3).

Skills	Low education		Secondary education		High education		All		
	Spain	All (Averange)	Spain	All (Averange)	Spain	All (Averange)	Averange	Min.	Max.
Planning	1.84	1.69	2.01	1.90	2.25	2.17	1.98	0.13	3.88
Reading	1.32	1.35	1.81	1.82	2.46	2.44	1.99	-1.16	6.19
Writing	1.51	1.49	1.99	1.86	2.34	2.29	2.00	-0.08	6.51
ICT	1.57	1.45	2.01	1.79	2.19	2.24	2.00	-0.34	6.29
Mathematics	1.71	1.59	2.10	1.88	2.27	2.22	1.98	-0.10	6.73
Physical Effort	3.87	3.96	3.19	3.44	2.43	2.25	3.12	1.00	5.00

Fig. 2. Age Range [4].

On the other hand, the aim is not to invade the mental or family health integrity of the worker beyond the working day, which is 8 h. This is indicated by the study carried out in [5], which details that there is a significant relationship between work shifts and the perception of conflict in which the work/family conciliation culture does not seem

Task and associated skills	
Skills	**Task**
Planning	Planning one's own activities and the activities of the others, organizing time.
Reading	Read documents (addresses, instructions, letters, reports, e-mails, articles, books, manuals, invoices, diagrams, maps).
Writing	Writing documents (letters, reports, e-mails, articles, forms).
ICT	Use e-mail, internet, spreadsheets, dosuments, processors, programming languajes, perform online transactions, participate in conversations (transfers, chats).
Numerical	Calculating prices, costs or budgets; using caculators; preparing graphs or tables; (algebra or formulas; advanced mathematics (calculus, trigonometry, regression).
Physycal effort	Frequency of physical work for long periods

Fig. 3. Tasks and associated skills [4].

to have a significant moderating effect. With this in mind, it is proposed that the tool should be, without sounding exhausting, useful within the shift, avoiding its use outside the working day at all.

3 Methodology

One of the most notorious efforts that are presented for the realization of this work is the channeling and obtaining of data, which with the help of the digital tool can be collected in a clear and effective way to be able to make use of them for any necessary action.

In this context, to avoid the waste of dead time that the idea of managing information through the use of cell phones in favor of labor productivity focused on the problem of control area and technologies, in order to obtain benefits in terms of performance and labor utilization, and once this has been obtained, to demonstrate whether it is also possible to obtain financial benefits, in addition to other factors as a consequence, such as:

- Increased productivity utilization per work shift (being 8 h).
- Time reduction in the registration process of administrative procedures.
- Possible economic savings based on overtime and consumable resources such as paper, toner and electricity.
- Information is always organized, updated and available 24 h a day.

For the innovation proposal related to the area of electric routing specifically, which seeks to estimate how many additional kilometers a unit could travel before needing a new 100% charge, this with the help of the data, kilometers of public transport units, as input and core, in order to guide the artificial intelligence model based on linear regression models and regression based on decision trees. Which is proposed here below.

1. Data collection.

In this section we capture the data generated daily in the electric route area, specifically, we will focus on the dates, unit identifier and the kilometers traveled in each load accumulated by each unit.

Once the aforementioned data have been collected, they will be grouped by type, and then sorted consecutively from the oldest to the most recent data, thus concluding the first phase.

2. Data processing.

The idea of implementing a Machine Learning model is proposed, to be integrated to the digital tool, considering the use of linear regression models or regression models based on decision trees.

Especially focused on the Neural Networks model, with characteristics of being a Unicapa trained model, this analyzed previously to the work done in [9] which details how it "addresses the temperature prediction of lithium-ion batteries in a metal foam thermal management system using artificial neural network models. The study compares the performance of different neural network models, including backpropagation neural network (BP-NN), radial basis function neural network (RBF-NN) and Elman neural networks (Elman-NN)." Since The results show that the Elman-NN model in this single-layer case has better adaptability and generalization ability for temperature prediction in the metal foam thermal management system.

Based on the above, although the paper presents advances in lithium-ion battery temperature prediction using neural network models, it also highlights the need for further research and improvements in this field to address existing challenges and limitations.

Therefore, in order to create a trained model, certain factors must be taken into account for the proposal in question, which are detailed below.

To predict the autonomy of the units based on factors such as initial charge percentage, kilometers traveled, and kilowatt-hours consumed. These regression models can help the electric area system estimate how many additional kilometers a unit could run before needing a new 100% charge, and thus optimize operating times both in terms of bus use and operator schedules.

As mentioned in [6] "For electric bus fleet management, research has been developed in areas such as electric bus system planning, charger location planning, and charging scheduling. These topics are identified as the main lines of research in electric bus fleet operation." Predicting this data will help SITEUR to know based on mileage how much a truck can perform on either of the two routes on which this line operates, avoiding incomplete turns, and avoiding service disruption. Meanwhile, the predicted data based on the KWh would serve as a tabulator to know the cost that generates the system load the bus and see if there are deficiencies in the economic performance generated by each electric bus and detect in time deficiencies or affectations that this can generate.

These models can also be used to estimate the autonomy based on different scenarios, which will allow us to make good decisions about the management of the units, to prevent them from arriving at the charging area with less than 20% charge. It is very common for buses to arrive in these conditions, being forbidden, which is very worrying since as mentioned in [7] the recommendation by battery manufacturers and professionals in the energy field is to avoid discharging the battery more than 20%, because deeper discharge levels can accelerate battery degradation and reduce its useful life. This is because operating the battery at higher discharge levels increases the cyclic and calendrical degradation rates, which can negatively affect the long-term capacity of the battery (Figs. 4 and 5).

A data analysis model was defined thanks to the collection of data in real time, this is of vital importance since it would allow us to analyze in depth the optimization of processes and with the correct data processing methods, the early detection of problems.

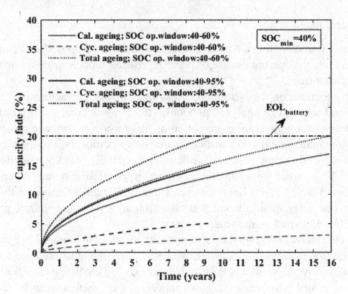

Fig. 4. Discharge slope [7].

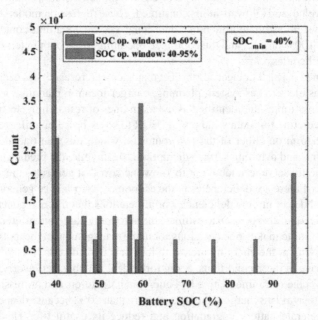

Fig. 5. Average load start. [7].

With this data in mind, an analysis of the data that has been accumulated historically over a period of 3 years was started, only in the area of the electric route, giving a total of more than 30,000 records, which are divided into groups, as shown in the following figure (Fig. 6):

ID CHARGER	UNIT	DATE	START	END	% LEVEL START	% LEVEL END	DURATION	KM	Kw/h	START USER CHARGE	END USER CHARGE
1036767901	C98 - UE-037	14/7/2021	12:50	15:22	42%	100%	02:32	2,931	0	icardona	icardona
1036766501	C98 - UE-019	14/7/2021	12:47	14:00	40%	100%	01:13	2,818	0	icardona	icardona
1036766501	C98 - UE-026	14/7/2021	11:25	12:00	71%	100%	00:35	2,796	0	icardona	icardona

Fig. 6. Database table short view. Self-Creation.

The data had to take different order structures and arrangements in order to be used with statistical tools to find patterns of relationship between them. To give an example, the dispersion analysis, applied to the data sets; % LEVEL START and DURATION.

Using statistical tools such as Minitab, a session is created where a graph is generated between the data to see if there is a relationship of any kind. The following is one of several graphs with which the data analysis is interpreted (Figs. 7 and 8).

↓	C1	C2-D
	% LEVEL START	**DURATION**
2	58.00%	00:52:00.000
3	57.00%	01:16:00.000
4	55.00%	01:05:00.000
5	55.00%	01:00:00.000
6	55.00%	00:59:00.000
7	54.00%	00:57:00.000
8	53.00%	00:58:00.000
9	52.00%	01:14:00.000
10	52.00%	01:02:00.000
11	51.00%	01:06:00.000
12	50.00%	01:11:00.000
13	50.00%	01:06:00.000
14	50.00%	01:06:00.000

Fig. 7. Table with data entered into Minitab. Self-Creation.

In the previous research where the work of [7] was referenced, the data obtained were analyzed to identify if we had the presence of bus load interruption incidents before they reach 100%, with the help of data filtering, the following information was obtained (Fig. 9).

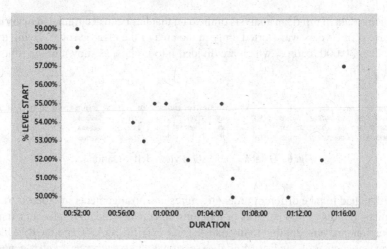

Fig. 8. Scatter plot showing that there is no relationship between data for attrition with irregular variation. Self-Creation.

↓	C1-T	C2
	Colaborator	No. of incomplete loads
1	icardona	30
2	vcastro	35
3	isalcido	0
4	jmurguia	81
5	oruiz	18
6	jrodriguez	3
7	faguilera	1
8	eleon	1

Fig. 9. Table of data with records of incidents interrupted bus loads by contributor. Self-Creation.

Taking into account these data in the Minitab tool, a statistical procedure was initiated with a focus on quality tools, using the Pareto diagram as a procedure, in order to prioritize problems that this exercise highlights, as well as to identify opportunities for improvement and to provide an effective control and follow-up process. The following figure will give us an overview of the data shown above (Fig. 10).

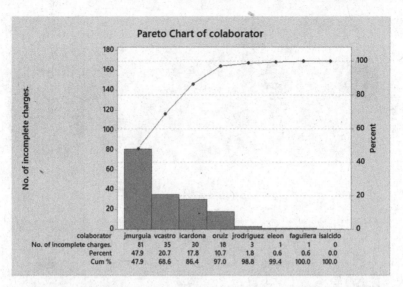

Fig. 10. Pareto chart for bus load interruption incidents. Self-Creation.

In the graph, we can see how the bars are presented in descending order of importance, with the most important cause in this case being jmurguia, the collaborator with the most incidents of this type, and with the greatest relative magnitude after vcatro and last in hierarchy with icardona, this is of utmost importance to prevent these events from occurring and can be addressed before they cause a major problem.

Once the optimization proposal has been identified, as well as its feasibility, based on the literature repository, a series of stages are processed to create a series of stages which will be crucial to carry out in order to carry out the artificial intelligence proposal with the aforementioned methods.

Once the optimization proposal has been identified, as well as its feasibility, based on the literature repository [8], a series of steps will be crucial to carry out in order to realize the artificial intelligence proposal with the aforementioned methods (Fig. 11).

Making use of Information and Communication Technologies (ICT'S) and digital tools for the development of the proposal such as:

1- PHP (Hypertext Preprocessor).
2- HTML (HyperText Markup Language).
3- CSS (Cascading Style Sheets).
4- Code editors (Visual Studio Code, Sublime Text, Atom).
5- Frameworks CSS (Bootstrap, Foundation y Bulma).
6- Design tools (Adobe XD, Figma, Sketch or even online tools such as Canva).
7- Version control (Git together with platforms such as GitHub or GitLab).
8- Testing and debugging (Chrome DevTools and Fire-fox tools).

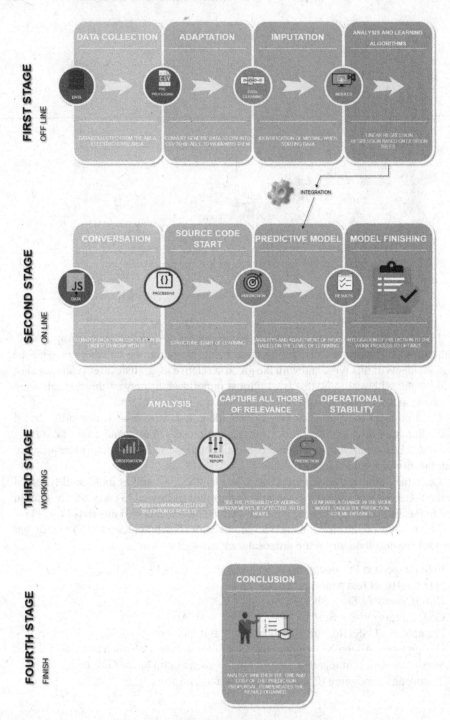

Fig. 11. Working Methodology. Self-Creation.

The creation and implementation of a digital application for the management of assets and the excessive use of untimely labor time is denoted, the indicators of use, noticeable savings margins and feasibility of use within the work area will be denoted with data obtained prior to the implementation of the digital tool, based on results obtained after its application.

Using sampling tools, a survey was conducted among the employees of the technology department, in which they answered 5 questions to find out their point of view about the implementation of this digital tool in their work activities.

Questions were as follows:

1- What type of operating system do you use?
2- What cellular company do you use?
3- What rate plan do you use?
4- Would you be willing to use your own mobile networks, using your own phone, for work, in order to reduce the use of paper formats and speed up the work in your area? Taking into account that you will not spend more than 300 mb per month of your data.
5- Does your phone constantly have bugs? (Crashes or closes apps out of the blue).

The results obtained by the department's employees are shown graphically below (Figs. 12, 13, 14, 15 and 16).

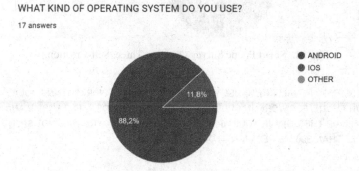

WHAT KIND OF OPERATING SYSTEM DO YOU USE?

17 answers

- ANDROID
- IOS
- OTHER

11,8%

88,2%

Fig. 12. Smart Phone Survey Question One. Self-Creation.

In this aspect [9] defines the research method "as a collective strategy that facilitates the development of the work, since it expands the investigative actions towards a conclusive end of methodological aspects.

WHICH CELLULAR COMPANY DO YOU USE?

17 answers

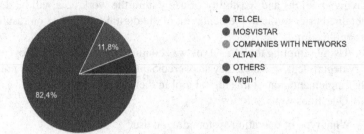

Fig. 13. Smart Phone Survey Question Two. Self-Creation.

WHICH RATE PLAN DO YOU USE?

17 answers

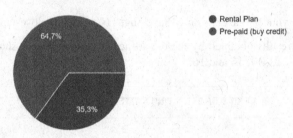

Fig. 14. Smart Phone Survey Question Three. Self-Creation.

WOULD YOU BE WILLING TO USE YOUR OWN MOBILE NETWORKS, USING YOUR
OWN PHONE, FOR WORK, IN ORDER TO REDUCE THE USE OF PAPER FORMS AND
SPEED UP THE WORK IN YOUR AREA? CONSIDERING THAT YOU WILL NOT SPEND
MORE THAN 300 MB PER MONTH OF YOUR DATA.

16 answers

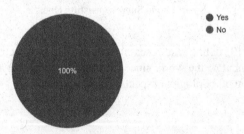

Fig. 15. Smart Phone Survey Question Four. Self-Creation.

The article has multiple factors to be able to carry out the proposed objective, the
digital tool for the capture of assets, as well as its outstanding functionalities that will

DOES YOUR PHONE CONSTANTLY HAVE BUGS? (FREEZES OR CLOSES APPS OUT OF THE BLUE)

17 answers

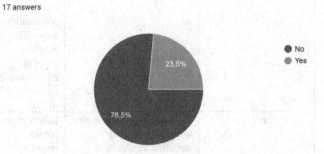

Fig. 16. Smart Phone Survey Question Five. Self-Creation.

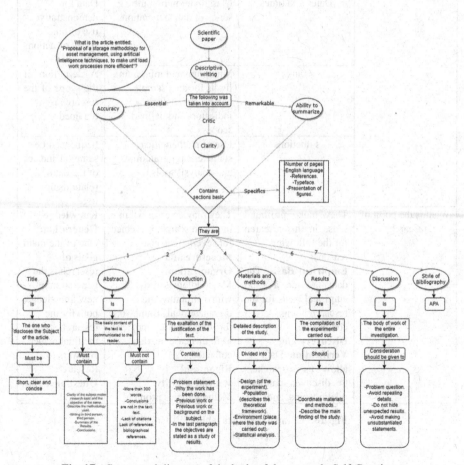

Fig. 17. Conceptual diagram of the body of the research. Self-Creation.

	Datum	Methodology	Results
Elementos clave	Text	Information search	Collection of background information on a specific problem without discriminating against any type of data.
	Numbers	Ordinance and grouping to enter them into statistical methods necessary for the research	Information to identify important factors to attack in the investigation.
	Dates and times	Classification by time series of data generation.	Data for demonstrative use of time series of actions presented.
	Ranks	Maximum and minimum limits for data sample collection, analysis indicators and defined scopes.	A definition of the scope of the research is obtained.
	Functions	Use of mathematical, statistical, programming and analysis tools.	Sequenced de-sampled indices of the data relationship used.
What is the point of research?	They make profound sense in this research for the following reasons: **Empirical Basis**: The data provide a solid empirical basis for the research being conducted. **Discovery and Verification**: Data allow us researchers to discover patterns, relationships, and	They play an essential role in the research presented for several reasons: **Structure and Organization**: Methodologies provide us with a structure and organizational framework for the research process, as in this case the data collected. **Rigor and Reproducibility:** Methodologies promote	**Knowledge Generation:** One of the main goals of research is the generation of new knowledge, both for the researcher and the final reader. Answers to Research **Questions:** The data and

Fig. 18. Comparative data table. Self-Creation.

	Datum	Methodology	Results
	trends in the information collected. **Objective Measurement**: Data provide an objective, quantitative measure of phenomena and variables. **Comparison and Contrast**: Data allow researchers to compare and contrast different situations, groups, or conditions.	scientific rigor by establishing standards and guidelines for data collection and analysis. **Coherence and Consistency**: Methodologies help maintain coherence and consistency in data collection and analysis throughout a research project. **Variables Assessment**: Methodologies help researchers define and measure variables accurately, which is essential for understanding the phenomena being studied.	methodologies help provide answers to the research questions posed at the beginning of the study, in this case it helps flesh out the theoretical framework that was generated. **Pattern and Trend Discovery:** Data analysis can reveal patterns and trends in the information collected, in this case we have statistical graphs and in a future work will be the probability with the correctness of the information. **Validation and Corroboration:** Data support and validate the claims made in an investigation. The empirical evidence provided by the data increases the credibility and reliability of the results.

Fig. 18. (*continued*)

integrate Machine Learning makes this work a multivariate perspective of what asset management is all about, integrating new technologies (AI) with rudimentary but accurate tools such as probability and statistical methods, to treat the data presented here in a comprehensive way with results that will lead to provide solutions not only to the system to which it is addressed but to any problem that needs it, knowing how to adapt its data and planning techniques for it.

With the results of this work, we will analyze the option of implementing it in other areas with similar problems that the control and technology department is facing so far, we are looking for an integral working model, friendly with processes of the same nature of asset capture and that is very useful at the time of integration immediately, without many adjustments, easy to use and friendly to employees, thus providing a methodology that adds value to the work environment of the SYSTEM highlighting other models of public transport and the use of the system in the same way (Figs. 17 and 18).

References

1. OECD. Part-time employment rate (indicator) (2023). https://doi.org/10.1787/f2ad596c-en. Accessed 22 July 2023
2. De Protección Al Salario, C. N. M. (s. f.). Derechos laborales de los trabajadores. gob.mx. https://www.gob.mx/conampros/acciones-y-programas/derechos-laborales-de-los-tra bajadores
3. OECD. Hours worked (indicator) (2023). https://doi.org/10.1787/47be1c78-en. Accessed 21 July 2023
4. Anghel, B., Lacuesta, A.: Envejecimiento, productividad y situación laboral. Boletín económico - Banco de España **1**, 9 (2020). ISSN 0210–3737
5. Pérez Rodríguez, V., Palací Descals, F.J., Topa Cantisano, G.: Cultura de conciliación y conflicto trabajo/familia en trabajadores con turnos laborales. Acción psicológica **14**(2), 193–210 (2017). ISSN 1578–908X
6. Manzolli, J.A., Trovão, J.P., Antunes, C.H.: A review of electric bus vehicles research topics – methods and trends (2022). ISSN 1364-0321. https://doi.org/10.1016/j.rser.2022.112211
7. Shabani, M., Wallin, F., Dahlquist, E., Yan, J.: The impact of battery operating management strategies on life cycle cost assessment in real power market for a grid-connected residential battery application (2023). https://doi.org/10.1016/j.energy.2023.126829
8. https://1drv.ms/x/s!AkyZ8AFM25mBgmBmOULGfIt9808_?e=irUg0r
9. Revista Científica Mundo de la Investigación y el Conocimiento, vol. 3, no. 3, pp. 1155–1176 (2019). ISSN: 2588–073X

Smart Geo-Reference System for the Prediction of the Mercalli Scale for Hurricanes in Los Cabos Area an Approximation of a Risk Atlas

Luis Fernando Bernal Sánchez[1,2] (iD), José Ismael Ojeda Campaña[1,2(✉)] (iD),
and Alberto Ochoa Zezzatti[3] (iD)

[1] Tecnológico Nacional de México, Matamoros, Mexico
jismael.oc@loscabos.tecnm.mx
[2] ITES de Los Cabos, Saltillo, Mexico
[3] Universidad Autónoma de Ciudad Juárez, Chihuahua, Mexico

Abstract. In this work, a smart system was developed to estimate the damage caused by hurricanes in Cabo San Lucas, based on physical and social vulnerability indicators obtained from the "Atlas of Risk and Vulnerability to Climate Change of the Municipality of Los Cabos, BCS". The system uses specific criteria to evaluate the potential destructive effects of hurricanes, considering the availability of relevant information and data. The values obtained were normalized to allow comparisons and weights assigned according to their relevance to the calculations. The integration of these indicators and their precise weighting provides a complete estimate of impact.

In addition to providing a damage estimation method, an intuitive and accessible visual Geo-Referencing approach was developed. The Mercalli scale was adapted and a graphic environment was used to facilitate understanding of the implications and consequences of natural events. This approach offers an effective tool to more clearly assess and explains the consequences of hurricanes.

In summary, the system combines vulnerability indicators, information from the "Atlas of Risk and Vulnerability to Climate Change of the Municipality of Los Cabos, BCS" and a visual representation based on a georeferenced system and the proposal to adapt the Mercalli scale to estimate damage caused by hurricanes. The resulting tool provides a detailed and understandable assessment of the impact and potential consequences of these natural events. With this approach, greater precision in damage estimation and a more complete understanding of its effects is achieved. The developed system constitutes a valuable contribution to addressing the challenges of managing and mitigating the impacts of hurricanes in the region.

Keywords: Smart Geo-Reference System · Mercalli scale · vulnerability indicators

1 Introduction

In recent decades, extreme natural events, such as hurricanes, have been a growing concern for coastal communities around the world. These devastating events can cause significant damage to infrastructure, the economy, and the lives of people in their path.

© The Author(s), under exclusive license to Springer Nature Switzerland AG 2024
H. Calvo et al. (Eds.): MICAI 2023 Workshops, LNAI 14502, pp. 259–273, 2024.
https://doi.org/10.1007/978-3-031-51940-6_19

[1]. The city of Cabo San Lucas, located in a region vulnerable to the arrival of hurricanes, has not been immune to the impacts of these powerful natural phenomena as shown in the following Fig. 1.

Municipality	Declaration type	Classification of the phenomenon	Type of phenomenon	Publication date	Date start	Date end	observations
Los Cabos	Disaster	Hydrometeorological	tropical cyclone	02/10/2003	09/21/2003	09/22/2003	Hurricane Marty
Los Cabos	Disaster	Hydrometeorological	tropical cyclone	09/25/2006	09/01/2006	09/03/2006	Hurricane John
Los Cabos	Disaster	Hydrometeorological	tropical cyclone	10/04/2007	09/04/2007	09/04/2007	Hurricane Henriette
Los Cabos	Disaster	Hydrometeorological	Drought	02/21/2012	05/01/2011	11/30/2011	No observations
Los Cabos	Disaster	Hydrometeorological	Rains	10/25/2012	10/16/2012	10/17/2012	Severe rain caused by Hurricane Paul
Los Cabos	Disaster	Hydrometeorological	Rains	09/05/2014	08/23/2014	08/27/2014	Severe rain and pluvial and river flooding
Los Cabos	Disaster	Hydrometeorological	Rains	09/15/2014	09/03/2014	09/06/2014	Severe rain caused by Hurricane Norbert
Los Cabos	Disaster	Hydrometeorological	tropical cyclone	09/22/2014	09/14/2014	09/15/2014	Hurricane "Odile"
Los Cabos	Climatological Contingency	Hydrometeorological	tropical cyclone	10/09/2014	09/14/2014	09/16/2014	SAGARPA. Natural disaster Cyclone "Odile"
Los Cabos	Disaster	Hydrometeorological	tropical cyclone	09/19/2016	09/06/2016	09/06/2016	Hurricane Newton
Los Cabos	Disaster	Hydrometeorological	tropical cyclone	09/08/2017	08/31/2017	08/31/2017	Tropical Storm Lidia
Los Cabos	Disaster	Hydrometeorological	tropical cyclone	10/01/2019	09/20/2019	09/20/2019	Hurricane Lorraine. Severe rain and strong winds.
Los Cabos	Disaster	Hydrometeorological	Rains	10/23/2019	10/13/2019	10/13/2019	Severe rain and river and pluvial flooding. Low pressure system.

Fig. 1. Natural disasters related to hurricanes in the last 20 years in the municipality of Los Cabos, Baja California Sur, México. **Source:** self-made

These devastating events can have far-reaching repercussions, affecting people's lives and well-being, as well as the economy and sustainable development of the region. According to the infographic of the National Center for Disaster Prevention of Mexico [2] can be seen how disasters caused by hydrometeorological phenomena have had a significant impact on the country as shown in Fig. 2. These events, combined with unplanned urban growth, poverty and environmental deterioration, have exacerbated desertification in the north of the country and caused increased flooding in coastal areas of the Gulf of Mexico [3].

National Water Commission prepares (CONAGUA, by Spanish acronym) a climate report of Mexico were details global climate conditions, precipitation, temperature and notable events [4].

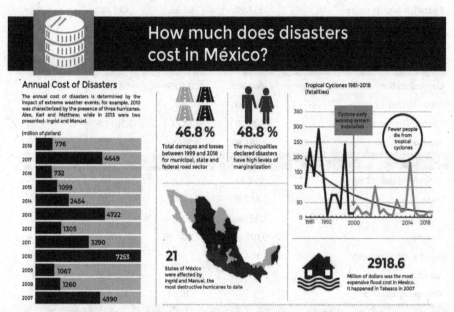

Fig. 2. Natural disasters in Mexico social and economic impact. **Source:** National Center for Disaster Prevention of Mexico [2]

The importance of addressing this problem lies into developing a comprehensive system that allows the consequences derived from damage caused by hurricanes to be adequately quantified. This approach is crucial to understand the real impact that these phenomena can have on physical infrastructure and local communities [5–7]. Furthermore, accurate damage estimation will allow for efficient resource distribution in emergency situations and in the post-hurricane recovery stage [8–10].

Hurricane Odile was one of the most intense hurricanes to affect the Los Cabos region in recent decades. Odile formed in 10 September 2014 in the eastern Pacific and rapidly intensified to become a Category 4 hurricane on the Saffir-Simpson scale. The hurricane made landfall on the Baja California peninsula on September 14, 2014, with sustained winds of up to 215 km/h. Odile caused significant damage in the Los Cabos region, including flooding, landslides, power and telecommunications outages, and damage to tourism infrastructure. It is estimated that the hurricane affected more than 200,000 people and caused more than $1 billion worth of property damage as shown in the Fig. 3.

Concept	Direct dagame	Losses	Total	Percentage of total
Social Infrastructure				
Living place	1,171.84	8.24	1,180.08	4.9
Education infrastructure	337.59	91.18	428.78	1.8
Hydraulic infrastructure	2,094.06	37.90	2,131.96	8.8
Subtotal	3,740.43	141.44	3,881.89	16.10
Economic infrastructure				
Communications and transportation	864.90	118.25	983.15	4.1
Urban Infrastructure	907.00	-	907.00	3.8
Electrical Infrastructure	520.00	1,647.00	2,167.00	9
Subtotal	2,291.90	1,765.25	4,057.15	16.90
Productive Sectors				
Agriculture and fishing	610.20	-	610.20	2.5
Trade	1,627.30	20.30	1,647.60	6.8
Industry	1,961.05	-	1,961.05	8.1
Tourism	9,075.86	1,064.16	10,140.02	42
Other guaranteed spins	1,504.66	-	1,504.66	6.2
Subtotal	14,779.07	1,084.46	15,863.53	65.60
Emergency attention		330.54	330.54	1.4
Grand Total	20,811.40	3,321.69	24,133.11	100

Fig. 3. Summary of damages caused by Hurricane Odile (2014) in the State of Baja California Sur, México. **Source:** National Center for Disaster Prevention of Mexico [11]

Hurricane Odile highlighted the importance of hurricane preparedness and response in the Los Cabos region. Hurricane Odile also had a significant impact on the region's economy, especially the tourism sector. Many hotels and restaurants were damaged and had to temporarily close, affecting the local economy and jobs. Additionally, the region's image as a tourist destination was affected by the damage caused by the hurricane.

2 Objective

The relevance of addressing this problem lies in the need to develop an intelligent Geo-Reference system that allows accurately quantifying the consequences derived from the damage caused by hurricanes in Cabo San Lucas. This approach is essential to understand the true impact that these phenomena can have on local infrastructure and communities.

This work aims to address this problem through the development of a system that combines indicators of physical and social vulnerability, based on information from the "Atlas of Risk and Vulnerability to Climate Change of the Municipality of Los Cabos, BCS" [12]. In addition, a methodology is used that combines specific criteria to evaluate the potential destructive effects of hurricanes, considering the relevance of each factor in the damage estimated. Likewise, an intuitive visual representation of Geo-Reference and the proposal of a scale based on the adaptation of the Mercalli scale have been implemented, which facilitates the understanding of the implications and possible consequences of each natural event.

3 Study Area

The geographical location of Los Cabos, on the Pacific coast is describe in Fig. 4, makes it vulnerable to hurricanes and tropical storms that form in the ocean. In recent years, Los Cabos has experienced rapid population and urban growth, which has placed pressures on infrastructure and public services. The region's population is estimated at around 300,000, but can increase significantly during the tourist season. The region also faces challenges in terms of access to drinking water, waste management and environmental protection.

Fig. 4. Satellite location of the city of Cabo San Lucas.

These hydrometeorological phenomena can have a significant impact on the economy and population of the region, especially in the tourism sector, which is one of the main sources of income. Therefore, it is important that the Los Cabos region is prepared to face the risks associated with hurricanes and tropical storms, and that prevention and mitigation measures are implemented to reduce the vulnerability of the exposed population and infrastructure.

4   Methodology

This study addresses the problem through the development of an intelligent Geo-Reference system that integrates several vulnerability indicators. The selected indicators evaluate different dimensions which are the following: the intensity of the hurricanes, the inherent vulnerability of the area, the capacity to respond to disasters and the local perception of risk. The implemented methodology combines specific criteria to evaluate the potential destructive effects of hurricanes, considering the relevance of each factor in estimating damage. The Fig. 5 shows a conceptual diagram of the system.

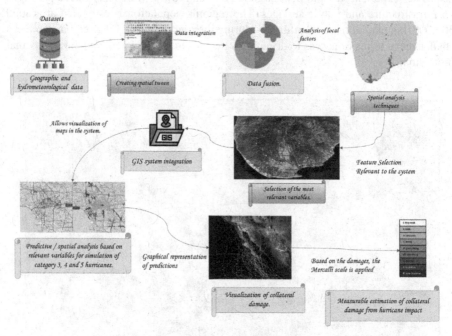

Fig. 5. Conceptual diagram on the applied methodology.

To obtain a comprehensive view of the physical and social vulnerability of the city, information from the "Atlas of Risk and Vulnerability to Climate Change of the Municipality of Los Cabos, BCS" was used [12]. This atlas was prepared through technical studies and analysis, and it contains valuable data on urban infrastructure, land use, population density and other factors relevant to damage estimation.

The vulnerability indicators selected in this study include factors related to the intensity of the hurricane, the inherent vulnerability of the area, the capacity to respond to disasters, and the local perception of risk. These indicators were selected because each addresses a critical aspect of hurricane vulnerability. Hurricane intensity, for example, is a direct indicator of a hurricane's destructive potential. The inherent vulnerability of the area, on the other hand, reflects the physical and socio-economic characteristics that can make an area more susceptible to hurricane damage. Disaster response capacity and local risk perception, meanwhile, provide insight into how local communities may be able to prepare for and respond to a hurricane.

4.1 Selected Indicators and Weight Assignment

To assign weights to each indicator, we used a method based on a comprehensive analysis of the relevant scientific literature and consultations with experts in the field of hurricane damage assessment. This approach allowed us to assign weights that reflect the relative importance of each indicator in estimating hurricane damage in Cabo San Lucas. Each weight assigned to an indicator represents its relative importance in estimating damage. An indicator with a higher weight has a greater impact on the final damage estimate.

In the hurricane hazard assessment study (Table 1) is presented the various components and their corresponding units of measurement in detail. These indicators have been defined to capture the fundamental physical characteristics of the event, allowing us to understand and model the nature and destructive potential of a hurricane. To evaluate the risk of hurricanes, different indicators are used that measure the risk components. Physical indicators measure the intensity of the hurricane, such as wind speed, atmospheric pressure, precipitation, and sea level.

Table 1. Components and indicators of hurricane danger.

Component	Indicator	Weight	V. minimum	V. maximum	Measurement units
Wind	Hurricane Category	0.5	1	5	Saffir Simpson Scale
Wind	Wind speed	0.3	100	300	km/h
Wind	Central pressure	0.2	800	1050	Mbar
Ocean Wave	Amplitude	0.4			m
Ocean Wave	Height	0.5			m
Ocean Wave	Wavelength	0.1			m
Rain	Intensity	0.3			Precipitation index: 0.0-20 (1/Practically constant); 20-40 (2/Weakly variable); 40-60 (3/Variable); 60-80 (4/Moderately variable); 80-100 (5/ Very variable)
Rain	Duration	0.3			Hours

Table 1 details all the components that are considered indicators of the potential risks associated with hurricanes. In this case, based on an exhaustive analysis of each component, a respective weight has been assigned to each of them. These weights represent the relative relevance and priority of each component in the calculations for estimating possible damage caused by a hurricane where we can realize that one of most relevant factors is the storm, which according to the analysis and expert opinions is the relevant factor at the time of the impact of a hurricane since it is the one that causes the greatest impact in terms of damage to infrastructure such as houses and public services. It is crucial that the sum of the weights assigned to each component results in a value of 1.

The evaluation of the hurricane prevention and response capacity in a specific geographic region is addressed. The objective is to understand and model the nature and destructive potential of hurricanes. A set of key indicators have been identified as shown in Table 2 allows measuring the capacity for prevention and response to these extreme climate events. These indicators cover different essential aspects to understanding and modeling the nature and destructive potential of hurricanes, as well as to strengthen the resilience of communities in the face of these challenges.

Table 2. Components and indicators of hurricane prevention and response capacity

Component	Indicator	Weight	Minimum value	Maximum value
Prevention and response capacity	Prevention capacity	0.5	0	4
	Response capacity	0.5	0	6
Social perception of risk	Local perception of risk	0.4	0	8
	Prevention and care capacity	0.5	0	10
	Resilience capacity	0.1	0	6

Table 2 was designed to evaluate indicators of the hurricane prevention and response capacity. These indicators allow us to measure the capacity of a community or region to anticipate, respond and recover effectively against these natural phenomena. The understanding and measurement of these aspects are fundamental to improve planning and reduce the impact of hurricanes in the affected areas where the greatest weights are inclined to the prevention and response capacity, followed by the attention capacity which are crucial to define how the hurricane will impact before, during and after said phenomenon already mentioned.

4.2 Transformation and Normalization of Indicators

The first step in building the model was the normalization of the selected indicators. This is a common practice in statistics that is used to ensure that all variables are on the same scale, thus preventing variables with larger ranges from dominating the damage scores. To achieve this, we use the Min-Max normalization technique. Mathematically, this technique is expressed as Eq. 1, to perform normalization, we use the Min-Max technique.

$$X_{norm} = \frac{X - X_{min}}{X_{max} - X_{min}} \tag{1}$$

In this case, we apply to all the indicators mentioned in Table 1, Table 2 and Table 3. In this formula, X represents the original value of the variable, X_{min} is the minimum value observed in the variable and X_{max} is the maximum observed value. By applying this formula, normalized values are obtained in a range from 0 to 1, where 0 corresponds to the minimum observed value and 1 to the maximum observed value.

The interaction of multiple variables of an integrated system is essential to evaluate the estimated damage that a hurricane can cause. To achieve this, it is necessary to normalize all these variables and bring them to a common scale, which facilitates the necessary calculations and analysis where the process of normalization of the variables allows combining different relevant aspects. By unifying these variables on a common scale, a reference framework is created to evaluate and estimate the possible impact and damage that a hurricane may generate in a certain geographic area.

With the normalized and transformed variables, we calculate the damage score for each hurricane instance as the weighted sum of the variables. Mathematically, this is expressed as the following (Eq. 2):

$$D = \sum_{a=1}^{3} W_a \cdot I_a + \sum_{b=1}^{3} O_b \cdot I_b + \sum_{c=1}^{3} R_c \cdot I_c + \sum_{d=1}^{2} P_d \cdot I_d + \sum_{e=1}^{3} S_e \cdot I_e \quad (2)$$

where D is hurricane damage estimation index calculated as the weighted sum of physical (W_a, O_b, R_c) and social indicators (P_d, S_e). W_a is for wind component, W_1 is for hurricane category (1–5); W_2 is for wind speed (100–300 km/h); W_3 is for central pressure (800–1050 Mbar) and I_a are normalized wind indicators of hurricane result of apply Eq. (1) for wind component of hurricane. The second component is ocean waves (O_b) where O_1 is amplitude, O_2 is height, O_3 is wavelength, measured in meters; I_b are normalized ocean waves indicators. The R_c is for rain component where R_1 is precipitation index; R_2 is duration of hurricane in hours; R_3 is pluvial in millimeters, I_c are normalized rain indicators. P_d are P_1: : Prevention component (0–4) and P_2 : Response capacity (0–6), I_d are normalized prevention and response capacity component; S_e indicates social perception of risk where S_1: : Local perception of risk (0–8); S_2: : Prevention and care capacity (0–10); S_3 : Resilience capacity (0–6), I_e are normalized social perception indicators. The weight of each physical's indicators is described in table 1 and for social indicators are in Table 2. Depending on the value of hurricane damage estimation index, a level will be assigned on the Mercalli scale in Fig. 6.

4.3 Adaptation of the Mercalli Scale for Estimating Damage from a Hurricane

To adapt the Mercalli scale to our study of damage caused by hurricanes, we made certain significant modifications. First, we classified the damage scores into five damage categories: Very low, Low, Medium, High and Very High. We have decided to divide the range of damage scores into five equal intervals. This is known as "binning" or "discretization" in statistics and is often used to transform continuous variables into ordered categories.

Once we have the range, we divide it by five to get the size of each interval. We then assign each damage score the category corresponding to its interval. Like shown in the Fig. 6.

Risk level	Hurricane Damage Estimation Index	Color scale
Very high	80-100%	
High	60-80%	
Half	40-60%	
Low	20-40%	
Very low	0-20%	

Fig. 6. Adaptation of the Mercalli scale for estimating damage from a hurricane.

Since the Mercalli scale was originally designed to evaluate the intensity of earthquakes based on observed damage and reported experiences, its use in this study involved an adaptation process to allow an equivalent evaluation of the effects of hurricanes.

5 Methodology Application

Three events have been selected to apply the proposed methodology, Hurricane Odile, Tropical Storm Lidia and Hurricane Paul, which have different categories of hurricane. Although this research includes a deterministic approach, the proposed methodology can be used for future scenarios thanks to the simulation model. To define this value, official data from the National Meteorological Service of Mexico were used.

5.1 Hurricane Odile

For the first case, the application of Hurricane Odile was made, which had the following characteristics shown below in Fig. 7.

Hurricane Danger Indicators

Indicators of social perception of risk and response capacity.

Wind (w_a)		
W_1	Hurricane Category	0.5
W_2	Wind speed	0.3
W_3	central pressure	0.4
Ocean wave Rain (O_b)		
O_1	Amplitude	0.2
O_2	Height	0.2
O_3	Wavelength	0.5
Rain (R_c)		
R_1	Intensity	0.1
R_2	Duration	0.4
R_3	Pluvial	0.2

Prevention capacity (P_d)		
P_1	Prevention capacity	0.1
P_2	Answer's capacity	0.3
Social perception of risk (S_e)		
S_1	Local perception of risk	0.5
S_2	Prevention and care capacity	0.5
S_3	Resilience capacity	0.2

Fig. 7. Values for indicators for Hurricane Odile.

After processing the values shown in Fig. 7 in the system and assigning the damage estimate for all the 37 colonies of Cabo San Lucas, display the following results shown in Fig. 8.

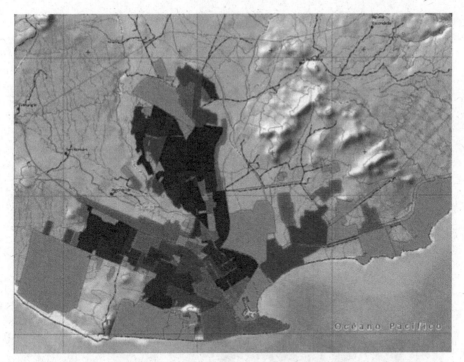

Fig. 8. Visual result on a Mercalli scale of the impact of Hurricane Odile on the city of Cabo San Lucas.

As we can see in Fig. 8, given the characteristics of Hurricane Odile, we can see that our damage estimation system indicates that all of the 37 neighborhoods of the city of Cabo San Lucas would suffer above the "medium" level of the scale of Mercalli proposed in Fig. 6 compared to reality since Odile has been the hurricane that has left the most damage in the city of Cabo San Lucas in the last 10 years. Odile caused serious damage in the state of Baja California Sur, according to CENAPRED, the damage was catastrophic for the entire region of Cabo San Lucas, as estimated by the proposed model, we can see an estimated view of what the damage from Hurricane Odile was like for the region of Cabo San Lucas where all the colonies were damaged in their entirety.

5.2 Tropical Storm Lydia

For the second case, the data from Tropical Storm Lidia will be used, which had a moderate level of impact in the city of Cabo San Lucas, but in this case the rainfall levels were extreme. Therefore, the neighborhoods with the highest risk of flooding were the most affected will use the following characteristics shown in Fig. 8.

It is processed in the system and the damage estimate assignment for all the 37 colonies of Cabo San Lucas returns the following results shown in Fig. 9.

Many colonies which in Hurricane Odile were not affected (Fig. 8), since in this hurricane according to CENAPRED the largest affected colonies were those that were

■ Hurricane Danger Indicators

░ Indicators of social perception of risk and response capacity.

Wind (w_a)		
W_1	Hurricane Category	0.3
W_2	Wind speed	0.1
W_3	Central pressure	0.3
Ocean wave Rain (O_b)		
O_1	Amplitude	0.2
O_2	Height	0.2
O_3	Wavelength	0.4
Rain (R_c)		
R_1	Intensity	0.1
R_2	Duration	0.3
R_3	Pluvial	0.7

Prevention capacity (P_d)		
P_1	Prevention capacity	0.1
P_2	Answer's capacity	0.3
Social perception of risk (S_e)		
S_1	Local perception of risk	0.5
S_2	Prevention and care capacity	0.5
S_3	Resilience capacity	0.2

Fig. 9. Values for the indicators for Hurricane Lidia.

Fig. 10. Visual result on a Mercalli scale of the impact of Tropical Storm Lidia on the city of Cabo San Lucas.

prone to flooding according to their location they lived in a risk area. According to the scale in Fig. 6 the highest level that predominated was "High", as shown in Fig. 10.

5.3 Hurricane Paul

For the third case, the application of Hurricane Paul was made, which had the following characteristics (Fig. 11). This hurricane had a similarity in characteristics to Hurricane Odile (Fig. 7) so it is expected to have similar results to Hurricane Odile.

Hurricane Danger Indicators		Indicators of social perception of risk and response capacity.		

Wind (w_a)		
W_1	Hurricane Category	0.3
W_2	Wind speed	0.4
W_3	Central pressure	0.3
Ocean wave Rain (O_b)		
O_1	Amplitude	0.2
O_2	Height	0.1
O_3	Wavelength	0.4
Rain (R_c)		
R_1	Intensity	0.3
R_2	Duration	0.3
R_3	Pluvial	0.1

Prevention capacity (P_d)		
P_1	Prevention capacity	0.1
P_2	Answer's capacity	0.3
Social perception of risk (S_e)		
S_1	Local perception of risk	0.5
S_2	Prevention and care capacity	0.5
S_3	Resilience capacity	0.2

Fig. 11. Values for the indicators for Hurricane Paul.

It is processed in the system and the damage estimated assignment for all the 37 colonies of Cabo San Lucas, results are shown in Fig. 12.

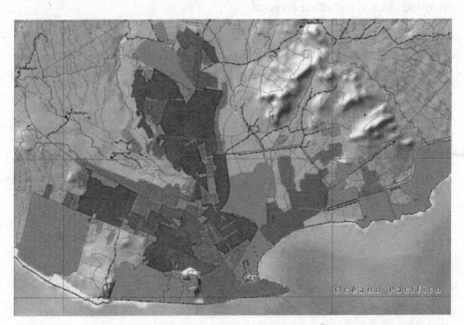

Fig. 12. Visual result on a Mercalli scale of the impact of Hurricane Paul on the city of Cabo San Lucas.

Hurricane Paul's was very similar impact to Hurricane Odile (Fig. 7), as we expected since the natural phenomena shared many similarities. According to the scale in Fig. 6, the highest level that predominated was "High" according to the results of the system.

6 Conclusions and Future Work

In this study, a methodology and system for the comprehensive and holistic estimation of damage caused by a hurricane was designed and implemented. The global requirements for new tools that support local governments and citizens is imperative. What is proposed is not limited to "revealing what can happen", but rather to provide critical information for the analysis and improvement of decisions before, during and after a hurricane.

The proposed system can be very useful in defining priorities according to the most affected areas in the city of Cabo San Lucas, offering a global vision of the possible magnitude of the impact. The indicators used to estimate damage were adjusted to the information available during the study period, but these can be the starting point to incorporate more indicators and thus strengthen the robustness of the system. Furthermore, the assigned weights can be modified according to the peculiarities of the case studied.

As a future line of research, the application of the Dijkstra's algorithm is contemplated to determine specific routes that allow social order to be restored after a hurricane. This aspect is critical, since, in addition to infrastructural damage, hurricanes can trigger social damage. An example of this, it is the looting that occurred during the disasters caused by hurricanes Odile and Lidia, where the population looted stores and supermarkets. This situation could have been avoided through the correct deployment of authorities, which is part of our proposal.

Figure 13 describes a proposal of the system, allowing it to be more reactive and provide support before, during and after a hurricane. This reflects the system's potential for growth and adaptation, so that it can respond more effectively to the changing demands and needs related to natural disaster management.

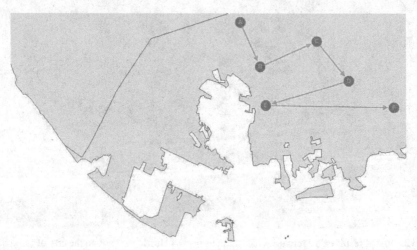

Fig. 13. System proposal applying the Dijkstra's algorithm for route suggestion to establish social order.

This system can be a tool for the government in decision making and being able to help in humanitarian logistics, where it is important to increase the resilience of society to these natural phenomena.

References

1. Aragón Durand, F.: Adaptación al cambio climático y gestión del riesgo de desastres en méxico: obstáculos y posibilidades de articulación. In: Graizbord, Mercado, A., Few, R., (eds.) Cambio climático amenazas naturales y salud en México, 1st edn. El Colegio de México, Centro de Estudios Demográficos, Urbanos y Ambientales Centro de Estudios Económicos, pp. 131–158 (2011)
2. Centro Nacional de Prevención de Desastres (CENAPRED). Los principales eventos registrados en el país durante 2000–2020 (2021). https://www.gob.mx/cms/uploads/attachment/file/673220/Presentaci_n_Impacto_Socioeconomico_compressed.pdf
3. Ruiz Rivera, N., Casado Izquierdo, J.M., Sánchez Salazar, M.T.: Los Atlas de Riesgo municipales en México como instrumentos de ordenamiento territorial. Investigaciones Geográficas. **88**, 146–162 (2015)
4. Comisión Nacional del Agua (CONAGUA). Reporte del clima en México. Septiembre 2016. Año 6. Número 9. CONAGUA y Servicio de Meteorología Nacional, pp. 131–158 (2016)
5. Cavazos Pérez, M.T.: Conviviendo con la Naturaleza El problema de los desastres asociados a fenómenos hidrometeorológicos y climáticos en México. In CONACYT RdDAaFHyC (eds), Conviviendo con la naturaleza. Primera ed. Tijuana: Ediciones ILCSA S.A. de C.V.C, pp. 1–45 (2015)
6. Hernández Aguilar, M.T.: Evaluación del riesgo y vulnerabilidad ante la amenaza de huracanes en zonas costeras del Caribe Mexicano: Chetumal y Mahahual. [Tesis Doctoral]. Chetumal 2014 (2023). http://hdl.handle.net/20.500.12249/98
7. Wilkinson, E.: Reducción de riesgos de desastres: marcos institucionales, politicas y tendencias. In: Graizbord. Cambio climático, amenazas naturales y salud en México. Primera ed. Ciudad de México: El Colegio de México Dirección de Publicaciones, pp. 33–98 (2011)
8. Garza Salinas, M.: Breve historia de la protección civil en México. In: Garza Salinas, M., Rodriguez, D. (eds.) Los desastres en México. Una perspectiva multidisciplinaria. México: Universidad Iberoamericana/UNAM, pp. 249–287 (1998)
9. Instituto Nacional de Estadística y Geografía (INEGI). Censo Nacional de Gobiernos Municipales y Delegacionales 2011: INEGI (2011). http://www.inegi.org.mx/est/contenidos/proyectos/censosgobierno/cng2011gmd/default.a spx
10. Ordóñez, A., Trujillo, M., Hernández, R.: Mapeo de Riesgos y Vulnerabilidad en Centroamérica y México Estudio de Capacidades Locales para Trabajar en Situaciones de Emergencia. OXFAM, Managua, Nicaragua (1999)
11. Secretaria de Gobernación (SEGOB)/Centro Nacional de Prevención de Desastres (CENAPRED). Diagnóstico de peligros e identificación de riesgos de desastres en México. Atlas Nacional de Riesgos de la Republica Mexicana Zepeda Ramos, González Martínez, editors. (CENAPRED), México (2014)
12. Instituto Municipal de Planeación Los Cabos (IMPLAN). Atlas de Riesgo y Vulnerabilidad al Cambio Climático del Municipio de Los Cabos, B.C.S.. IMPLAN, Los Cabos (2022). http://rmgir.proyectomesoamerica.org/AtlasMunPDF/2022/03008_LOS_CABOS_2022.pdf

Intelligent Applications for the Inclusion of People with Hearing Disabilities in the Communication

Marco Antonio Martínez, Julio Cesar Ponce Gallegos⬤, Alejandro Padilla Diaz(✉)⬤, and Francisco Javier Álvarez Rodríguez⬤

Universidad Autónoma de Aguascalientes, Aguascalientes, México
julio.ponce@edu.uaa.mx, apadillarobot@gmail.com

Abstract. The present research focuses on the development of an innovative application aimed at promoting the inclusion of individuals with hearing disabilities. The primary objective is to facilitate real-time communication and access to information through a voice recognition and translation system. The application aims to eliminate the linguistic barriers faced by people with sensorial disabilities focus to deaf people, enabling them to fully participate in various social contexts. Artificial intelligence will play a crucial role throughout the application's development, as will the efficient management of the database that handles the internal translation processes. To ensure usability and accessibility, comprehensive testing will be conducted to individuals with hearing disabilities, as well as with experts in the fields of inclusion and assistive technology.

Keywords: hearing disability · voice recognition · linguistic barriers · artificial intelligence · database · usability · accessibility

1 Introduction

Communication is fundamental for the social development of human beings. When, for any reason, speech is impeded, the possibility of achieving true social fulfillment is considerably reduced. As a means of socialization and compensatory mechanism, deaf individuals have developed their own language, sign language. In the face of the inability of deaf individuals to use spoken words and as an act of solidarity with this sector of the population, society must commit to promoting the dissemination of sign language among hearing individuals. Nowadays, society advances, and new technologies are created to assist certain segments of the population, but the question is: Why, with so much technology available, is it not being used to create a more effective communication system between those with hearing disabilities and those without? Current systems are slow and costly, leading to significant inclusion challenges for individuals with such disabilities in social, professional, and other contexts.

Actually is estimated 1.3 billion people experience significant disability. This represents 16% of the world's population, or 1 in 6 of people. The persons with disabilities,

H. Calvo et al. (Eds.): MICAI 2023 Workshops, LNAI 14502, pp. 274–284, 2024.
https://doi.org/10.1007/978-3-031-51940-6_20

suffer to stigma, discrimination, poverty, exclusion from education and employment, and barriers faced in the systems itself [1].

In Mexico 7.6% of the population lives with hearing disabilities. It is estimated that, in Mexico, about 2.3 million people suffer from hearing disabilities, of which more than 50% are over 60 years old, just over 30% are between 30 and 59 years old, and about 2% are girls and children, according to data from the Ministry of Health [2].

Artificial intelligence is an area of science with great growth where there are different emerging technologies. Some of them are being used to support the inclusion of people with sensorial disabilities, an example is the use of Augmented Reality for people with visual disabilities [3], Tangible Interfaces, Gamification, Extended Reality for Blind and Autistic People [4] to mention a few, this work shows the emerging technology related to voice recognition to people whit visual disability [5].

In response to this need, an innovative application has emerged, which is a voice recognition to sign language translator. This application combines powerful voice recognition technology with a deep understanding of sign language, providing a revolutionary tool to facilitate seamless communication between hearing and deaf individuals. It should be noted that sign language can vary in certain aspects between different regions or communities. Therefore, the application must consider these variations and generate translations adapted to the specific characteristics and conventions of each context. This application has the potential to facilitate real-time communication, allowing deaf individuals to interact more effectively and fully participate in various environments.

A series of applications have been developed with the aim of addressing the inclusion issues of people with hearing disabilities. However, they are not sufficiently accurate in tackling this problem since they only translate word by word and take time to display the entire sentence that the user wants to understand. Additionally, glove sensor-based systems have been explored, but their accessibility and costs make them less viable for the general population. The designed application seeks to solve the issues of other systems, as well as to innovate and offer highly precise and high-quality natural language processing. To achieve this, a series of important concepts were needed to address the development of the translation application effectively.

To do this, it is necessary to understand what artificial intelligence (AI) is, which, according to Oracle, Mexico [6], refers to a discipline of computer science that focuses on designing and developing systems capable of performing tasks that generally require human intelligence. AI is based on the idea that machines can learn from experience and improve their performance in specific tasks over time. For this purpose, a branch of AI focusing on speech recognition was used. As stated by Rodríguez, J. L. O. [7] in his article, computer speech recognition is a complex task involving pattern recognition and biometric systems. Typically, the speech signal is sampled in a range between 8 and 16 kHz. In the experiments reported in this work, a sampling frequency of 11025 Hz was used. The speech signal needs to be analyzed to extract relevant information once it has been digitized.

According to MDN [8], the Web Speech API is the application programming interface (API) developed by the World Wide Web Consortium (W3C) that allows website and web application developers to integrate speech recognition and speech synthesis functionalities into their products [9]. It is important that the application has a high level of effective language, where the most well-supported idea comes from Zendesk [10], defining effective communication as one in which a message is shared, received, and understood without altering its ultimate purpose. In other words, the sender and the receiver interpret the same meaning. This way, doubts and confusion are avoided while meeting expectations regarding what has been conveyed.

It is known that the database is a very important technology for processing all the information provided by the Web Speech API. Therefore, Oracle mentions that it is an organized collection of structured information or data, usually stored electronically in a computer system [11]. Typically, a database is managed by a database management system (DBMS). Undoubtedly, the database has a close relationship with natural language processing. According to Jurafsky and Martin [12], natural language processing is a branch of artificial intelligence that focuses on the study of interaction between computers and human language. The goal of NLP is to enable computers to understand, process, and generate human language naturally, just as a person would. Speaking about the front-end or visual interface of the application, it is necessary to consider usability and accessibility.

The International Organization for Standardization mentions that usability refers to the extent to which a product or system can be effectively, efficiently, and satisfactorily use d by its users to achieve specific goals in a given context. It is a discipline that focuses on designing products and systems that are easy to use. Likewise, the International Organization for Standardization states that accessibility refers to the ability of a product, service, or environment to be used by all people, regardless of their physical, cognitive, or sensory abilities [13].

2 Methodology

The methodology employed in this work is based on experimental research, which consists of several stages: data collection, data analysis, and measurement. To carry out this process, it was broken down into various procedures that can be observed in Fig. 1. First, an inductive course on Mexican Sign Language (LSM) was conducted with the purpose of understanding how deaf individuals interpret information, in order to perform effective natural language processing.

Subsequently, semantic fields were defined based on vocabulary, identifying the most frequently used fields and words. Once this step was completed, each LSM sign was created, totaling approximately 650 signs [14]. Next, speech recognition was implemented using the Web Speech API to integrate it with the database creation and application design.

After all these elements were defined, the most complex phase was tackled: natural language processing, which allowed for the adaptation of information for deaf individuals to comprehend. Following this, a sequential search in the database was conducted to display each corresponding sign to the user. This process culminated successfully with an accuracy of over 90% and a processing time of less than 2 s.

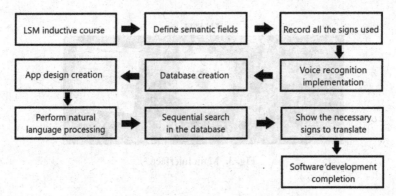

Fig. 1. "Experimental Research" Methodology

3 Results

The methodological process previously outlined has been carried out. First, the selection of semantic fields was undertaken, as can be seen in Fig. 2. It is worth noting that a total of 650 signs were recorded [15], a sufficient quantity to establish basic and intermediate dialogues. Once these semantic fields were defined, the conception and design of the application, as detailed in Fig. 3, took place, and the voice recognition functionality was integrated. When the corresponding button is activated, as shown in Fig. 4, the application begins to transcribe and process the user's auditory input.

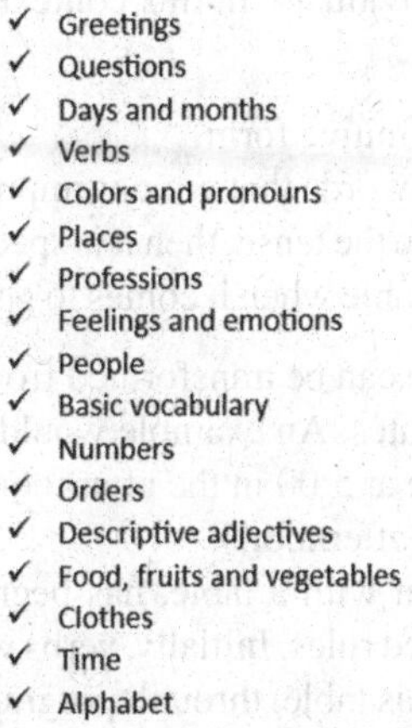

✓ Greetings
✓ Questions
✓ Days and months
✓ Verbs
✓ Colors and pronouns
✓ Places
✓ Professions
✓ Feelings and emotions
✓ People
✓ Basic vocabulary
✓ Numbers
✓ Orders
✓ Descriptive adjectives
✓ Food, fruits and vegetables
✓ Clothes
✓ Time
✓ Alphabet

Fig. 2. Semantic Fields in Spanish

Once these interfaces are defined and when the Web Speech API transmits the information obtained from speech recognition, the next step is to process that information in order to generate the corresponding signs for the message intended for people with hearing impairments. To carry out natural language processing with this information,

Fig. 3. Main Interface

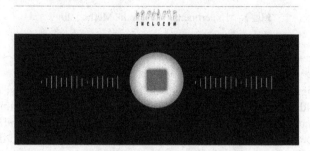

Fig. 4. Voice Recognition Interface

the process began by understanding the difference in communication for deaf individuals as compared to hearing individuals. In this context, three fundamental rules were established, which are:

- They handle verbs in their infinitive form.
- They remove connectors and words that are not important to convey a message.
- If the sentence doesn't indicate the tense, then it is specified whether it is past, present, or future. (This is more applicable when it comes to gestures rather than written form)

With these 3 rules, a message can be transformed from how a listener communicates to how a deaf person communicates. An example would be:

Listener: "The movie will be at 5:00 in the afternoon."

Deaf: "movie future be 5:00 afternoon."

An algorithm, in conjunction with a table, has been developed for the purpose of applying the previously mentioned rules. Initially, verbs were processed in their infinitive form using the table in Fig. 5. This table, through programming, analyzes the text string containing the message and detects the verbs present in it. When a verb is identified, it is replaced with its corresponding infinitive form. Additionally, in this process, the third rule is implemented, which involves determining the tense of the message. To do this, a numerical value is assigned in the position before the verb: 0 for past, 1 for present, and 2 for future.

Next, the algorithm proceeds to the second phase, which involves the removal of connectors and terms that do not contribute to the understanding of the message. For

			id_verb	verb_trad	verb_infinit	tiempo_verb
☐	✏ Edit ⬚ Copy ⊘ Delete		10	estar	estar	1
☐	✏ Edit ⬚ Copy ⊘ Delete		11	estoy	estar	1
☐	✏ Edit ⬚ Copy ⊘ Delete		12	estamos	estar	1
☐	✏ Edit ⬚ Copy ⊘ Delete		13	estan	estar	1
☐	✏ Edit ⬚ Copy ⊘ Delete		14	esta	estar	1
☐	✏ Edit ⬚ Copy ⊘ Delete		15	estado	estar	0
☐	✏ Edit ⬚ Copy ⊘ Delete		16	estaba	estar	0
☐	✏ Edit ⬚ Copy ⊘ Delete		17	estabas	estar	0
☐	✏ Edit ⬚ Copy ⊘ Delete		18	estabamos	estar	0
☐	✏ Edit ⬚ Copy ⊘ Delete		19	estaban	estar	0
☐	✏ Edit ⬚ Copy ⊘ Delete		20	estuve	estar	0
☐	✏ Edit ⬚ Copy ⊘ Delete		21	estuviste	estar	0

Fig. 5. Synthesizer (conversion of verbs to their infinitive form)

this purpose, the table described in Fig. 6 was used, in which the algorithm carries out an exhaustive search to identify if the message contains terms listed in the table, proceeding to eliminate them if affirmative.

			id_eli	nom_eli
☐	✏ Edit ⬚ Copy ⊘ Delete		1	es
☐	✏ Edit ⬚ Copy ⊘ Delete		2	la
☐	✏ Edit ⬚ Copy ⊘ Delete		3	las
☐	✏ Edit ⬚ Copy ⊘ Delete		4	los
☐	✏ Edit ⬚ Copy ⊘ Delete		5	o
☐	✏ Edit ⬚ Copy ⊘ Delete		6	:

Fig. 6. Eliminator (Removes unnecessary words and connectors)

Once the message has been adapted for the understanding of people with hearing disabilities, the search for each required linguistic signal to convey that message is initiated. To carry out this process, the approach described in Fig. 7 is employed, where the algorithm performs a thorough search in a predefined table for the corresponding signal. If the signal is not found in the table, the system breaks down the word in question into its individual components, generating a distinctive signal for each letter. Essentially, a phonetic enumeration of the word not found in the table is carried out to ensure that any term can be translated effectively and accurately.

Once the natural language processing process is completed, the subsequent action consists of presenting the identified gestures in the last step, which are displayed on the interface, as illustrated in Fig. 8.

Once the message has been presented in sign language to the user, the interface intended for the hearing-impaired user to input the message they wish to convey to

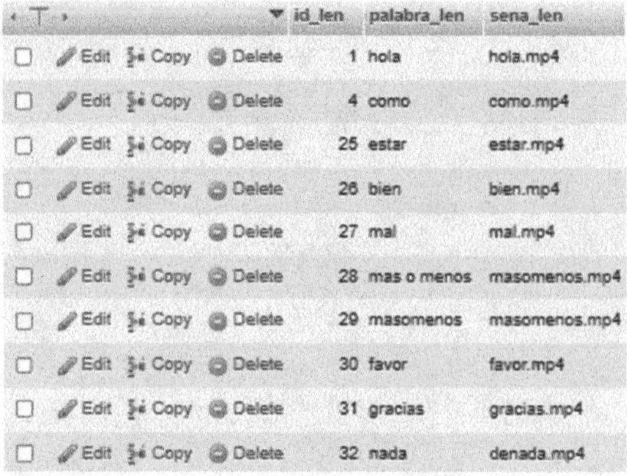

				id_len	palabra_len	sena_len
☐	✎ Edit	⫶ Copy	● Delete	1	hola	hola.mp4
☐	✎ Edit	⫶ Copy	● Delete	4	como	como.mp4
☐	✎ Edit	⫶ Copy	● Delete	25	estar	estar.mp4
☐	✎ Edit	⫶ Copy	● Delete	26	bien	bien.mp4
☐	✎ Edit	⫶ Copy	● Delete	27	mal	mal.mp4
☐	✎ Edit	⫶ Copy	● Delete	28	mas o menos	masomenos.mp4
☐	✎ Edit	⫶ Copy	● Delete	29	masomenos	masomenos.mp4
☐	✎ Edit	⫶ Copy	● Delete	30	favor	favor.mp4
☐	✎ Edit	⫶ Copy	● Delete	31	gracias	gracias.mp4
☐	✎ Edit	⫶ Copy	● Delete	32	nada	denada.mp4

Fig. 7. Language (Sign Search)

Fig. 8. Sign Language Display Interface

the hearing user is displayed, as illustrated in Fig. 9. Clicking on the speaker icon will activate voice transmission, thus establishing high-quality communication between both users, free from temporary issues, signal quality degradation, and other limitations.

Fig. 9. Text-to-Speech Translation Interface

Once the application was completed, it underwent evaluation by several professionals specialized in the field. Subsequently, a presentation was held at the Aguascalientes

Special Education Institute, as illustrated in Fig. 10, where an invitation was extended to teachers from the institute to evaluate the application. These educators, both men and women, used the application among themselves, and they were provided with detailed information about the objectives and concepts presented in this article.

Fig. 10. Application Evaluation

In total, three presentations were conducted during the technical council meetings. In addition to involving individuals with hearing disabilities, their family members were also included in the process. Once each presentation was concluded and after several participants had the opportunity to test the application, a questionnaire was provided to them. Figure 11 shows question 1 of the questionnaire, where the majority of the respondents were education professionals, including teachers. Other types of users, such as social workers, students, managers, and administrative staff, also participated.

What type of user are you?
35 answers

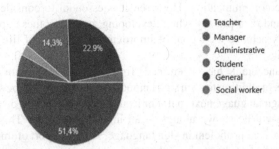

Fig. 11. Question 1 of the assessment.

Below, in Fig. 12, the results of question 2 are presented, which proved to be very useful as it highlights the diversity of users in terms of age. This is of particular importance when specifying the applicability of the tool, as age can sometimes correlate with experience in using Mexican Sign Language (LSM).

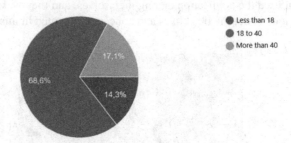

Fig. 12. Question 2 of the assessment.

Then, they were asked for their opinion about the application, and after analyzing all the responses, a significant acceptance from the users was observed. Many of them expressed a genuine desire for the application to be available in production, which will undoubtedly contribute significantly to improving inclusion in society. Finally, they were inquired about possible improvements or suggestions for the ongoing development of the application, and detailed tracking of these comments has been carried out, as detailed in the Discussion section of this article.

4 Discussion

In an increasingly digital and connected world, smart applications are playing a fundamental role in the inclusion of people with hearing disabilities. These tools not only facilitate everyday communication but also open new opportunities for participation in society. From voice recognition apps and automatic sign language translation to connected assistive hearing devices, technology is paving the way for more accessible and equitable communication. However, it is essential to consider aspects such as accessibility and data privacy when developing and using these applications, ensuring that they truly achieve their goal of improving the quality of life for people with hearing disabilities.

Advancements have been made for future work that is in development. In Fig. 7's interface, in addition to each sign language phrase, there will be a graphical representation of that sign language next to it. For example, if you want to display the sign for a house, the sign will be shown alongside an image of a house. This is because not all deaf individuals are proficient in sign language, so the support of images will aid in a quicker understanding of the message the speaker wants to convey to the deaf person.

Additionally, efforts are being made to expand the sign language vocabulary, creating signs for different semantic fields such as medicine, engineering, education, sports, etc. This way, more technical communication can be established without the algorithm having to spell out words extensively because certain signs are not in the database.

The final improvement will be to make the application accessible throughout Mexico. This is because sign languages differ in some states or regions, so signs will be created and separated depending on the region, and the translation will be based on that. This ensures that no matter where in the country the application is used, it will always work with 100% precision and efficiency.

5 Conclusions

The significant impact that an artificial intelligence-developed application can have on helping society is interesting. In the case of this research, it is concluded that the development of an intelligent application for inclusion, along with the created and implemented form, was of great assistance. Based on experimentation and testing, it was evident that people were quite interested in the application. This led to the conclusion of how necessary the application is and what sets it apart from other applications in the market.

The development of an application that translates speech into Mexican Sign Language represents a significant advancement in inclusion and accessibility for individuals with hearing impairments in Mexico. This application has the potential to break down existing communication barriers, enabling a smoother and more natural interaction between people who use different forms of communication.

By providing real-time translation from speech to Mexican Sign Language, the application facilitates communication in various settings, such as education, employment, healthcare, and social interactions. Deaf individuals will be able to access real-time information, actively participate in conversations, learn about various subjects, and express their thoughts and feelings more effectively.

It is crucial to emphasize that the inclusion of people with hearing impairments requires a collective effort from society, institutions, teachers, administrators, social workers, students, and individuals. Adequate policies and legislation are needed, along with the promotion of accessible environments and tailored resources.

References

1. World Health Organization. Disability (2023). https://www.who.int/news-room/fact-sheets/detail/disability-and-health
2. Santiago, R.: El 7.6% de población mexicana vive con discapacidad auditiva, Oceano Medicina (2023). https://mx.oceanomedicina.com/nota/actualidad-ec/el-7-6-de-poblacion-mexicana-vive-con-discapacidad-auditiva/
3. Ponce Gallegos, J.C., Montes Rivera, M., Ornelas Zapata, F.J., Padilla Díaz, A.: Augmented reality as a tool to support the inclusion of colorblind people. In: Stephanidis, C., Antona, M., Gao, Q., Zhou, J. (eds.) HCII 2020. LNCS, vol. 12426, pp. 306–317. Springer, Cham (2020). https://doi.org/10.1007/978-3-030-60149-2_24
4. Ramos Aguiar, L.R., et al.: Implementing gamification for blind and autistic people with tangible interfaces, extended reality, and universal design for learning: two case studies. Appl. Sci. 13(5), 3159 (2023)
5. Cerón, C., Archundia, E., Fernández, J., Flores, C.: Design of a web system of learning evaluation of students with visual disability by voice recognition. In: Nguyen, N.T., Chbeir, R., Exposito, E., Aniorté, P., Trawiński, B. (eds.) ICCCI 2019. LNCS (LNAI), vol. 11684, pp. 592–602. Springer, Cham (2019). https://doi.org/10.1007/978-3-030-28374-2_51

6. Oracle, ¿Qué es la inteligencia artificial (IA)?. https://www.oracle.com/mx/artificial-intellige nce/what-is-ai/. Accessed 29 Apr 2023
7. W3C, Web Speech API. https://dvcs.w3.org/hg/speech-api/rawfile/tip/speechapi.html. Accessed 29 Apr 2023
8. MDN. Web Speech API. https://developer.mozilla.org/es/docs/Web/API/Web_Speech_API. Accessed 29 Apr 2023
9. World Wide Web Consortium. About the World Wide Web Consortium (W3C). https://www. w3.org/Consortium/. Accessed 1 Dec 2022
10. Zendesk. ZendeskSuit, https://support.zendesk.com/hc/es. Accessed 29 Apr 2023
11. Oracle. ¿Qué es una base de datos?. https://www.oracle.com/mx/database/what-is-database/. Accessed 3 May 2023
12. Jurafsky, D., Martin, J.H.: Speech and Language Processing, 3rd edn. Pearson Education Inc, London (2019)
13. ISO, ISO 9241–11:2018, Usability: Definitions and concepts. https://webstore.ansi.org/pre view-pages/ISO/preview_ISO+9241-11-2018.pdf. Accessed 1 Dec 2022
14. Valenzuela, T.E., Gutiérrez, C.E., Hernández, A.E.: Manual de lengua de señas mexicana. Colegio de México (2014)
15. CONAPRED: Diccionario básico de Lengua de Señas Mexicana: Manos que hablan, voz que comunica (2007). https://www.conapred.org.mx/documentos_cedoc/DiccioSenas_Man osVoz_ACCSS.pdf

TICCAD: A Resource for Higher Education During and After Sars-CoV2

María Esmeralda Arreola Marín[1]([⊠]), Mariela Chávez Marcial[1],
José Iraic Alcantar Alcantar[1], and Edgar Gonzalo Cossio Franco[2]

[1] Tecnológico Nacional de México/ITS de Ciudad Hidalgo, Hidalgo, México
{marreola,jiraic}@cdhidalgo.tecnm.mx
[2] Instituto de Información Estadística y Geográfica de Jalisco, Zapopan, México
edgar.cossio@iieg.gob.mx

Abstract. The purpose of this research is to generate knowledge about the teaching-learning process that is currently carried out inside and outside the classroom through the use of TICCAD, in the educational sector as media and digital tools for information and communication, responding to the way in which the school community; teachers and students adapted to the new normality due to the SARS-CoV-2 pandemic in Mexico (period March 2020 to January 2022).

Information, Communication, Knowledge, and Digital Learning Technologies, also known as TICCAD, are fundamental tools used mainly to strengthen the work of teachers in digital skills issues. Currently, the process of evolution that has occurred in the education sector has made it possible to analyze and detect the benefits and areas of opportunity that need to be considered for the development of the contents of the study plans and programs established by the Ministry of Public Education (SEP) [21] and thus ensure a learning environment inside and outside the classroom because they have become invaluable allies.

Keywords: TICCAD · Teaching · SARS-Cov2 · Learning · Technology and Competencies

1 Introduction

As of March 2020 Mexico is declared in a health emergency due to the spread of a virus better known as SARS-Cov2 which has significantly affected the population worldwide, in response to concerns for public health, certain preventive measures have been recommended to have certain preventive measures through the World Health Organization [WHO] and in the education sector by the Secretary of Public Education [SEP] [21] it was agreed that classes were suspended in person looking for a number of alternatives for children and adolescents to continue with their education.

In the educational sector as digital information and communication media and tools, responding to the way in which the school community; Teachers and students adapted to the new normal due to the SARS-CoV-2 pandemic in Mexico [9]. Information, Communication, Knowledge and Digital Learning Technologies, also known as TICCAD, are

H. Calvo et al. (Eds.): MICAI 2023 Workshops, LNAI 14502, pp. 285–297, 2024.
https://doi.org/10.1007/978-3-031-51940-6_21

fundamental tools used mainly to strengthen the work of teachers in digital skills issues. Currently, the process of evolution that has occurred in the educational sector has made it possible to analyze and detect the benefits and areas of opportunity that must be considered for the development of the contents of the study plans and programs established by the Ministry of Public Education (SEP) [21] and in this way ensure a learning environment inside and outside the classroom because they have become invaluable allies [5].

In agreement with Luna, J. [16] notes, TICCAD allows you to fully exploit a series of resources, platforms, devices and tools that society currently has in order to strengthen and expand the knowledge, as well as skills that human beings acquire in the area of digital technology to potentialize their creativity as well as the development of a motivational environment in educational institutions and in this way it is possible to contribute positively to the education of girls, boys and adolescents in the country.

TICCAD has revolutionized the way of communication, the way in which the student acquires knowledge and the way in which the teacher transmits it to his pupils and even the capacity that these two social actors manage the information, we must consider that talking about Technology in the educational field is not a new topic, however, it is a topic that is constantly involved in a series of social, cultural and economic changes.

The implementation of TICCAD allows to continue with the educational plans and programs, with the purpose of integrating the knowledge from year to year to achieve that these digital tools are a substantial resource in the learning and teaching process in the face of the "new normality" that young people, teachers, and parents have faced with the return to classes in a face-to-face way.

In this way it is clearly expressed the fact that it is a study and inquiry into the reality of a social phenomenon involving the educational system which allows to analyze, explore, examine to understand a series of problems, the formulation of hypotheses or suggested solutions, the collection, organization as well as the evaluation of data, the formulation of deductions the scope of consequences, the presentation of conclusions to determine and evaluate the use of TICCAD in education in the face of a health emergency of SARS-CoV-2.

Torres, P. [22] says, "the incorporation of technology in the educational process requires teachers to acquire new skills for the exploration and construction of knowledge, allowing them to have physical tools that make students open their minds to new ways of learning to work, facing the need to experiment with new models to carry out the teaching and learning process" [p. 74]. Among its many positive aspects, the National Education Agreement [21], proposes two primary objectives:

To offer an excellent, inclusive, and equitable education to the children, adolescents, and young people of our country and to grant teachers of the National Education System (NES) [21] the right to better training and constant updating. To achieve such purposes, today we have powerful allies: information, communication, knowledge, and digital learning technologies [p. 6].

They are considered as didactic material with a level of importance as appendices in the educational process, part of its objective is to generate a potential of pedagogical character that should be incorporated by its structure in the study plans and programs, being a strategy for the generation of useful knowledge for different educational levels.

The author Cordovez, C. [6], states: "Education is adapting to this pace and becoming more dynamic by using the operational properties of technology" [p. 3]. The incursion of technology has significantly impacted education, in the sense of whether it is bringing about a profound change in the prevailing educational paradigms, in the way learning, teaching, and assessment take place.

In this context, García, P. [21], points out that "The change from face-to-face classes to distance education occurred suddenly. Neither teachers nor students had time to prepare, so they adapted with the resources they had available." Even the closing of schools due to the health emergency has no historical comparison, students stop going to the institutions and start learning from home with the support of teachers who found it necessary to seek and incorporate means of communication and digital tools to continue with the distance teaching and learning process.

To the autor Escudero, X. [9], the use of TICCAD during the SARS-CoV-2 health crisis made it possible to carry out activities that strengthen students' learning and performance, considering them as a key element to overcome the contradictions and contextual problems that SARS-CoV-2 has generated for education, but, above all, that make it possible to think about the possibility of moving forward.

The TICCAD incorporated in the pandemic time were platforms for videoconferencing, virtual classrooms (we worked on the design of activities and adjustment of study programs), use of browsers and academic search engines, educational apps, tools for the development of digital resources, with this it is possible to develop collaborative work and project-based learning, which to date [2023], have remained in the classroom [2, 7]. Giving way to new forms of study such as dual education, inverted classroom, among others; as well as a new learning called: autonomous learning, for Manrique, L. [17], Autonomous learning is "a new attitude towards learning and knowledge construction. It provides collaborative and meaningful learning". Also Margalef, L. [18] says "autonomous learning and is asynchronous and synchronous as it overcomes time and space limits" [p. 37]. The main purposes of this learning are the development of intellectual, personal, and social autonomy, independent work.

This article arises from the investigation of a postgraduate thesis (Master of Pedagogy), whose main research question for the development of the project was generated by investigating how TICCAD impacts the teaching and learning process in higher education before and after the presence of a pandemic; SARS-Cov2? To clearly identify the problem, it is essential to pinpoint the underlying factors in order to provide a starting point for the research process. The study is based on a psychopedagogical framework with a mixed methodology. It adopts a socio-critical paradigm and employs descriptive and correlational techniques, incorporating elements of documentary and field research. The applied methods include observation, interview scripts, questionnaires, and sociograms.

2   Methods

For the development of this inquiry, it is considered to be answered the research question that arises in time of pandemic: How do TICCAD impact on the teaching and learning process in higher education during the presence of a pandemic; SARS-Cov2? based on a psycho-pedagogical line, with a mixed approach, whose paradigm is Socio-critical, its scope is descriptive and correlational, a type of documentary and field research, the techniques applied were: observation, interview, survey and its instruments: case studies, field diary, interview script, questionnaire, and sociogram, respectively [20]. Therefore, the following purposes are proposed for the development of this research:

To analyze the educational impact generated by TICCAD in the teaching and learning process during and after the SARS Cov2 pandemic in higher education students. Specific objectives: to know the use that teachers and students make of TICCAD for their education. To examine the impact caused by SARS-Cov2 in the Ricardo Flores Magón and to observe the mastery and work done by teachers and students at the higher level. To support the thesis work, two variables were structured with the objective of identifying the key elements that make up the research data to answer the research questions and even identify how they are related to each other. The dependent variable indicates: The use of TICCAD as an educational resource and the independent variable: SARS CoV-2 [9] and the teaching and learning process, for this reason they have quantitative and qualitative characteristics and properties of the studied phenomenon that acquire values that allow the relationship to be observed. Between both variables [20]. Table 1 shows the categories of analysis, referring to the variables involved in the research.

Table 1. Categories of analysis

VARIABLE X *TICCAD*	*X1. The use of TICCAD as an educational resource.* *X2. Digital Tools* *X3. Educational platforms*
VARIABLE Y *Students in time of SARS-Cov2*	Y1. SARS-Cov2 Y2. Teaching and learning process. Y3. Educational impact

Within this research, the hypothesis to be tested is as follows: The implementation of TICCAD, as the primary resource designed to analyze, apply, and evaluate the teaching and learning processes during a health emergency caused by SAR-Cov2, benefits many adolescents in continuing their higher education by providing an efficient interaction system between teacher and student. The type of study of this research is correlational since it allows the researcher to measure the degree of relationship that exists between two or more concepts or variables; since it is possible to identify the relationships that exist between two or more concepts, categories, or variables in a particular context. These correlations are based on hypotheses subject to testing [20], the fundamental part of this scope is presented when analyzing how a variable can behave by focusing on the behavior of the others which are linked [19]. Table 2 presents the hypotheses generated from their relationship with the independent variable (X) TICCAD and the dependent variable (Y) Students in SARS-Cov2 time.

Table 2. Working hypothesis

Hypothesis	Description
H1	The implementation of TICCAD (X) as a main resource designed to analyze, apply, and evaluate the teaching and learning processes in time of health emergency by SAR-Cov2 (Y), benefits many adolescents to continue with their higher education being an efficient interaction system between teacher and student.
H2	With the use of TICCAD (X), student learning is achieved (Y).
H3	Teachers integrating TICCAD (X) achieve the teaching of the subjects of study (Y).
H4	TICCADs (X) are fundamental tools for teaching and learning for students and teachers at the higher education level during and after SAR-Cov2 (Y).

Faced with the pandemic that occurred, some public and private educational institutions saw the need to use and incorporate TICCAD in order to continue with the education of young people in higher education with the aim of achieving the expected learning set by the Secretary of Public Education (SEP) [21], however, it must be considered that not all social sectors have the economic and geographical resources necessary to have access to such means, which has been a constraint.

3 Results

The research techniques implemented for data collection were selected according to their level of scope for the collection of data and their veracity for the analysis of the results; observation, interview and surveys were used, with the support of a series of instruments such as case studies, field diary, interview script, questionnaire, and sociogram.

To collect data, we utilized four techniques: documentary analysis, fieldwork employing the ethnographic method, observation, and surveys. This required essential tools such as bibliographic files, a computer, storage units, a field diary, an observation guide, and the questionnaire. It's worth noting that some of this data was gathered through traditional techniques during fieldwork, involving engagement with the interest group and visits to libraries for source consultation. The results obtained from the application of 14 questionnaires to students aged 13, 14 and 15 years old, who were third grade secondary school students and whose educational evolution in the use of TICCAD has been observed throughout their stay at the "Ricardo Flores Magón" educational institution, when the SARS CoV-2 pandemic reached Mexico, to the community of Santiago Tilapa and the students had to take classes online, until their return to the classrooms.

Additionally, a questionnaire was administered to 12 teachers, including 7 female and 5 male teachers, responsible for the educational process of the students who participated in the questionnaire. They taught various subjects such as mathematics, biology, physics, chemistry, Spanish, arts, English, geography, history, civic and ethical training, workshop, physical education, tutoring, healthy living, and gender equality. This spanned from first grade in the 2020–2021 school year, through second grade in 2021–2022, to third grade in 2022–2023. The objective was to observe, analyze, and interpret their teaching and learning process through the utilization of TICCAD as an educational resource during the period of confinement.

The study population was the educational institution with a registered enrollment of 890 students as the space from which the population will be delimited, which corresponds to group "A" currently in third grade made up of 27 enrolled students; of which 16 are female students and represent 59.2% and the remaining 40.8% are identified by 11 male students. The sample that was considered in total is 14 students who represent 51.8% of the total sample, this is the result of half of the total established population, the questionnaire was applied to 7 female and 7 male students.

It should be noted that during this process, it was crucial to also observe, comprehend, and analyze the work conducted by teachers from various disciplines who attended to the group of students representing the sample. This approach enabled the monitoring of their learning process through the use of TICCAD. The teacher enrollment comprised a total of 45 teachers between the morning and afternoon shifts, with 24 being female, representing 53%, and 21 male, representing 47%. Therefore, the sample consisted of 14 teachers who supervised their academic training.

The unit of analysis is the main environment and the sample that is studied within the research work, considering the determination, type, calculation, and identification of the elements of the population that are immersed in said unit of study being students and teachers since they comply with the sampling parameters and that in the development of chapter four the information will be presented.

As a result, it is obtained that the TICCAD that were incorporated during the SAR-Cov2 time were: the use of cloud computing (Google Drive and Drop box), integration of platforms for videoconferencing being Meet, Microsoft Teams, and Zoom as the main technological media used by young people, virtual classrooms such as Google Classroom and Moodle, design of activities through Genially, Kahoot, Educaplay and Canva as online support tools for the creation of interactive content and browsers such as Google Chrome, Mozilla Firefox and Internet Explorer, which corresponds to the independent variable (x) of the thesis, see Fig. 1.

As it can be seen in Fig. 2, most used mobile devices, a considerable percentage uses the cell phone as the main device, representing 92.9% of the total sample, which indicates that there are 13 students. In second place, the desktop computer is shown as a technological tool that 9 students have at home, representing 64.3%.

It is essential to identify why students used these platforms or educational tools, which were: submitting tasks and for communication, accessing virtual meetings and receiving documents in PDF and JPG format with 92.9%, entering virtual platforms represented by 78.6%. % and finally with 64.3% is product creation. See Fig. 3. Purpose of using platforms.

According to new data from INEGI [14], there were 342 thousand people trained and working in the field of information and communication technologies [TIC] in Mexico. 75.6% of the Mexican population (88.6 million people) used the Internet, according to

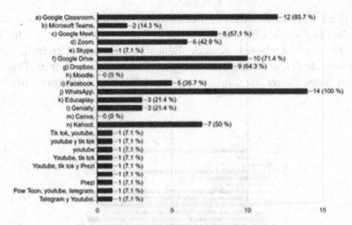

Fig. 1. Platforms used during the pandemic.

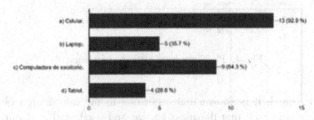

Fig. 2. Most used mobile device

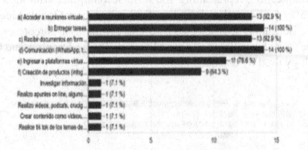

Fig. 3. Purpose of using platforms.

the Encuesta Nacional sobre Disponibilidad y Uso de Tecnologías de la Información en los Hogares [ENDUTIH][1] [14]. Therefore, the TICCADs that continued to be used after SAR-Cov2 are multimedia content presentations: PowerPoint, Canva, Genially, Slideshare, Prezi, Padlet, Emaze, Jamboard, Powtoon, Piktochart, ZohoShow, wordwall, among others.

Collaborative tools: forums, Blogs, wikis, Webquest, and Padlet. Cloud computing: DropBox, Drive, Box, Box, iCloud, OneDrive, Nextcloud, Mega, Adrive, Claro-Drive, AmazonDrive. Messaging and social networks: Skype, WhatsApp, Google Chat,

SnapChat, Telegram, Matrix, Allo, Twitter and email. Online platforms and campuses: GoCongr, Moodle, Google Classroom, Microsoft Teams, and TedEd. Tools to dynamize virtual and face-to-face classes: Kahoot, EducaPlay, Quizizz, Cerebriti, AhaSlides, Cerebriti, Brainscape, Quizlet, Peardeck, Jigsaw Planet, Pixton, Kubbu, Paper.li, Popplet, Random Name Picker, mentimeter, appsorteo. See Fig. 4. Platforms used after the pandemic.

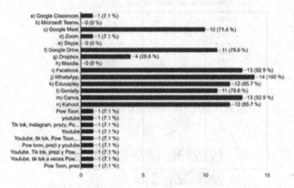

Fig. 4. Platforms used after the pandemic.

On the other hand, it is shown that according to the information obtained from the teachers, it is analyzed that the most known and used platforms during the time of confinement were: Google Classroom as a virtual classroom and WhatsApp as a means of communication, representing 100%. In second place with 83.3% is Kahoot, subsequently 75% of the sample selected applications such as: Google Meet and Google Drive, therefore Educaplay was selected by 33.3%, while 25% is represented by Zoom as an alternative for video calls, 16.7% are Microsoft Teams, Dropbox and Genially. See Fig. 5. Technology used by teachers.

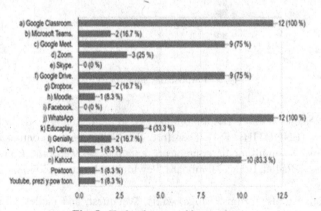

Fig. 5. Technology used by teachers

It is important to know if TICCAD after the confinement period and with the return to the classrooms are still useful. With this item it is possible to analyze and interpret that, in fact, the majority of the teachers surveyed make use of different platforms, applications and technological tools. For the digital learning of their students, information that can be compared and verified in the same way with the information obtained in the previous point.

The applications and platforms with the highest demand for use are: Google Drive, WhatsApp, and Educaplay, all with a demand of 91.7%. Following closely with 83.3% demand are Genially and Kahoot. Notably, before the pandemic, Genially did not hold significant relevance, but it is now widely used by teachers. Dropbox and Canva represent a case of 75%, a different scenario from the analysis in item number 9. Meanwhile, Facebook is utilized by 66.7% of respondents. Among the results with the least demand are: Google Meet with 33.3%, Google Classroom and Moodle with 8.3%. Additionally, we have included answers provided in another section: Telegram, YouTube, TikTok, and PowToon.

Knowing the results of the research, the teachers consider the importance of taking a course or training, since 100% mention that they have interest and willingness to enroll in a course on TICCAD, which has already aroused their interest and commitment. With their performance as people at the service of education. See Fig. 6.

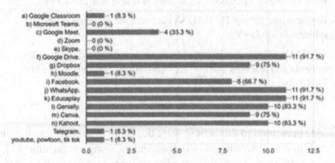

Fig. 6. Recommended platforms.

This data yields quantifiable insights, emphasizing the need to delve into what motivates teachers to pursue technological courses. This is driven by the imperative to acquire new knowledge and their personal interest in furthering their education, particularly in specialized courses focusing on TICCAD. It is evident that they are already somewhat acquainted with TIC and TAC, and they recognize the importance of staying updated.

Consequently, they emphasize the crucial necessity for teachers to continually engage in their training process. There is still ample terrain to explore, especially considering that TICCAD resources were in existence even prior to the emergence of the social phenomenon of SARS CoV-2. Society will seek an educational resource to adapt to the teaching and learning process, a resource that was not widely known before. Now presents an opportune moment to delve further into them. This involves not only acquiring proficiency in managing platforms but also recognizing the paramount importance of using these media responsibly to extract the maximum benefit from these educational technological resources.

Through the analysis and interpretation of data collected from the sample population, the veracity of the proposed hypotheses is confirmed. Over the course of three academic cycles, we observed and studied the social phenomenon. This confirmed that both teachers and students in the institution adopted the use of TICCAD as an educational resource for various academic activities during the period of confinement due to the pandemic. This significantly impacted the teaching and learning process for both parties involved. Table 3, showcasing the contrast of the hypotheses tested in this research, is provided below.

Table 3. Hypothesis testing

Hypothesis	Description	State
H1	The implementation of TICCAD (X) as a main resource designed to analyze, apply, and evaluate the teaching and learning processes in time of health emergency by SAR-Cov2 (Y), benefits many adolescents to continue with their higher education being an efficient interaction system between teacher and student	Accepted
		Accepted
		Accepted
H2	With the use of TICCAD (X), student learning is achieved (Y)	Accepted
H3	Teachers integrating TICCAD (X) achieve the teaching of the subjects of study (Y)	Accepted
H4	TICCADs (X) are fundamental tools for teaching and learning for students and teachers at the higher education level during and after SAR-Cov2 (Y)	Accepted

The TICCAD presented in this article were designed to support teaching by facilitating content development, fostering autonomous and meaningful learning [11], promoting self-esteem, and cultivating a positive attitude towards technological advancements at a higher level. This empowers students to pursue their studies with greater confidence and proficiency.

As a result of this exploration, students were able to identify errors during the handling of the activities, generate problem-solving guides, develop their logical thinking through technological applications, manipulate it easily due to its easy access interface, measure its scope and generate self-evaluation, greater skill in the use of TICCAD. Improved communication, application of new teaching-learning methods, use of different digital tools, development of competencies, personalized and collaborative work.

4   Discussion

According to the results acquired, it becomes evident that TICCAD in education and society demonstrate a remarkable degree of flexibility and adaptability in an increasingly dynamic environment. Before 2020, the integration of technology in education was a gradual process. However, with the emergence of the SARS-CoV-2 pandemic, it became imperative. Currently, it has transitioned to a voluntary but essential aspect of education, driven by the recognition of the substantial benefits that arise from the correct implementation of TICCAD [13].

The mixed approach developed in this research process has allowed the analysis of two perspectives, on one hand, quantitative data obtained through surveys and on the other hand, qualitative data collected through an interview process and sociograms, which have allowed substantiating information, as well as raising new perspectives that were not foreseen at the beginning of this research process. As a result of the quantitative analysis, employing both descriptive and correlational statistical methods [1], it became evident that 71.4% of the surveyed teachers found the transition from classroom to virtual mode to be challenging. Additionally, the same percentage indicated a lack of familiarity with TICCAD tools. Consequently, 100% of those surveyed affirmed the importance of implementing a training plan focused on the use of technological tools. Such training proved to be of substantial assistance in adapting to the virtual mode.

For the qualitative analysis of the information obtained in the interviews applied, it is stated that the transition from face-to-face to virtual mode at the beginning of the pandemic became complicated, since they did not have the necessary knowledge to perform adequately in these virtual environments, especially when the implemented training was not of a practical nature, which meant that the management of TICCAD and the adaptation to these new scenarios became a challenge [3, 4]. Based on this somewhat bitter experience, most of the interviewees agreed that the technological tools, beyond the difficulties presented, have resources and applications that lead to the strengthening of the students' knowledge, which is why it is important to have a training plan in the use of TICCAD, to help them take advantage of the potential that technologies have to offer [13].

It is proposed to implement a structured training plan for the use of TICCAD. This plan aims to establish a clear roadmap including the following steps: identify a TICCAD need, establish objectives, define the contents according to the need, choose a trainer who may be a person from the same institution or an external agent expert in the area, select the teachers who will receive the training [8], establish the teaching modality, which can be virtual or face-to-face, develop the theory and subsequent practice of the topics addressed, to finally follow up on what was covered in the training, in order to adjust certain eventualities that arise in daily practice and continue to be used to promote student learning in this digital age, the information age.

5   Conclusions

The integration of TICCAD in higher education during the pandemic generated a diversity of opportunities to innovate teaching practice. In particular, it facilitated the merging of virtual environments with real-world elements at various levels of interaction, effectively adapting to the new reality of that time. The level of acceptance was such that, in the process of developing digital educational resources, many teachers turned to technologies focused on augmented reality and virtual reality, provided their context and knowledge permitted. For its implementation at the higher level, the creation of digital resources involved a systematic process based on pedagogical and methodological approaches, as well as technological guidelines that promote autonomous [7] and meaningful learning [11] for interactivity with the main object of study (students).

Once left behind the fears and uncertainty of how to integrate TICCAD in the classroom, the results were satisfactory; so it can be determined with the results obtained that TICCAD came to stay, since January 2022 that returned to the classroom in a face-to-face manner continued to be used; various digital tools have emerged and innovated for the creation of content, apps, browsers, search engines, virtual classrooms, simulators, computer clouds, artificial intelligence is increasingly present in the educational environment, collaborative work, autonomous, meaningful and project-based learning continues to be promoted; even the curricula have been updated, based on the results obtained after the SAR-Cov2 experience and its relationship with technology focused on the educational field.

With this research, we successfully validated the hypothesis and addressed the research question. Most importantly, we have tackled a problem. The acquired results not only affirm the current impact but also underscore the need for continuous development and updates. This encourages the creation of new tools and platforms, opening up fresh alternatives for study, teaching, and learning, ensuring that these advancements are here to stay.

Finally, the efficiency and quality of the resources applied by teachers and used by students is examined, considering indicators valued by students such as usability, accessibility and impact on their learning [15]. TICCAD, as technological tools, have increased the degree of significance and educational conception, establishing new models of communication, in addition to generating spaces for training, information, debate, reflection, among others; breaking with the barriers of traditionalism in the classroom [10, 12]. It should be taken into account that the teaching-learning process in the classroom, making use of TICCAD, requires a set of competencies that the teacher must acquire with the logic of adding a methodology capable of taking advantage of technological tools, where teacher training should be considered one of the first options before facing new educational challenges.

References

1. Avila, H.: Introducción a la metodología de la investigación. Eumed, México (2006)
2. Baca, M., et al.: El aprendizaje autónomo: Una competencia ineludible en la sociedad del conocimiento. Universidad de Guanajuato, Guanajuato (2016)
3. Barajas, J.: La clasificación de los medios tecnológicos en la educación a distancia. Un referente para su selección y uso. Apertura, Guadalajara (2009)
4. Belloch, C.: Las tecnologías de la información y comunicación (T.I.C.). Unidad de tecnología educativa, Valencia (2021). https://www.uv.es/~bellochc/pdf/pwtic1.pdf
5. Bodero, H.: El impacto de la calidad educativa. Revista Apuntes de Ciencias Sociales, México (2014)
6. Cordovez, C.: La utilización de las tecnologías de información y comunicaciones (TIC) en la enseñanza de la optometría. Universidad de la Salle, México (2020)
7. Crispín, M.L., et al.: Aprendizaje Autónomo. Orientaciones para la docencia. Universidad Iberoamericana, México (2011)

8. Cuartas, A.: Conceptualización de ambientes virtuales de aprendizaje. Bogotá: Areandina (2017)
9. Escudero, X.: La pandemia de coronavirus SARS-COV-2 (COVID-19): Situación actual e implicaciones para México. Revista Cardiovascular, México (2020)
10. Esteban, M.: Las estrategias de aprendizaje en el entorno de la Educación a Distancia (EaD). Consideraciones para la reflexión y el debate. Introducción al estudio de las estrategias de aprendizaje y estilos de aprendizaje, en Revista de Educación a Distancia. Murcia. Número 7 (2003a). http://www.um.es/ead/red/6/documento6.pdf. Accepted 12 Feb 2010
11. García, F.: Influencia de las TIC en el aprendizaje significativo. UNIR, España (2011)
12. García, F.: La tecnología su concepción y algunas reflexiones con respecto a sus efectos. CIECAS-IPN, México (2010)
13. García, P.: Educación en Pandemia: Los riesgos de las clases a distancia. IMCO, México (2021). https://imco.org.mx/wp-content/uploads/2021/06/20210602_Educacio%CC%81n-en-pandemia_Documento.pdf
14. Instituto Nacional de Estadística y Geografía [INEGI]. Encuesta Nacional sobre Disponibilidad y Uso de Tecnologías de la Información en los Hogares (ENDUTIH) (2022). https://www.inegi.org.mx/contenidos/saladeprensa/boletines/2023/ENDUTIH/ENDUTIH_22.pdf
15. Khvilon, E.: Las tecnologías de la información y la comunicación en la formación docente. UNESCO, Francia (2004)
16. Luna, J.M.: Tecnologías de Información, Comunicación, y Conocimiento para el aprendizaje digital en tiempos de pandemia: un balance crítico desde los imaginarios de la sostenibilidad. UNAM, México (2021)
17. Manrique Villavicencio, L.: "El aprendizaje autónomo en la educación a distancia" LatinEduca2004.com Primer Congreso Virtual Latinoamericano de Educación a Distancia – Perú (2004)
18. Margalef García, L., Pareja Roblin, N.:. "¿Qué aprendemos del aprendizaje autónomo?" Red Estatal de Docencia Universitaria (REDU). Seminario Internacional 2-07: El desarrollo de la autonomía en el aprendizaje (2007)
19. Ramos, C.: Los alcances de una investigación. CienciAmérica, Ecuador (2020)
20. Sampieri, R., Fernández, C., Baptista, P.: Metodología de la investigación. Mc Graw Hill, México (2006)
21. SEP. Agenda Digital de Educación. SEP, México (2020)
22. Torres, P., et al.: Tecnología educativa y su papel en el logro de los fines de la educación. Educare, Venezuela (2017)

Implementation of Time Series to Determine Purchase and Use of Electric Cars in a Smart City Considering Generation Z as Target Population

Shaban Mousavi Ghasemlou[1]([✉]), Alberto Ochoa-Zezzatti[1,2], Vianey Torres[1], Erwin Martinez[1], and Victor Lopez[2]

[1] Doctorado en Tecnología, UACJ, Ciudad Juárez, Mexico
al237855@alumnos.uacj.mx

[2] CADIT, Posgrados Anahuac, Universidad Anahuac Campus Norte, Naucalpan de Juárez, Mexico

Abstract. One of the biggest issues facing the globe today is greenhouse gas (GHG) emissions. In the transportation industry, the development of electric vehicles has recently been the main focus of research globally to satisfy the permissible GHG limitations where vehicles that operate on oil contribute a significant amount of GHG. In recent years, there has been a great increase in the study of electric vehicles (EVs). This integrative evaluation is an attempt to close the gap in assessments that evaluate and show the demand and development of EVs in great detail. This study provides numerous thought-provoking insights on specific events, such as the rise in popularity and desire for electric vehicles on a worldwide scale, the demand for power, and the role of government in the development of smart cities. This study gained a new perspective with the addition of the concept of a smart city created by EV implementation. The overview would be helpful for policy-makers as well as academics. The findings of this study provide a summary of the arguments made by investors, regulators, and members of Generation Z in favor of electric vehicles.

Keywords: electric vehicle · Generation Z · Smart City · Time Series Models

H. Calvo et al. (Eds.): MICAI 2023 Workshops, LNAI 14502, pp. 298–312, 2024.
https://doi.org/10.1007/978-3-031-51940-6_22

Graphical Abstract:

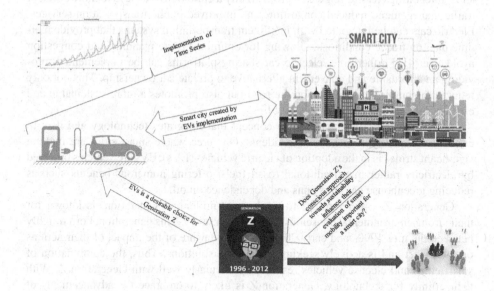

1 Introduction

The convergence of two major technological advancements, namely the concept of "smart cities" and the rising popularity of electric cars, has the potential to revolutionize urban landscapes and transportation systems. A smart city refers to a city that utilizes advanced technology and data-driven solutions to enhance the quality of life for its residents. Electric cars, on the other hand, are vehicles powered by electricity instead of traditional fossil fuels, offering a greener and more sustainable mode of transportation. Integrating electric cars into the infrastructure of a smart city holds tremendous promise in terms of reducing pollution, enhancing energy efficiency, and improving overall mobility. One of the key benefits of electric cars in the context of a smart city is the significant reduction in carbon emissions. Transportation is a major contributor to greenhouse gas emissions, and the widespread adoption of electric cars could play a crucial role in combating climate change [1].

Electric vehicles produce zero tailpipe emissions as they are powered by electricity stored in rechargeable batteries. This not only helps to improve air quality, but it also reduces the reliance on fossil fuels, making cities more sustainable and environmentally friendly [2]. Moreover, when electric cars are integrated into smart city infrastructure, they become part of a larger ecosystem that optimizes energy consumption and management. Smart charging stations can be strategically placed throughout the city, equipped with advanced monitoring systems to ensure efficient power distribution. These charging stations can potentially be powered by renewable energy sources such as solar or wind, further reducing the carbon footprint of electric vehicles. Additionally, smart grids can be implemented to balance the electricity demand from electric cars, saving costs and reducing strain on the existing power infrastructure [3]. Another significant advantage of electric cars in a smart city is the potential for improved mobility and transportation systems. Integrating electric vehicles with smart city technologies can lead to more efficient traffic management, reduced congestion, and improved public transportation services. Electric cars can be integrated with intelligent transportation systems that provide real-time data on traffic conditions, allowing for optimized route planning and congestion avoidance [4]. Furthermore, electric car-sharing platforms can be implemented, providing an affordable and convenient alternative to private car ownership. This not only reduces the number of vehicles on the road but also promotes a more sustainable and efficient urban transportation system [5].

A smart city is a visionary urban concept that integrates technology and data to enhance the quality of life for its residents. One area where smart cities are making significant strides is in the adoption of electric vehicles (EVs). EVs are vehicles powered by electricity rather than traditional fossil fuels, offering numerous benefits such as reducing greenhouse gas emissions and dependence on oil.

Generation Z, often referred to as the post-millennial generation, is known for their strong environmental consciousness and tech savvy. This generation, born roughly between the mid-1990s and early 2010s, is highly aware of the impact of their actions on the planet and is actively seeking sustainable solutions. Thus, the combination of smart cities and electric vehicles resonates particularly well with Generation Z. With their affinity for technology, Generation Z is likely to embrace the advancements of smart cities that provide EV infrastructure, such as charging stations, and implement policies that promote the adoption of electric vehicles. They understand the importance of sustainable transportation and are likely to prioritize environmentally friendly options when it comes to commuting and travel [6]. Moreover, Generation Z has grown up in a digital age where connectivity and data are central to their [7]. Smart cities, with their emphasis on connectivity and data-driven decision-making, align with the preferences and expectations of this generation. They value convenience, efficiency, and innovative solutions, all of which are key aspects of smart cities.

The integration of electric cars into the infrastructure of a smart city presents a promising opportunity to address pressing environmental concerns and enhance overall mobility. By reducing carbon emissions, optimizing energy consumption, and improving transportation systems, the combination of smart cities and electric cars has the potential to create cleaner, greener, and more livable urban environments [8]. However, it is important for policymakers, urban planners, and technology developers to collaborate and invest in the necessary infrastructure to maximize the potential benefits of this technology convergence.

This study will offer important insights into the variables affecting the rise of EVs and smart cities in their development and early baseline. The research findings can help urban planners, academics, and policymakers make educated choices, encourage teamwork, and promote sustainable mobility options in smart cities. In addition, the following parameters are also examined in this study:

1. The effects of generation Z on smart cities
2. The different types of renewable energy used in smart cities and how they affect the environment.
3. Impact on Stakeholders (Organizations, Government, and Citizens).

2   Literature Review

The data for the monthly home energy consumption time series E, for a period of six years, was considered by Abdel et al. The UBJ model is built using data from the first five years, and the last year is seaside for model evaluation. The main characteristics of this timeseries are (i) a clear seasonal pattern within each year, seasonality of length 12, with maximum consumption around July and minimum around April; and (ii) a variance that is nearly constant with time, suggesting that no additional processing (e.g. use of log transform) is necessary to compress data [9].

Pegah et al. [10] studied that predicts the total power usage of a population of air conditioners. In this study, a control approach is proposed for modifying the thermostat settings of the air conditioners so that their overall aggregated power consumption would follow a desirable trajectory. First, the controller arranges the ACs into groups of similar patterns by using their unchecked power usage. Similarity in room temperature and thermal properties is implied by similarity in power usage. As a result, the controller will have better resources to improve performance by adjusting the set points of similar rooms.

Climate change (global warming), renewable energy (reducing reliance on oil), stopping pollution (reducing emissions, eliminating pesticides), recycling and reducing waste, protecting wildlife, and optimizing resource use are the top environmental concerns for them, according to research by Monika et al. [11].

The necessities and requirements of the energy consumption forecast model are to be addressed, according to Jin et al. [12]. They introduced a unique deep learning-based approach to further construct latent space by embedding electric information independently because many prior methods have limitations on explanatory power. They modeled not only performs predictions with state-of-the-art accuracy, but it also examines the correlation between input values and anticipated values to offer a higher degree of justification.

According to James et al. [13], there are various ways to usefully model a certain process. Different models of dynamic motion, including Newtonian, relativistic, quantum, statistical, and continuum mechanics, each add to our understanding of the same phenomenon in a different way. They needed a way to compare the relative efficiency and accuracy of different models because they might compete in this situation.

Wenjing et al. [14], reviewed the literature-proposed economy-driven EV charging management strategies. Although study on the subject has been extremely active in recent years, to our knowledge no work has been published that provides a thorough summary of the many techniques taken. The goal of this work is to classify the available models, emphasize their key hypotheses and findings, and compare them in order to pinpoint the most promising mechanisms and future prospects.

A hybrid neural network-based load forecasting system made up of linear and non-linear neural networks was taken into consideration by Reinschmidt et al. [15]. They are functional and prepared for use in businesses. Additionally, they discussed two subsystems that would be added to the system to handle various odd situations [15].

3 Development

EVs have emerged as a crucial strategy to cut carbon emissions and advance a cleaner, more sustainable future in the current era, which is defined by environmental consciousness and the demand for sustainable transportation. The adoption of EVs is a desirable choice for Generation Z, who were born between the middle of the 1990s and the middle of the 2000s and are particularly interested in technology and the environment. In this paper, we will investigate how to forecast EVs demand in a smart city with 1,500,000 residents from 2023 to 2050, given a starting base of 7,000 EVs.

3.1 Time Series Models (ARIMA)

Time series models are a class of potent statistical methods for analyzing and predicting data that changes over time, such as the auto-regressive integrated moving average (ARIMA). Numerous disciplines, including finance, economics, epidemiology, and environmental research, have found extensive use for ARIMA models. They are able to recognize and model complicated temporal patterns in data because they are based on the principles of autoregression, differencing, and moving averages. ARIMA models give researchers and analysts the ability to make educated forecasts about future values by spotting patterns, seasonality, and abnormalities within time series, offering crucial insights for decision-making and planning. The benefits and drawbacks of utilizing this are:

1. ARIMA models are a well-known and dependable technique for time series forecasting since they are based on sound statistical concepts. They offer a well-organized framework for creating models and studying time series data [16].
2. Simplicity: Compared to some more complex machine learning models, ARIMA models are comparatively easy to comprehend and use. As a result, they are usable by a variety of users, even those who lack in-depth machine learning knowledge [17].
3. Interpretability: By providing coefficients that can be used to analyze how previous values of a time series affect future values, ARIMA models provide information that may be used to gain understanding of the underlying dynamics of the data.
4. Effective for Trend and Seasonality: ARIMA models are especially good at identifying and predicting time series data that exhibit distinct trends and seasonal patterns. They can manage seasonal changes when coupled with seasonal ARIMA (SARIMA) models.
5. Data Efficiency: To generate accurate forecasts, ARIMA models don't need a lot of data. They are effective with small and brief time series datasets.

   And about the disadvantages:

1. Sensitivity to Model Selection: When using ARIMA models, model orders (p, d, and q) must be carefully chosen in accordance with the properties of the time series data. Poor model performance might result from making the improper ordering decisions [18].
2. ARIMA models make the assumption that the time series data is stationary, which means that its statistical characteristics do not alter over time. The data may need to

be differed in order to get stationarity, and choosing the right amount of differencing is not always simple.

3. Lack of Seasonality Handling: Time series data with significant seasonal trends are difficult for ARIMA models to handle. SARIMA (Seasonal ARIMA) models were developed to address this problem; however, they could still need to be carefully tuned.

4. Limited Explanatory Power: Since ARIMA models are primarily employed for time series forecasting, they might not be able to shed light on the underlying factors that underlie the observed data patterns. They don't work as well for causal analysis.

3.2 The Status of Electric Vehicles (EVs)

Hybrid EVs (HEV) and various types of all-electric vehicles (AEV) are two subcategories of EVs. An AEV just has a motor that is powered by the power supply. Battery EVs (BEV) and EVs powered by fuel cells (FCEV) are further categories of AEVs. An energy storage system (ESS) and a power control unit (PCU) are components of a BEV. The PCU's connection to a hydrogen tank (HT) and fuel cells (FCs) distinguishes BEVs from FCEVs. The FCEV does not need an external charging mechanism. However, BEV only uses the network's external power source to load a storage device. A plug-in hybrid electric vehicle (PHEV) is a type of HEV that can use the grid for electricity. Figure 1 shows the many types of EVs and the sources of power for their wheels.

Specific details about the low-cost EVs produced by various manufacturers are shown in Fig. 2. According to several charging theories, the figure also depicts the projected charging time needed to charge the car from 0% to 80%. Here, the first stage of charging is equivalent to 110–120 V, the second stage is equivalent to 22–240 V, and the third stage, also known as DC fast charging (DCFC), is equivalent to 200–800 V. It is clear that an electric vehicle's range depends on the battery charge. However, after roughly 100 km, the battery only lasts 200 to 400 km in some car types and some other models. However, in China, the majority of the current EV models have a range of over 400 km [19].

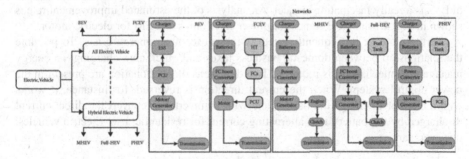

Fig. 1. Classifications of EVs.

3.3 The Demand of EVs

The first stage in determining the basic metals for future energy-based transportation is to set up a scenario where the quantity of EVs and anticipated future demand for the following metals may be calculated. With regard to historical (2022) and future (2050) year scenarios, such as baseline (BS), moderate (MS), and stringent (SS) results, Fig. 3 depicts the annual growth of three various types of EVs (BEV, PHEV, and HEV). The integrated model to analyze greenhouse impacts (IMAGE), which was created for the database of the shared socioeconomic pathways (SSP), provided the data necessary to make improvements. An SSP is a persistent problem that enters the network as a result of environmental changes. They are dependent on five distinct accounts, which equates to the absolute number of drivers in the basic scenario is predicted to increase from 1.13 billion in 2011 to 2.6 billion in 2050 based on the results of improving the situation. By 2050, there will probably be 2.25 billion and 2.55 billion station wagons worldwide, respectively, depending on economic conditions. Figure 3 demonstrates that in these three examples, the supply of three EVs rose year over year, despite the challenging circumstances [20].

3.4 EV- Related Technology

Experts, organizations, and strategic developers from various nations have expressed a great deal of interest in the creation of EVs. EVs coordinate all kinds of individual accomplishments and categorize the entire EV field into several crucial regions, which can offer more crucial point-by-point information [21]. EVs can help with the decarbonization of transportation due to their beneficial usage characteristics and low pollution levels, and the expansion of low-carbon emission urban areas has thus become one of the models to boost the enthusiasm of the automotive industry.

In any case, innovation is crucial to the future success of the electric car industry [22]. Politicians are seriously considering reform in the area of EVs in many nations, including Sweden, China, Malaysia, and South Korea [23]. They are also making plans to encourage EV technology progress. However, technological advancement in the area of EVs is a really fascinating subject. An analysis of the estimated improvement rate is shown in Fig. 4, where PE stands for power electronics and EM for electric motor.

Furthermore, it is frequently cited as the cause of an energy crisis. To promote the relationship between domestic business taxes and quick auto charging, an energy management mechanism is required. Two directions of fractionation are present in the power control system. When the current direction is reversed, for instance, it works with a rectifier to charge the battery and with an inverter to convert the direct current discharged by the battery into alternating current for residential use. Electric vehicles, however, can:

1. The charging station has large-scale loads compared to residential loads.
2. In this situation, the transmission capacity has better response speeds.
3. The charging points are available and have highflexibility.

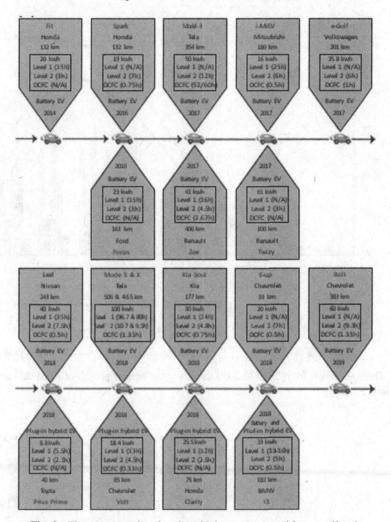

Fig. 2. The most popular electric vehicle currency positions are listed.

4 Results and Discussion

Smart cities have become an increasingly prominent topic of discussion in recent years, driven by the need for sustainable and efficient urban development. One key aspect of smart cities is the integration of EVs as a means of transportation. As the world moves towards a carbon-neutral future, EVs are seen as an essential component of reducing greenhouse gas emissions. The expanding significance of smart cities in the modern world is highlighted in this sentence. It states that the necessity for sustainable and effective urban development has caused smart cities to become a hot topic of discussion in recent years. The use of EVs as a mode of transportation is a crucial component of smart cities. EVs are viewed as essential in lowering greenhouse gas emissions as the world strives to achieve a carbon-neutral future [25].

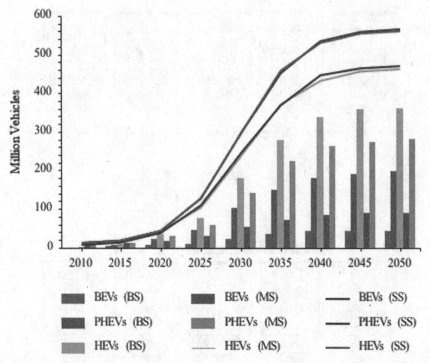

Fig. 3. Three scenarios from 2010 to 2050 that show the annual growth of three different EV stock kinds.

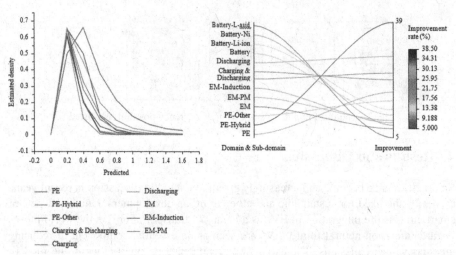

Fig. 4. The estimated technological improvement rates of domains and subdomains. Redrawn and taken permission from Elsevier [21].

Generation Z, the demographic cohort born between the mid-1990s and mid-2000s, plays a crucial role in shaping the adoption and development of EVs in smart cities. As digital natives, Generation Z individuals are well-versed in technology and have a high level of environmental awareness. This generation has grown up with smartphones and other digital tools that have shaped their values and preferences. The demographic cohort born between the mid-1990s and mid-2000s plays a role in influencing the adoption and development of electric vehicles (EVs) in smart cities. As digital natives, Gen Z individuals are highly familiar with technology and possess a strong environmental consciousness. Growing up with smartphones and other digital tools has shaped their values and preferences, making them more inclined towards sustainable transportation options like EVs [26]. The mobility choices of Generation Z and their evolving views toward car ownership are highlighted in a McKinsey report [27] and represented in our research as in Fig. 5.

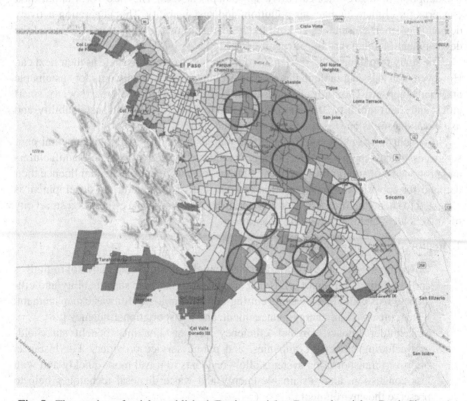

Fig. 5. The number of articles published (Review articles, Research articles, Book Chapters)

Along with personal vehicles, Gen Z supports sustainable and shared transportation. Since becoming widely adopted in city centers, public transportation and micro mobility choices like bike and e-scooter rentals have greatly increased in popularity among younger people. Nearly 55% of Gen Z consumers are interested in cooperative mobility solutions and are open to sharing their personal vehicles.

For many young people in Generation Z, sustainability and environmental consciousness are vital considerations in their choices and actions. They are more likely to prioritize eco-friendly solutions, making them a prime target audience for EV manufacturers and proponents of smart cities. As these young individuals become drivers and decision-makers, their demand for sustainable transportation options like EVs is expected to increase.

Moreover, Generation Z's technological mindset aligns well with the smart city concept. They are accustomed to using mobile apps, connected devices, and data-driven solutions. EV infrastructure, such as charging stations and intelligent transportation systems, can be seamlessly integrated into their tech-driven lifestyles. Generation Z is likely to embrace the convenience and efficiency of EVs while also appreciating the environmental benefits they offer.

As Generation Z becomes more influential in society, their support and advocacy for EV adoption in smart cities can drive substantial changes. They are vocal about their concerns for climate change, air pollution, and sustainability. This generation actively demands greener alternatives and are likely to champion the transition to EVs and the development of smart city infrastructure.

The study found that 50% of Gen Z would prefer an electric vehicle as their next car. However, Gen Z is adaptable in their mobility choices and typically opts for sustainable and shared options. The study also adds to the body of knowledge on smart society, smart cities, and smart mobility. These new trends aim to support long-term sustainability, and using car sharing is one approach to create smart cities.

In conclusion Generation Z's emphasis on sustainability and environmental consciousness makes them a significant target audience for electric vehicle manufacturers and proponents of smart cities. As they continue to grow in numbers and influence their demand for eco-friendly transportation options and sustainable urban development is expected to shape the future of mobility and urban planning. The effects of a smart city on stakeholders and the role of the government:

1. Improved Efficiency and Sustainability:

   - Government Role: The government is essential to the implementation of initiatives and regulations for smart cities that are meant to increase sustainability and efficiency. This could involve using cutting-edge technologies for waste management, water conservation, energy management, and transportation efficiency [28].
   - Stakeholder Impact: Greater efficiency and sustainability benefit stakeholders including citizens, companies, and public service providers. For instance, improved transportation systems allow residents to travel more quickly and with less congestion, and sustainable energy and waste disposal techniques help to preserve the environment.

2. Enhanced Quality of Life:

   - Government Role: Governments can leverage smart city technologies to enhance the quality of life for their citizens. This may involve implementing smart infrastructure, including intelligent lighting, public Wi-Fi networks, and smart healthcare systems, to provide better services and amenities.

- Stakeholder Impact: Citizens experience an improved quality of life through enhanced connectivity, access to public services, and better healthcare facilities. Stakeholders such as healthcare providers and businesses also benefit from increased connectivity and improved infrastructure, leading to better service delivery and economic growth.

3. Economic Development and Innovation:

- Government Role: By enacting laws that encourage investment, encourage entrepreneurship, and advance R&D, governments can boost economic growth and innovation in smart cities. They can enact beneficial regulatory frameworks and offer incentives for companies to use cutting-edge technology.
- Stakeholder Impact: A smart city ecosystem offers chances for businesses, entrepreneurs, and startups. Innovation, economic growth, and job creation are fueled by the availability of cutting-edge infrastructure, data access, and supportive policies.

4. Data Privacy and Security:

- Government Role: Governments have a responsibility to establish robust data privacy and security frameworks in smart cities. They must enact legislation and regulations to protect citizens' personal information and ensure the secure handling of data collected from various sources.
- Stakeholder Impact: Stakeholders, particularly citizens, rely on the government to safeguard their privacy and protect them from cybersecurity threats. Building trust through transparent data governance practices and strong security measures is crucial to ensure stakeholder confidence in smart city initiatives.

In a nutshell, the government's job in a smart city is to provide the leadership, oversight, and infrastructure required for putting smart initiatives into action and monitoring them. Governments influence stakeholders through their actions through increasing productivity, enhancing quality of life, encouraging citizen involvement, fostering economic development, and protecting data privacy and security.

5 Conclusion and Future Recommendation

Conclusion. It is anticipated that advancements in EV development and contributions to the overall infrastructure and resources of renewable energy will enhance the image of electric vehicles on a worldwide scale. To ensure the greatest benefits from EVs with circulation, additional technological advancements are essential. These include reasonable and appropriate charging rules, smart cities, strong adaptive frameworks, business structures, policy, CO_2 emission reduction, and measures to measure the impact on the environment, health, and power grid. In addition, Energy Internet will develop into a cutting-edge network in the future and fully compute the energy framework using the most recent energy frame panel. This study provides an introduction to every component of the electric car development framework. Electric vehicles must be well-known on the market after implementing EV principles and their adoption globally. The issues of smart cities, electric vehicles, and generation Z are hot topics that researchers are particularly interested in exploring, as can be seen in the image below (Fig. 6).

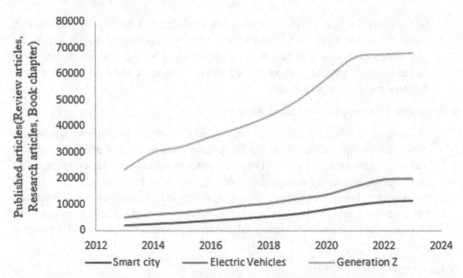

Fig. 6. Segmentation by distribution pole of the location of 10,000 electric vehicles associated with a Smart City and their need for electrical energy through a Smart Grid.

This study examines each of the components of the electric vehicle development structure, the role of the government in the smart city, and the advantages and disadvantages of using renewable energy in the smart city. Electric vehicles should be recognized in the market after the implementation of EV principles and their global acceptance. In addition, many existing architectural components that load EV and communication networks, including their strengths, adaptability, control, and coordination power, as well as their advantages and disadvantages, are thoroughly evaluated and improved. The report provides recommendations for further future research into tidal containment.

Future Suggestion, it is believed that the unique ideas can be helpful in overcome the barriers to EV development after analyzing the most recent research on the status of electric vehicles (EVs). Additionally, discussing all the relevance in one study is intolerable. The research needs some future suggestions for illuminating its usefulness for further enhancement, as provided below:

1. Information and communication in EV– smart cities with the development of renewable energy sources should be more advanced. We need to gather additional research materials or data from online surveys in order to make the best decision, and the concept may come from EV-developed nations.
2. Create fresh business and policy plans for EVs for customers' goods and services.
3. Grid-connected EVs battery charging still has drawbacks. It might take some time for these impacts to diminish, which will improve the likelihood of integrating EBs with renewable energy sources.
4. Recycling the components used in EVs batteries is difficult. Therefore, a new energy storage technique is required.

References

1. Kirimtat, A., Krejcar, O., Kertesz, A., Tasgetiren, M.F.: Future trends and current state of smart city concepts: a survey. IEEE Access **8**, 86448–86467 (2020)
2. Larminie, J., Lowry, J.: Electric Vehicle Technology Explained. Wiley, Hoboken (2012)
3. Tan, K.M., Ramachandaramurthy, V.K., Yong, J.Y.: Integration of electric vehicles in smart grid: a review on vehicle to grid technologies and optimization techniques. Renew. Sustain. Energy Rev. **53**, 720–732 (2016)
4. Zamboulis, A., et al.: Chitosan and its derivatives for ocular delivery formulations: Recent advances and developments. Polymers **12**(7), 1519 (2020)
5. Kamargianni, M., Li, W., Matyas, M., Schäfer, A.: A critical review of new mobility services for urban transport. Transp. Res. Procedia **14**, 3294–3303 (2016)
6. Singh, B., Roy, P., Spiess, T., Venkatesh, B.: Sustainable integrated urban & energy planning, the evolving electrical grid and urban energy transition. The Centre for Urban Energy. Ryerson University (2015)
7. Katz, R., Ogilvie, S., Shaw, J., Woodhead, L.: Gen Z, Explained: The Art of Living in a Digital Age. University of Chicago Press (2022)
8. Razmjoo, A., Kaigutha, L.G., Rad, M.V., Marzband, M., Davarpanah, A., Denai, M.: A technical analysis investigating energy sustainability utilizing reliable renewable energy sources to reduce CO_2 emissions in a high potential area. Renew. Energy **164**, 46–57 (2021)
9. Abdel-Aal, R., Al-Garni, A.Z.: Forecasting monthly electric energy consumption in eastern Saudi Arabia using univariate time-series analysis. Energy **22**(11), 1059–1069 (1997)
10. Yazdkhasti, P., Ray, S., Diduch, C.P., Chang, L.: Using a cluster-based method for controlling the aggregated power consumption of air conditioners in a demand-side management program. In: 2018 International Conference on Smart Energy Systems and Technologies (SEST), pp. 1–6. IEEE (2018)
11. Wawer, M., Grzesiuk, K., Jegorow, D.: Smart mobility in a Smart City in the context of Generation Z sustainability, use of ICT, and participation. Energies **15**(13), 4651 (2022)
12. Kim, J.-Y., Cho, S.-B.: Electric energy demand forecasting with explainable time-series modelling. In: 2020 International Conference on Data Mining Workshops (ICDMW), pp. 711–716. IEEE (2020)
13. Bezdek, J.C.: Pattern Recognition with Fuzzy Objective Function Algorithms. AAPR, Springer, Boston (1981). https://doi.org/10.1007/978-1-4757-0450-1
14. Weigend, A.S.: Time series analysis and prediction using gated experts with application to energy demand forecasts. Appl. Artif. Intell. **10**(6), 583–624 (1996)
15. Reinschmidt, K., Ling, B.: Artificial neural networks in short term load forecasting. In: Proceedings of International Conference on Control Applications, pp. 209–214. IEEE (1995)
16. Praveen, P., Shravani, S., Srija, R., Tajuddin, M.: A model to stock price prediction using deep learning. In: 2023 International Conference on Sustainable Computing and Smart Systems (ICSCSS), pp. 242–252. IEEE (2023)
17. Muhasina, K., et al.: 7 machine data mining learning and. Artificial Intelligence in Bioinformatics and Chemoinformatics, p. 117 (2023)
18. Dumitru, C., Maria, V.: Advantages and Disadvantages of Using Neural Networks for Predictions. Ovidius University Annals, Series Economic Sciences, vol. 13, no. 1 (2013)
19. Said, D., Cherkaoui, S., Khoukhi, L.: Queuing model for EVs charging at public supply stations. In: 2013 9th International Wireless Communications and Mobile Computing Conference (IWCMC), pp. 65–70. IEEE (2013)
20. Habib, K., Hansdóttir, S.T., Habib, H.: Critical metals for electromobility: global demand scenarios for passenger vehicles, 2015–2050. Resour. Conserv. Recycl. **154**, 104603 (2020)

21. Feng, S., Magee, C.L.: Technological development of key domains in electric vehicles: improvement rates, technology trajectories and key assignees. Appl. Energy **260**, 114264 (2020)

22. Correa, G., Muñoz, P., Falaguerra, T., Rodriguez, C.: Performance comparison of conventional, hybrid, hydrogen and electric urban buses using well to wheel analysis. Energy **141**, 537–549 (2017)

23. Brady, J., O'Mahony, M.: Travel to work in Dublin. the potential impacts of electric vehicles on climate change and urban air quality. Transp. Res. Part D: Transp. Environ. **16**(2), 188–193 (2011)

24. Barhagh, S.S., Mohammadi-Ivatloo, B., Anvari-Moghaddam, A., Asadi, S.: Risk-involved participation of electric vehicle aggregator in energy markets with robust decision-making approach. J. Clean. Prod. **239**, 118076 (2019)

25. Lnenicka, M., et al.: Transparency of open data ecosystems in smart cities: definition and assessment of the maturity of transparency in 22 smart cities. Sustain. Cities Soc. **82**, 103906 (2022)

26. Lalić, D., Stanković, J., Bošković, D., Milić, B.: Career expectations of generation Z. In: Anisic, Z., Lalic, B., Gracanin, D. (eds.) IJCIEOM 2019. LNMIE, pp. 52–59. Springer, Cham (2020). https://doi.org/10.1007/978-3-030-43616-2_6

27. Varma, I.G., Chanana, B., Lavuri, R., Kaur, J.: Impact of spirituality on the conspicuous consumption of fashion consumers of generation Z: moderating role of dispositional positive emotions. Int. J. Emerg. Mark. (2022)

28. Eremia, M., Toma, L., Sanduleac, M.: The smart city concept in the 21st century. Procedia Eng. **181**, 12–19 (2017)

CIAPP 2023

Blood Cell Image Segmentation Using Convolutional Decision Trees and Differential Evolution

Adriana-Laura López-Lobato$^{(\boxtimes)}$ [ID], Héctor-Gabriel Acosta-Mesa[ID], and Efrén Mezura-Montes[ID]

Artificial Intelligence Research Institute, University of Veracruz, Campus Sur, Calle Paseo Lote II, Sección Segunda No. 112, Nuevo Xalapa, 91097 Xalapa-Enríquez, Veracruz, México
adrilau17@gmail.com, {heacosta,emezura}@uv.mx
https://www.uv.mx/iiia

Abstract. Semantic segmentation is an important process in computer vision that assigns labels to the pixels of an image to divide it into regions of interest. The most used machine learning model for this problem is the Convolutional Neural Network (CNN), in which high-performance results are obtained, however, they are difficult to understand and explain, which is not very useful in fields where explainability is fundamental, as in medicine. As an alternative, there are Convolutional Decision Trees (CDT), a tool that is easy to interpret due to its intuitive and user-friendly graphic structure. In this article, a method is proposed to induce an optimized CDT with different kernel sizes using the Differential Evolution algorithm, obtaining F1-scores greater than 0.92 on a set of blood cell images for erythrocyte segmentation, a relevant task for doctors and laboratory technicians.

Keywords: Semantic segmentation · Convolutional decision tree · Differential evolution

1 Introduction

The study of blood tissue (blood) is of vital importance for doctors and laboratory technicians, since its main function is to transport oxygen, carbon dioxide and the essential nutrients for life, as well as waste products [5]. Blood is composed of erythrocytes, thrombocytes, and leukocytes, each one fulfills specific tasks for the proper functioning of the human body, so the identification and characterization of each of these elements is essential for the diagnosis of blood-related diseases [7]. To observe the characteristics of blood samples, the Dark-Field Microscopy (DFM) technique has been used since it is a powerful optical tool that improves the visibility of the samples by balancing the components of the blood and the background of the images [6,17]. However, although these images are helpful to experts, their analysis can be a tedious, imprecise and subjective process, which makes it necessary to use segmentation techniques with automated algorithms that require minimal user interaction [16,19].

H. Calvo et al. (Eds.): MICAI 2023 Workshops, LNAI 14502, pp. 315–325, 2024.
https://doi.org/10.1007/978-3-031-51940-6_23

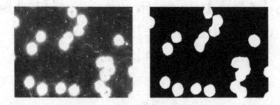

Fig. 1. Semantic segmentation technique applied to DFM images to identify erythrocytes in blood tissue.

This project proposes the use of a semantic segmentation technique applied to DFM images that highlights the blood features of interest. Semantic segmentation is a process that consists of assigning labels to the pixels of an image to distinguish between objects of interest and the background of the image; see Fig. 1. There are several methods that have been used to solve the image segmentation problem [9,12]; however, Convolutional Neural Networks (CNN) are the most used [10]. High-performance results are obtained with CNN; however, they are often called black boxes because it can be difficult to understand and explain the process they follow to obtain results, so their use in critical contexts, such as in the medical area, must be thoroughly studied [3,11].

An alternative to CNN are Decision Trees (DT), since they are an easy tool to understand due to their intuitive and user-friendly graphical structure [2]. For the application in image segmentation, Convolutional Decision Trees (CDT) [8] were proposed, constructed with convolution kernels obtained by maximizing the information gain.

To address optimization problems in the DT induction process, different metaheuristics have been used, such as the Differential Evolution (DE) algorithm [13,14]. DE is one of the most popular metaheuristic search strategies for solving optimization problems; it incorporates stochastic elements and parameters that enhance its ability to explore the problem domain, even when the parameters must be adjusted to the specific problem [4]. In particular, using the DE algorithm to induce CDT, as opposed to the traditional maximization process detailed in [8], has the benefit of being able to modify the objective function (information gain) and select non-continuous functions.

In [1], a global search approach is introduced for inducing CDT with DE. Multiple CDTs are induced using this technique, and the most effective one is chosen for segmentation. Nevertheless, this process can be very complex in terms of computational time, since the classification accuracy of each instance (pixel) of the training set must be assessed on each tree created with the DE process.

In this research study, a method is proposed for conducting a local search using DE to identify the optimal convolutional kernel size and the corresponding convolutional kernel for each CDT node. Furthermore, the method is compared to a global search approach, since previous experiments in [1] shows that the use of a global search technique results in F1-scores below 0.6.

To build the CDT with the proposed method, the user first provides a list of kernel sizes and, after applying the learning process with these kernel sizes, the kernel with the best F1-score at each partition node is selected. This results in a single CDT with convolutional kernels of different sizes, which is an interpretable supervised learning model that can be used for image segmentation.

2   Methods

2.1   Differential Evolution Algorithm

The DE algorithm is one of the most popular metaheuristics search strategies for solving optimization problems. It uses a population, which represents potential solutions to the problem coded according to its needs, and gets a highly competitive solution by generating new populations through an iterative process.

Mutation and crossover operators are applied to each individual in the population (parent) to obtain a new individual (child) that maintains some characteristics of the predecessor. These operators involve two parameters, a scaling factor F and a crossover rate CR, set by the user. Afterwards, a selection operator is used, which preserves or removes the ancestor by comparing its fit with that of the child. This process is carried out with each of the individuals in the population until forming a new population. This procedure is repeated for a certain number of iterations, denoted as generations, and the individual with the best fit in the population of the last generation is considered the optimal solution.

There are various evolutionary strategies for the DE process, however the two most popular are DE/rand/1/bin and DE/best/1/bin. The most used is the DE/rand/1/bin strategy, proposed by Storn and Price [15]. This strategy uses a random individual from the current population to obtain children in the mutation operator. The DE/best/1/bin strategy uses the individual from the current population with the best fit to obtain children in the mutation operator. The main difference between these strategies is convergence. The DE/rand/1/bin strategy may converge more slowly than the DE/best/1/bin strategy [18].

After carrying out several experiments with these strategies, and confirming similar results to those shown in [1], better results were obtained with the DE/best/1/bin strategy, and in less time, so this variant was selected for this project.

2.2   Proposed Model (DE-CDT-BKS)

In this work, a local search strategy is proposed, denoted as DE-CDT-BKS (BKS = Best Kernel Size), in which the DE algorithm is applied to find the size of the kernels and the kernels of the nodes of a CDT of given depth. The information of each pixel (instance) of the image is encoded as the vector with the shades of gray of the pixels in the neighborhood of size $s \times s$ that surround it, adding the value 1 for the bias, see Fig. 2.

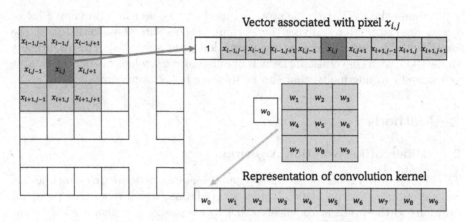

Fig. 2. Coding an instance associated with a pixel and a convolution kernel.

The value of s is taken from a given list S of kernel sizes, with odd values. For the DE process, the variant DE/best/1/bin is considered where each individual in the population is a vector of size $s^2 + 1$ corresponding to the weights of the kernel that we want to find an the value for the bias, see Fig. 2. The population size, the scaling factor F, the crossing rate CR and the number of generations are parameters given by the user. To determine which branch of the tree is taken when classifying an instance, a perceptron structure is used, where the product of the values of the coded instance and those of the kernel is passed through an activation function (sigmoid) that returns a label 1 or 0, see Fig. 3.

The fit value is determined by the F1-score metric and the purpose is to maximize it with the DE algorithm. Other metrics such as cross-entropy and information gain were used to measure the fit value, but the best results were

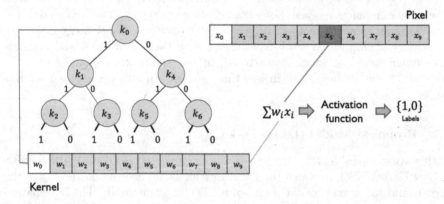

Fig. 3. Codification of a CDT and processing an instance associated with a pixel.

obtained with the F1-score metric, so this metric was used to measure the fitness of individuals.

The DE process is carried out with each of the values of the set S. For each value s, the individual with the best fitness is found and, among them, the one with the maximum F1-score is selected as the best solution.

The CDT structure classifies the instances through the convolution process established with each kernel found, until reaching the depth given by the user. For the root node kernel, all instances are considered for the DE process, and for the other kernels, only those tagged with the class corresponding to the branch in which the kernel is located. If there are more than 20 instances belonging to different classes and at least 5% of them belong to one class, a new kernel is created. Otherwise, DE-CDT-BKS does not find a new kernel. Thus, with the DE-CDT-BKS method, an optimized CDT with different kernel sizes is obtained, without the use of a pruning process.

3   Experiments and Results

In this section, some of the results obtained by inducing a CDT with the DE-CDT-BKS process are reported using the last 12 images from the database "Blood detection in dark-field microscopy images"[1], where 365 images obtained using the dark field microscopy technique are shown to observe and segment erythrocytes. The last 12 images were selected to observe the behaviour of the proposed model, as a first approximation, since they are in the same dimension conditions (384×288).

Controlled experiments were carried out to calibrate the parameter values of the DE algorithm, by taking fixed training and test sets with 8 and 4 images, respectively, and reducing the number of pixels with a process of resizing an image in half. The following values were considered for the parameters of the DE-CDT-BKS process: depths from 1 to 5 for the tree, population size and number of generations set to 50, $S = \{3, 5, 7\}$, and for the values of F and CR the different combinations of 0.1, 0.5, 0.7 and 0.9 were considered. After performing several experiments with these characteristics, the highest F1-scores are obtained with $F = 0.7$ and $CR = 0.5$, so these two values were maintained in the following experiments.

Table 1 shows the results obtained by modifying the resizing of the images (Redim), the depth of the tree (Depth), the population size (PopSize) and the number of generations (NumGen), keeping the training and test sets fixed, and the parameters $S = \{3, 5, 7\}$, $F = 0.7$ and $CR = 0.5$.

In experiments 1–8, the resize is 0.5 and trees of depth 3 and 5 are induced, varying the population size and the number of generations with combinations of values 100 and 200. The best results are obtained with 100 individuals and 200 generations, showing that an increase in the population size doubles the execution time, with worse results. The best result is obtained in experiment 2 with a tree of depth 3, see Fig. 4(a).

---

[1] https://github.com/PerceptiLabs/bacteria/tree/main.

Table 1. Results obtained with the proposed method for CDT induction. The F1-score and Accuracy measures were calculated with the actual and predicted labels of the 4 images in the test set. The best results by image resizing are highlighted with numbers in italics. The best result when resizing the images to scale 0.5 is obtained in experiment 2, the best result when resizing to scale 0.75 is obtained in experiment 10 and the best result when using the full images is obtained in experiment 15.

Exp	Redim	Depth	PopSize	NumGen	Time (hr)	F1-score	Accuracy
1	0.5	3	100	100	3.41	0.88558	0.94885
2	0.5	3	100	200	6.63	0.92973	0.96819
3	0.5	3	200	100	6.88	0.88525	0.94789
4	0.5	3	200	200	13.22	0.92931	0.96809
5	0.5	5	100	100	3.88	0.90254	0.95680
6	0.5	5	100	200	6.71	0.92891	0.96791
7	0.5	5	200	100	6.97	0.89605	0.95643
8	0.5	5	200	200	13.18	0.92848	0.96779
9	0.75	2	100	100	6.62	0.89188	0.95409
10	0.75	2	100	200	13.33	0.92900	0.96933
11	0.75	3	100	100	7.73	0.90625	0.96005
12	0.75	3	100	200	15.39	0.92832	0.96915
13	0.75	5	100	100	9.98	0.9101	0.96166
14	0.75	5	100	200	18.58	0.92515	0.96718
15	1	1	100	200	10.33	0.93207	0.97145
16	1	2	100	200	12.33	0.93147	0.97129
17	1	3	100	200	14.28	0.93088	0.97113

In experiments 9–14 the resize is 0.75 and trees of depths 2, 3 and 5 are induced, with a population of 100 individuals and varying the number of generations between 100 and 200. The best results for depth are obtained with 200 generations; and the best result is obtained in experiment 10 with the tree of depth 2, see Fig. 4(b).

In the last three experiments with the complete images, trees of depths 1, 2 and 3 are induced with populations of 100 individuals and 200 generations. Again, the best result is obtained with the tree with the lowest depth, corresponding to experiment 15, see Fig. 4(c).

The results of these experiments indicate that inducing a CDT with the proposed method produces better results with shallow trees and, in the context of DTs, shallow trees are easier to represent and interpret.

The increase in population size does not represent an improvement for the process, but the increase in the number of generations does. Furthermore, the execution time seems to be proportional with respect to these two parameters.

The best result was obtained by inducing a CDT of depth 1, but when comparing the results of experiments 2, 10 and 15, a difference of 2 to 3 thousandths

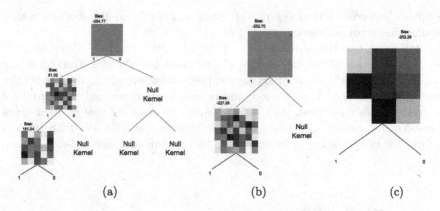

Fig. 4. CDTs induced with the CDT-DE-BKS method corresponding to experiments 2 (a), 10 (b) and 15 (c). Note: the root nodes in figures (a) and (b) are kernels of size 3 × 3.

in the F1-score and large differences in execution times are observed, so the number of instances of the corresponding training sets generated by the resizing process affects the execution time.

Table 2 shows the best and worst results obtained with the CDT of experiment 15 in the test set images and 3 graphs; the original image of the erythrocytes, the original image with the segmentation overlay, in blue, and the image of the real labels (ground truth) with the segmentation overlay.

With the CDT, a precise distinction is obtained between pixels of light tone (class 1) and dark tone (class 0), however, the ground truth of the images

Table 2. Best result and worst result, per image, obtained with the CDT induced in experiment 15. The pixels corresponding to class 1 predicted by the CDT, corresponding to erythrocytes, are shown in blue.

Image	Graphical comparisons		
355 F1-score: 0.9544 Accuracy: 0.9838			
366 F1-score: 0.8892 Accuracy: 0.9565			

classifies the pixels within the erythrocytes as class 1, so the method makes classification errors in these pixels.

With respect to the explainability of the CDT, its structure allows an analysis of each kernel of the tree and how they carry out the classification of the image pixels. In the context of erythrocyte segmentation, in the first kernel the classification obtained is very good, and in the lower kernels only the profile of the erythrocytes is detailed, see Fig. 5, however it is expected that in other types of images and contexts, the kernels obtained with the DE-CDT-BKS process perform the classification of images pixels by patterns and shapes, see Fig. 6.

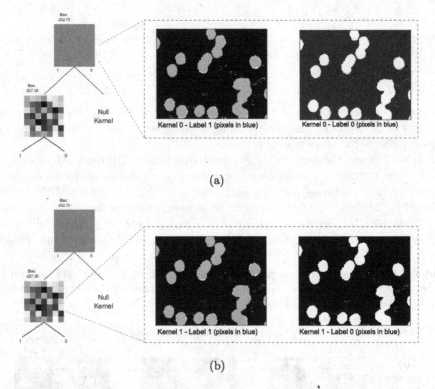

Fig. 5. Analysis of the segmentation process of image 355 for each kernel of the CDT induced in experiment 10. Kernel 0 in figure (a) and kernel 1 in figure (b). Label 1 corresponds to erythrocyte pixels and label 0 to the background of the image.

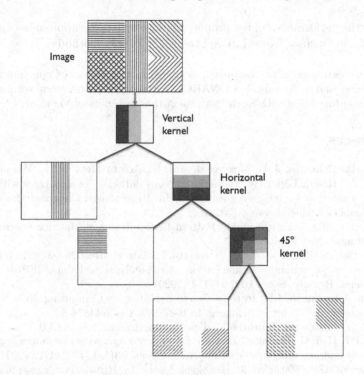

Fig. 6. Explainability in CDT

4 Conclusions and Future Work

In this project, a method is proposed for the induction of an optimized CDT, without the use of a pruning process, with different kernel sizes using the DE algorithm, which serves as an alternative to the approaches proposed in [1,8].

Generating the tree locally by finding the CDT kernels one by one allows to establish a process to compare different kernel sizes in each node, by carrying out a simple convolution on the corresponding pixels of the image to be segmented. The proposed method is the first approach in which different kernel sizes are considered.

With the proposed process, satisfactory results are obtained for the segmentation of images of erythrocytes obtained with the dark field microscopy technique, by inducing small and, consequently, explainable CDTs, with F1-scores greater than 0.92 when trained with few images and without any other additional process, so the use of the DE algorithm seems to be a favorable tool for the induction of CDTs. Furthermore, the proposed model can be used to segment other types of images by analyzing them per pixel.

As future work, we want to implement techniques to reduce the execution time of evaluating the individuals. Also, we want to analyze the capabilities of using the DE algorithm with self-adapted parameter values of F and CR, and compare the performance of the proposed model with other approaches for image segmentation, such as Convolutional Neural Networks methods.

Acknowledgements. The first author is funded by the National Council of Humanities, Sciences and Technologies (CONAHCyT), through a postdoctoral scholarship at the Artificial Intelligence Research Institute of the University of Veracruz.

References

1. Barradas Palmeros, J.A., Mezura Montes, E., Acosta Mesa, H.G., Márquez Grajales, A., Rivera López, R.: Induction of convolutional decision trees with differential evolution for image segmentation. In: Proceedings: Congreso Mexicano de Inteligencia Artificial, vol. 8 (2023)
2. Bishop, C.M., Nasrabadi, N.M.: Pattern Recognition and Machine Learning, vol. 4. Springer, New York (2006)
3. Diakogiannis, F.I., Waldner, F., Caccetta, P., Wu, C.: ResUNet-a: a deep learning framework for semantic segmentation of remotely sensed data. ISPRS J. Photogramm. Remote. Sens. **162**, 94–114 (2020)
4. Eiben, A.E., Smith, J.E.: Introduction to Evolutionary Computing. NCS, Springer, Heidelberg (2015). https://doi.org/10.1007/978-3-662-44874-8
5. Finn, G.: Histología. Editorial Medica Panamericana, España (2001)
6. Gao, P.F., Lei, G., Huang, C.Z.: Dark-field microscopy: recent advances in accurate analysis and emerging applications. Anal. Chem. **93**(11), 4707–4726 (2021)
7. Junqueira, L.C., Carneiro, J.: Histologia básica. In: Histologia básica, pp. 512–512 (1985)
8. Laptev, D., Buhmann, J.M.: Convolutional decision trees for feature learning and segmentation. In: Jiang, X., Hornegger, J., Koch, R. (eds.) GCPR 2014. LNCS, vol. 8753, pp. 95–106. Springer, Cham (2014). https://doi.org/10.1007/978-3-319-11752-2_8
9. Lateef, F., Ruichek, Y.: Survey on semantic segmentation using deep learning techniques. Neurocomputing **338**, 321–348 (2019)
10. Minaee, S., Boykov, Y., Porikli, F., Plaza, A., Kehtarnavaz, N., Terzopoulos, D.: Image segmentation using deep learning: a survey. IEEE Trans. Pattern Anal. Mach. Intell. **44**(7), 3523–3542 (2021)
11. Molnar, C.: Interpretable Machine Learning: A Guide for Making Black Box Models Explainable. 2nd edn (2022). https://christophm.github.io/interpretable-ml-book
12. Patil, D.D., Deore, S.G.: Medical image segmentation: a review. Int. J. Comput. Sci. Mob. Comput. **2**(1), 22–27 (2013)
13. Rivera-Lopez, R., Canul-Reich, J.: Construction of near-optimal axis-parallel decision trees using a differential-evolution-based approach. IEEE Access **6**, 5548–5563 (2018)
14. Rivera-Lopez, R., Canul-Reich, J., Mezura-Montes, E., Cruz-Chávez, M.A.: Induction of decision trees as classification models through metaheuristics. Swarm Evol. Comput. **69**, 101006 (2022)
15. Storn, R., Price, K.: Differential evolution-a simple and efficient heuristic for global optimization over continuous spaces. J. Global Optim. **11**, 341–359 (1997)
16. Usmani, U.A., Roy, A., Watada, J., Jaafar, J., Aziz, I.A.: Enhanced reinforcement learning model for extraction of objects in complex imaging. In: Arai, K. (ed.) Intelligent Computing. LNNS, vol. 283, pp. 946–964. Springer, Cham (2022). https://doi.org/10.1007/978-3-030-80119-9_63

17. Verebes, G.S., Melchiorre, M., Garcia-Leis, A., Ferreri, C., Marzetti, C., Torreggiani, A.: Hyperspectral enhanced dark field microscopy for imaging blood cells. J. Biophotonics **6**(11–12), 960–967 (2013)

18. Zhang, J., Sanderson, A.C.: JADE: adaptive differential evolution with optional external archive. IEEE Trans. Evol. Comput. **13**(5), 945–958 (2009)

19. Zhu, C., Ni, J., Li, Y., Gu, G.: General tendencies in segmentation of medical ultrasound images. In: 2009 Fourth International Conference on Internet Computing for Science and Engineering, pp. 113–117. IEEE (2009)

Multi-objective Evolutionary Algorithm Based on Decomposition to Solve the Bi-objective Internet Shopping Optimization Problem (MOEA/D-BIShOP)

Miguel A. García-Morales, José A. Brambila-Hernández[✉], Héctor J. Fraire-Huacuja, Juan Frausto-Solis, Laura Cruz-Reyes, Claudia Guadalupe Gómez-Santillan, Juan Martín Carpio Valadez, and Marco Antonio Aguirre-Lam

National Technological Institute of Mexico, Technological Institute of Ciudad Madero, Ciudad Madero, Tamaulipas, Mexico
alfredo.brambila@outlook.com

Abstract. The main contribution of this paper is to develop a new solution method applied to the bi-objective Internet shopping problem. The first objective of this problem considers minimizing the total cost of the shopping list, and the second objective is minimizing the shopping list delivery time. This solution consists of a Multi-objective Evolutionary Algorithm Based on Decomposition to Solve the Biobjective Internet Shopping Optimization Problem (MOEA/D-BIShOP). The proposed MOEA/D-BIShOP algorithm obtains an approximate Pareto optimal set for nine types of real-world instances classified according to their size into small, medium, and large. These instances have been obtained through web scraping, which consists of extracting information from some technology products on the Amazon site. The Biobjective Internet shopping optimization problem is different from the Internet shopping optimization problem because it considers the purchase cost and the delivery conditions of each product. This algorithm is compared to the state-of-the-art work by applying two heuristics to each objective respectively and obtaining an approximate Pareto Front (APF), transforming the second objective function into a constraint within an integer linear programming algorithm. The results demonstrate that the proposed algorithm (MOEA/D-BIShOP) has equal statistical performance compared to the only work from the state-of-the-art. Three metrics were used: Hypervolume, Generalized Dispersion, and Inverted Generational Distance. The non-parametric Wilcoxon and Friedman test is applied to validate the results obtained with a significance level of 5%.

Keywords: Internet Shopping Optimization Problem · Multi-objective Optimization · Web Scraping · Bi-objective

1 Introduction

The Internet is an extensive communication network that allows interaction of computing facilities over long distances. The use of the Internet in companies has changed the way of doing business since it allows the use of commercial strategic tools for marketing, sales,

H. Calvo et al. (Eds.): MICAI 2023 Workshops, LNAI 14502, pp. 326–336, 2024.
https://doi.org/10.1007/978-3-031-51940-6_24

and customer services. Due to the significant relevance of the Internet within companies, it has caused electronic commerce to contribute to most of a company's profits [1]. In addition, the Internet allows access to various smart devices such as sensors, cameras, smart cities, and decision support systems [2–4], which today is called Internet Shopping. The Internet shopping problem is a classic electronic commerce scenario that allows customers to purchase goods or services through the Internet [5]. The main advantage of online shopping is the variety of products and services offered, which allows no restrictions on time, place, and under any circumstances [1].

In some product purchases, it takes work to optimize the total cost when the price of the products, the shipping cost, and the delivery time of multiple online stores must be considered [1]. Typically, customers should select the store with the best total cost and delivery time [1]. These decisions aim to minimize effort and maximize customer benefit [1].

Chung [6] proposes for the first time an IShOP optimization model that involves two objectives, the total purchase cost and the delivery time, using a multi-objective optimization model. Chaerani et al. [7] assume that Chung's model is similar to the maximum flow problem with circular demand (MFP-CD) because it has multiple sources in several stores. The model proposed by Chung includes a delivery time decision variable. Chaerani et al. [7] transform that decision variable into an adjustable variable using the adjustable robust counterpart (ARC) method with maximum delivery time uncertainty. Chaerani et al. [1] consider the above and propose a Benders decomposition method to solve the Adjustable Robust Countparty Problem for Internet Shopping (ARC-ISOP).

This article proposes a new solution method by implementing a multi-objective optimization algorithm MOEA/D to solve the Bi-objective Internet Shopping Problem (MOEA/D-BIShOP). In the computational calculations, nine instances were used with real-world data of Amazon technological products obtained through a Web Scraping technique.

1.1 Definition of the Problem

Chung [6] first proposed the "Internet shopping optimization problem model" with two objectives. In this problem, a customer wants to buy a set of n products N online, which can be purchased in a set of m available stores M. Now, the set N_i contains the products available in store i, each product $j \in N_i$ has a cost of p_{ij}, a shipping cost f_j, and a delivery time d_{ij}. The shipping cost is charged if one or more products are purchased in the store i. The Bi-objective Internet Shopping Optimization Problem (BIShOP) consists of minimizing the total cost of purchasing all products N, considering the cost-plus shipping costs, and minimizing the delivery time.

The notation in Table 1 is used to define and describe the problem.

The model considers the purchase cost and delivery time limitations as objectives at the same time. This bi-objective model is described as follows:

$$Min \sum_i \sum_j p_{ij} x_{ij} + \sum_j f_j y_j \qquad (1)$$

$$Min\, max_{i,j}\left(d_{ij} x_{ij}\right) \qquad (2)$$

Table 1. Notation table.

Variable/Parameter	Description
M	Set of stores.
N	Set of products
I	Vector solution
m	Number of stores, $|M|$
n	Number of products, $|N|$
j	Store indicator
i	Product indicator
N_i	Container of products available in a store j
f_j	Shipping cost of all products in the store j
p_{ij}	Cost of product i in store j
d_{ij}	Delivery time of a product i in store j
x_{ij}	Binary variable that indicates whether product i is purchased in store j
y_j	Binary variable indicating whether to add the sipping cost of store j

$$s.t. \sum_j x_{ij} = 1, \forall i = 1, \ldots, n \tag{3}$$

$$\sum_i x_{ij} \le ny_j, j = 1, \ldots, m \tag{4}$$

$$x_{ij} = 0/1, y_j = 0/1 \tag{5}$$

The objective function (1) establishes that we want to minimize the purchase cost, including the price of the products and the shipping cost. Objective function (2) means that you want to minimize the delivery time of all products. Constraint (3) states that all products to be purchased must be selected from available stores, and constraint (4) means that a fixed shipping cost is incurred whenever there is some selection of products in the store. Constraint (5) establishes the binary decision variables.

2 General Structure of the MOEA/D-BIShOP Algorithm

This section describes all the elements that make up the general structure of the proposed multi-objective optimization algorithm MOEA/D-BIShOP. The algorithm uses a vector representation I of length N for each solution within the population. This vector contains each product in the store where it can be purchased. Equations 1 and 2 show the calculation of the two objective functions, respectively.

In this work, the crossover operator selects two parents at random and divides them in half, obtaining two children from the mixture between both parents, considering a crossover point. It selects one of the two children obtained and randomly decides which will advance to the mutation process. Consequently, the mutation operator loops through each element of the selected child and modifies the current store in the range $[1, m]$.

2.1 Crossover Operator

This operator randomly selects two solutions called $parent_1$ and $parent_2$ [8]. The solution $child_1$ is generated by taking the initial half of $parent_1$ and joining it with the second half of $parent_2$. Later, to form $child_2$, the initial half of $parent_2$ is joined with the second half of $parent_1$ [9]. Subsequently, a random number is generated; if this generated value is less than 0.5, the crossover operator selects $child_1$; otherwise, it takes $child_2$ to advance to the mutation process. The crossover operator uses $\lfloor N/2 \rfloor$ or $\lceil N/2 \rceil$ as the crossover point.

2.2 Mutation Operator

The mutation process takes the candidate solution selected by the crossover operator. It immediately positions itself on the first element of the solution and generates a random number; if this random value is less μ, the current element of the solution is replaced by a random value in the online stores range $[1, m]$. This process continues until all elements of the current solution have been traversed.

2.3 The Multi-objective Evolutionary Algorithm Based on Decomposition to Solve the IShOP Bi-objective Problem (MOEA/D-BIShOP)

The multi-objective evolutionary algorithm based on decomposition (MOEA/D) was developed by Zhang and Li [10–12] and serves as a reliable and robust alternative for working with MOPs. This algorithm shares some characteristics of the weighted sum and population-based approach. It starts by distributing a set of λ weight vectors in the objective functional space, then creates a matrix of T nearest vectors using the Euclidean distance between vectors, thus creating neighborhoods [13]. The basic MOEA/D algorithm uses the Tchebycheff decomposition defined in Eq. 6:

$$\min g^{te}\left(x|\lambda^j, z^*\right) = max_{1 \leq i \leq m} \frac{1}{\lambda_i^j}|f_i(x) - z_i^*| \tag{6}$$

The MOEA/D-BIShOP algorithm is represented in Algorithm 1. In steps 1 to 4, λ reference vectors are established, neighborhoods are created using the T nearest neighbor vectors as criteria, and the ideal Z point is calculated. The main loop runs through all individuals within the population. In step 7, two parents are chosen. These are taken from the neighborhoods created in $B(i)$. $B(i)$ is traversed, and two parents are chosen randomly; then, the crossover and mutation operators are applied to generate a single child. In the final part of the algorithm, the Z value is updated again. The aggregation values of the two are calculated using λ reference vectors; likewise, the aggregation value of the child y^i is replaced with a simple criterion: if the child y^i has an aggregation value less than one of the parents, it is replaced; otherwise, the parent remains, and the population is not modified.

Algorithm 1. MOEA/D-BIShOP Algorithm

Input: MOP − Bi-objective IShOP Problem
Pop − Population
nPop − Population size
fileSize − File size
Stopping criterion
N − the number of subproblems considered in MOEA/D-BIShOP
A uniform distribution of N weight vectors: $\lambda^1, ..., \lambda^N$
T − the number of weight vectors in the neighborhoods of each weight vector
Output: EP

1: $EP = \emptyset$
2: Compute the Euclidean distances between any two weight vectors and then compute the weight vectors T closets to each weight vector.
3: **for** $i \leftarrow 1$ **to** N **do**
4:     $B(i) = \{i_1, ..., i_T\}$ $\lambda^{i_1}, ..., \lambda^{i_T}$ are T nearest weight vectors λ^i
5: **end for**
6: Generate initial population $x^1, ..., x^N$ randomly.
7: $FV^i = F(x^i)$
8: Initialize $z = (z_1, ..., z_m)^N$ for the bi-objective IShOP
9: **while** stopping criterion not met **do**
10:    **for** $i \leftarrow 1$ **to** N **do**
11:        Randomly select two indices k, l from $B(i)$, and generate a new solution y from x^k and x^l using genetic operators
12:        Apply a problem-specific repair/improvement heuristic on y to produce y'
13:        **for** $j \leftarrow 1$ **to** m **do**
14:            **if** $z_j < f_j(y')$ **then**
15:                $z_j = f_j(y')$
16:            **end if**
17:        **end for**
18:        **foreach** index $j \in B(i)$ **do**
19:            **if** $g^{te}(y'|\lambda^j, z) \leq g^{te}(x^j|\lambda^j, z)$ **then**
20:                $x^j = y'$
21:                $FV^j = F(y')$
22:            **end if**
23:        **end foreach**
24:    **end for**
25:    remove from EP all solutions dominated by $F(y')$
26:    insert $F(y')$ in EP if there are no solutions in EP that dominate $F(y')$
27: **end while**

---

3 Computational Experiments

The names of instances determine their size, m represents the number of stores, and n is the number of products. For the experimental test, three sets of real-world instances of different sizes were used, and each subset contains 30 instances.

Table 2. Definition of instances.

Small	Medium	Large
$3n20m$	$5n240m$	$50n400m$
$4n20m$	$5n400m$	$100n240m$
$5n20m$	$50n240m$	$100n400m$

The designs are obtained from Web Scraping of multiple technological products (USB flash, Modem, RAM) that were carried out on Amazon's e-commerce website. In this process, approximately 8002 records containing product names, prices, suppliers, delivery time, and shipping costs were obtained.

Figure 1 shows the process of building the instances from real-world data described below: collect product and store information from the Amazon.com page. Build an application in the Python language that allows us to explore within the search engine and obtain information using the Web Scraping technique, using various keywords such as laptop, headphones, and speakers, among others. With a depth of 10 pages for each, the Beautiful Soup Python library is used to process the information.

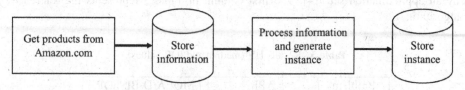

Fig. 1. General instance generation process.

A first version of the instances has been generated, and its construction is carried out by taking the products obtained with a defined price range and the stores are obtained. Shipping times are defined arbitrarily (randomly) with values between 1 and 5 days. For the shipping cost, four arbitrary values are used, which are assigned randomly. These values are 88, 99, 120, and 140. The types of instances generated are shown in Table 2.

3.1 Configuration of the Parameters

The configuration parameters of the proposed MOEA/D-BIShOP algorithm are shown below $pop = 100$, $pc = 0.6$, $pm = 0.01$, $maxIter = 1000$, and $\mu = 0.02$. The above configuration was determined based on related works found in the state-of-the-art. Modifying the values of the parameters can affect the behavior of the algorithm and, therefore, the quality of the solutions. The size of the population is important because it affects the diversity and convergence of the algorithm. A small population can lead to loss of performance, diversity, and early convergence. An inadequate number of generations can cause the algorithm to converge prematurely or have excessive resource consumption, and incorrect use of the crossover and mutation operators can lead to deadlocks or inefficient explorations of the solution space and the size of the neighborhood because it determines the number of neighboring solutions to explore contributes to the quality of the generated solutions. In the computational experiments, the 30 non-dominated fronts were obtained from each of the three sets of instances for each subset; subsequently, non-parametric tests were applied, and the p-value was obtained to determine if there were significant differences in favor of the implemented algorithm.

3.2 Results

The experiment results are presented in Tables 3, 5, and 7, organized by metric. Friedman and Wilcoxon non-parametric tests were used with a significance level of 5%. The first column of each table corresponds to the problem, while the second column corresponds to the algorithm results. The algorithm in the second column is the reference (APF). In the table, the symbol ▲ means that there is statistical significance in favor of the reference algorithm, the symbol ▼ means that there is statistical significance in favor of the algorithm that is compared with the reference algorithm (in favor of the current column), and the symbol == means there is no statistical significance.

Hypervolume
The hypervolume (HV) calculates the volume of the objective space weakly dominated by an approximation set [14]. The first column in Table 3 represents the reference algorithm.

Table 3. Results HV (median and IQR values).

Problem	APF	MOEA/D-BIShOP
$3n20m$	1.00e+00 0.00e+00	0.00e+00 1.00e+00 ▲
$4n20m$	1.00e+00 0.00e+00	0.00e+00 0.00e+00 ▲
$5n20m$	1.00e+00 0.00e+00	0.00e+00 1.00e+00 ▲
$5n240m$	1.00e+00 0.00e+00	0.00e+00 0.00e+00 ▲
$5n400m$	1.00e+00 0.00e+00	0.00e+00 0.00e+00 ▲
$50n240m$	1.00e+00 0.00e+00	0.00e+00 0.00e+00 ▲
$50n400m$	1.00e+00 0.00e+00	0.00e+00 0.00e+00 ▲
$100n240m$	1.00e+00 0.00e+00	0.00e+00 0.00e+00 ▲
$100n400m$	1.00e+00 0.00e+00	0.00e+00 0.00e+00 ▲

As can be seen, in the hypervolume metric, the reference algorithm is better in all nine problems.

Friedman test

The $p - value$ calculated with the Friedman test is 0.002699796063260207, so with a level of statistical significance of 5%, it is significant. Table 4 below shows the average ranks per algorithm obtained with the Friedman test.

Table 4. Average ranks for HV.

Algorithm	AVG Rank
APF	1
MOEA/D-BIShOP	2

Friedman's non-parametric test determines that the significant differences favor the state-of-the-art algorithm. The above shows that the algorithm obtains better approximate Pareto fronts for all the evaluated instances.

Generalized Spread

Generalized Spread (GS) evaluates the degree of dispersion and uniformity of the solutions identified. In Table 5, the first column is the reference algorithm.

Table 5. Results GS (median and IQR values).

Problem	APF	MOEA/D-BIShOP
$3n20m$	5.01e-01 7.39e-04	5.01e-01 6.04e-04 ▼
$4n20m$	4.36e-01 1.55e-02	4.38e-01 1.36e-02 ==
$5n20m$	0.00e+00 0.00e+00	0.00e+00 0.00e+00 ==
$5n240m$	5.12e-01 7.99e-03	5.12e-01 6.28e-03 ==
$5n400m$	0.00e+00 0.00e+00	0.00e+00 0.00e+00 ==
$50n240m$	4.16e-01 1.32e-02	4.17e-01 1.46e-02 ▲
$50n400m$	5.09e-01 7.24e-03	5.10e-01 1.04e-02 ▲
$100n240m$	5.03e-01 1.58e-03	5.03e-01 1.73e-03 ▲
$100n400m$	4.09e-01 5.55e-03	4.09e-01 6.36e-03 ==

As can be seen, in the hypervolume metric, the reference algorithm is better statistically in three of nine problems.

Friedman test

The $p - value$ calculated with the Friedman test is 0.09558070454562984, so with a level of statistical significance of 5%, it is significant. Table 6 below shows the average ranks per algorithm obtained with the Friedman test.

Friedman's non-parametric test determines no significant difference between both algorithms. Therefore, the approximate Pareto fronts obtained in both algorithms have similar performance.

Table 6. Average ranks for GS.

Algorithm	AVG Rank
APF	1.22
MOEA/D-BIShOP	1.77

Inverted Generational Distance

The inverted generation distance (IGD) gives the average distance between any point in the reference set and its nearest point in the approximation set [14]. In Table 7, the second column is considered as the reference algorithm.

Table 7. Results IGD (median and IQR values).

Problem	APF	MOEA/D-BIShOP
$3n20m$	1.42e+02 6.22e+01	1.44e+02 5.56e+01 ▲
$4n20m$	3.70e+00 1.43e+00	3.90e+00 1.08e+00 ▲
$5n20m$	1.34e+154 0.00e+00	1.34e+154 0.00e+00 ==
$5n240m$	1.14e+01 9.58e+00	1.33e+01 7.18e+00 ▲
$5n400m$	1.34e+154 0.00e+00	1.34e+154 0.00e+00 ==
$50n240m$	9.55e+00 5.84e+00	9.48e+00 5.82e+00 ==
$50n400m$	1.84e+01 1.05e+01	1.80e+01 1.08e+01 ▼
$100n240m$	4.62e+01 2.42e+01	4.53e+01 2.50e+01 ▼
$100n400m$	1.93e+01 1.26e+01	1.86e+01 1.31e+01 ==

As can be seen, in the hypervolume metric, the reference algorithm is better statistically in three of nine problems.

Friedman test

The $p-value$ calculated with the Friedman test is 0.7388826803635397, so with a level of statistical significance of 5%, it is not significant. Table 8 below shows the average ranks per algorithm obtained with the Friedman test.

Table 8. Average ranks for IGD

Algorithm	AVG Rank
MOEA/D-BIShOP	1.44
APF	1.55

Friedman's non-parametric test determines no significant difference between both algorithms. Therefore, the inverted generation distance metric indicates that the both algorithms find the best solution in fewer iterations.

4 Conclusions

Finally, with the results obtained, it is observed that the proposed MOEA/D-BIShOP algorithm shows statistically equal performance in two of the three metrics evaluated: generalized spread and inverted generational distance, which suggests that the algorithm has good dispersion in the solutions and has a convergence similar to the reference algorithm. Therefore, it is assumed that by using other genetic operators and including new elements, a competitive solution method can be developed concerning the state-of-the-art algorithms. This work proposes as future work to develop an NSGA-II algorithm and develop a method for adaptive selection of control parameters. The development of an algorithm that converges more quickly to optimal solutions, and the quality of these solutions is superior, would be very useful to be applied in online stores and search engines on the Internet, providing the user with tools that allow them to perform searches using more than one attribute at a time and therefore allow us to make good decisions.

References

1. Chaerani, D., Saksmilena, S., Irmansyah, A.Z., Hertini, E., Rusyaman, E., Paulus, E.: Benders decomposition method on adjustable robust counterpart optimization model for internet shopping online problem. Computation **11**(2), 37 (2023)
2. Zamir, M., et al.: Face detection & recognition from images & videos based on CNN & Raspberry Pi. Computation **10**(9), 148 (2022)
3. Afzal, K., Tariq, R., Aadil, F., Iqbal, Z., Ali, N., Sajid, M.: An optimized and efficient routing protocol application for IoV. Math. Probl. Eng. **2021**, 1–32 (2021)
4. Malik, U.M., Javed, M.A., Zeadally, S., ul Islam, S.: Energy-efficient fog computing for 6G-enabled massive IoT: recent trends and future opportunities. IEEE Internet Things J. **9**(16), 14572–14594 (2021)
5. Kumar, S.: Online shopping-a literature review. In: National Conference on Innovative Trends in Computational, pp. 129–131 (2015)
6. Chung, J.B.: Internet shopping optimization problem with delivery constraints. Distrib. Sci. Res. **15**(2), 15–20 (2017)
7. Chaerani, D., Rusyaman, E., Marcia, A., Fridayana, A.: Adjustable robust counterpart optimization model for internet shopping online problem. In: Journal of Physics: Conference Series, vol. 1722, no. 1, p. 012074. IOP Publishing (2021)
8. Holland, J.H.: Adaptation in Natural and Artificial Systems. University of Michigan Press, And Arbor (1975)
9. Umbakar, A.J., Sheth, P.D.: Crossover operators in genetic algorithms: a review. ICTACT J. Soft Comput. **6**(1) (2015)
10. Zhang, Q., Li, H.: MOEA/D: a multiobjective evolutionary algorithm based on decomposition. IEEE Trans. Evol. Comput. **11**(6), 712–731 (2007)
11. Eiben, A., Smith, J.: Introduction to Evolutionary Computing. Natural Computing Series, Springer, Heidelberg (2015). https://doi.org/10.1007/978-3-662-44874-8
12. Li, H., Zhang, Q.: Multiobjective optimization problems with complicated Pareto sets, MOEA/D and NSGA-II. IEEE Trans. Evol. Comput. **13**(2), 284–302 (2009). http://ieeexplore.ieee.org/document/4633340/
13. García, C.: A celullar Evolutionary Algorithm to Tackle Constrained Multiobjective Optimization Problems [Tesis de maestría, Instituto Nacional de Astrofísica, Óptica y Electrónica]. Repositorio institucional del INAOE (2020). https://inaoe.repositorioinstitucional.mx/jspui/bitstream/1009/2155/1/Mc_Thesis_Cosijopii.pdf

14. Brambila-Hernández, J.A., García-Morales, M.Á., Fraire-Huacuja, H.J., del Angel, A.B., Villegas-Huerta, E., Carbajal-López, R.: Experimental evaluation of adaptive operators selection methods for the dynamic multiobjective evolutionary algorithm based on decomposition (DMOEA/D). In: Castillo, O., Melin, P. (eds.) Hybrid Intelligent Systems Based on Extensions of Fuzzy Logic, Neural Networks and Metaheuristics. SCI, vol. 1096, pp. 307–330. Springer, Cham (2023). https://doi.org/10.1007/978-3-031-28999-6_20

A Surrogate-Assisted Differential Evolution Approach for the Optimization of Ben's Spiker Algorithm Parameters

Carlos-Alberto López-Herrera[✉][iD], Héctor-Gabriel Acosta-Mesa[iD], and Efrén Mezura-Montes[iD]

Artificial Intelligence Research Institute, University of Veracruz, Veracruz, Mexico
carlosalberto.lopezherrera91@gmail.com, {heacosta,emezura}@uv.mx
https://www.uv.mx/iiia/

Abstract. Spiking neural networks (SNNs) differentiate themselves from traditional artificial neural networks by modeling the behavior of neurons in a more biologically plausible manner. Consequently, they employ discrete spikes or events to communicate information. Therefore, codifying analog signals into spike trains is a fundamental pre-processing step in SNNs. Ben's Spiker Algorithm (BSA) has become one of the most used codification methods. Moreover, having optimal parameters allows for efficient and desirable performance. This paper contrasts two Kriging-Assisted Differential Evolution (KADE) approaches against Differential Evolution (DE) in said optimization task. The implementation is tested in a synthetic signal, and an electroencephalographic (EEG) signal to assess the consistency of the method. Furthermore, the Signal to Noise Ratio (SNR) metric was used to evaluate the performance of the implementations. Our findings demonstrate that KADE reduces the computational time of the implementation while achieving similar reconstructed signals. Specifically, the KADE approaches and DE implementation achieved a mean SNR of 9.04, 9.08, and 9.30, respectively, for the synthetic signal while reaching 10.27, 10.79, and 11.43, respectively, for the EEG signal.

Keywords: Ben's Spiker Algorithm · Differential Evolution · Kriging · Kriging- Assisted Differential Evolution · EEG signals

1 Introduction

In the brain, the interaction between neurons is done by transmitting action potentials (or spike trains) to other nearby neurons. Inspired by this, Spiking Neural Networks (SNNs) were proposed as a more biologically realistic approximation [1]. Nevertheless, all real-world signals are, in nature, both analog and temporal. Hence, it is crucial to have proper techniques to transform signals into spike trains while preserving sufficient information to leverage SNNs.

© The Author(s), under exclusive license to Springer Nature Switzerland AG 2024
H. Calvo et al. (Eds.): MICAI 2023 Workshops, LNAI 14502, pp. 337–348, 2024.
https://doi.org/10.1007/978-3-031-51940-6_25

In the specialized literature, there are two big groups of encoding methods [2]: Rate and Temporal coding schemes. On one hand, the rate coding strategy focuses on how information is encoded. On the other hand, temporal coding methods encode signals based on the timing of significant events. It has been found that rate coding suffers from wide latency periods between spikes, which may affect the performance of SNNs for some applications. Hence, temporal coding has been used in more recent works [2].

BSA, introduced by Scherauwen & Van Campehout (2003) [3], is one of the most prominent temporal coding schemes, where the main idea is that an analog signal can be derived from a spike train using the convolution and a Finite Impulse Response (FIR) filter [2]. From there, BSA uses a suitable filter to produce a spike train based on the comparison of two errors; the first error ($E1$) involves the sum of differences between the original signal and the FIR filter, while the second error ($E2$) represents the aggregated value of the signal. Therefore, a spike is produced whenever the first error is smaller than the weighted (by a *threshold* value) second error. Consequently, the decoding process is achieved by the convolution of the encoded spike train signal and the FIR filter. Moreover, the FIR filter's composition relies on two parameters: *filter size* and *cutoff frequency*. Thus, the filter size (Fs), the cutoff frequency (Cf), and the threshold (Th) are of great importance. For a more in-depth description of the BSA, Petro et al. (2020) [2] provides a comprehensive review of the subject. Some algorithms have been used to optimize the BSA parameters (2). One promising option is using Evolutionary Algorithms (EAs).

Over the years, there has been an essential consolidation of EAs as an optimization technique for complex problems. This group can be described as population-based algorithms that take inspiration from natural evolution. Generally, the objective function (or fitness value) of an individual (a member of a population) is obtained using an explicit function, a computational simulation, or an experiment [4]. Nevertheless, in some cases, these evaluations are too expensive (i.e., time-consuming or computationally costly) [5]. To overcome this, the combined usage of Surrogate-Assisted (SA) models and EAs have been proposed [6]. Such techniques are called Surrogate-Assisted Evolutionary Algorithms (SAEAs).

The concept of the surrogate model was driven by the idea of having a mathematical tool capable of imitating a real objective function as accurately as possible [7] while reducing computational cost/time in an expensive problem [4]. Among the most common models are the Radial Basis Function (RBF), Support Vector Machine (SVM), Polynomial Approximation Model (PAM), and Kriging model (KM). Regarding the latter, it has been shown that KM has a robust and efficient performance for a low number of variables (less than 15) [7], as is the case of BSA optimization.

Also known as Gaussian processes, KM was first introduced in 1963 [8] as a geo-statistical method for interpolation processes, and its formulation undergoes two principal components:

$$y(x) = f(x) + Z(x) \tag{1}$$

where $f(x)$ represents a linear model and $Z(x)$ is a systematic departure from the linear model that follows a Gaussian random function [9] under any sample point x. One of the most appealing properties of this model is that it provides a value of uncertainty. Furthermore, it has been established that a large degree of uncertainty at a given point can hint at regions poorly explored, indicating good candidates for finding better solutions [4]. This property has been taken further by using different model management criteria (infill sampling), such as the Probability of Improvement (PoI), the Lower Confidence Bound (LCB), and the Expected Improvement (EI).

In recent years, EI has become more notable than its counterparts [10]. This function balances local and global search [11] by measuring the value of improvement that a given point x is expected to get when compared with the best prediction so far y_{best}. Mathematically:

$$EI(x) = (y_{best} - \hat{y}(x)) \cdot \Phi\left(\frac{y_{best} - \hat{y}(x)}{\hat{s}(x)}\right) + \hat{s}(x) \cdot \phi\left(\frac{y_{best} - \hat{y}(x)}{\hat{s}(x)}\right) \quad (2)$$

where $\hat{y}(x)$ is the predicted value, $\hat{s}(x)$ is the uncertainty obtained, $\Phi(\cdot)$ and $\phi(\cdot)$ are the normal cumulative distribution functions and probability density function, respectively.

The main goal of this paper is to analyze the behavior of a Kriging-Assisted Differential Evolution (KADE) approach for optimizing BSA parameters. The rest of this paper is structured as follows: In Sect. 2, the problem statement is given. Section 3 provides a brief literature review on KM used as a surrogate technique in EAs. Section 4 describes the proposed approach's methodology. Section 5 lays out the experiments conducted and shows the results. Furthermore, in Sect. 6, a general discussion is made of the evidence observed. Finally, Sect. 7 consists of some conclusions attained as well as ideas for future work.

2   Problem Statement

In this work, the problem of interest is the optimization of BSA parameters. As demonstrated in [2], the search space in this optimization task is an irregular landscape, which justifies the usage of EAs to this end. Also, this paper is intended as a follow-up of [12], where it was found that Differential Evolution (DE) could be considered as an optimizer of BSA parameters. Furthermore, the primary purpose of this work is to show that it is possible to lower the computational time of the optimization in a synthetic signal and a more challenging signal such as an Electroencephalographic (EEG) signal. To measure the BSA performance, two metric criteria will be used. The first one will measure the influence of the parameters on the reconstruction process, and the second one will indicate how saturated the spike signal is. As such:

- **Signal to noise ratio (SNR):** Measures the relation involving the original signal power and the noise signal power. Noise is the difference between the original signal (s) and the decoded signal (r). Higher SNR values mean better results. It is defined as:

$$SNR = 10 \cdot \log_{10} \left(\frac{\sum_t^N s_t^2}{\sum_t^N (s_t - r_t)^2} \right) \tag{3}$$

- **Absolute firing rate (AFR):** Indicates the saturation of the spike train (sp). Lower AFR values mean a less saturated signal. It is defined as:

$$AFR = \frac{\sum_t^N |sp_t|}{N} \tag{4}$$

In both Eqs. 3 and 4, N means the total lenght of the signal. The SNR metric will be set as the objective function of the optimization, and therefore, the procedure is a maximization problem.

3   Related Work

In the literature, there is evidence of Kriging-Assisted (KA) implementations as surrogate models for EAs. Emmerich et al. [13] proposed a KA Evolutionary Strategy (ES) using an exponential kernel as a Gaussian process and LCB as an infill strategy for a set of benchmarks. Zhou et al. [14] used a KA in a Genetic Algorithm (GA) as a global surrogate to screen promising individuals, employing PoI as an infill sampler. Later, Tian et al. [15] made a comparative analysis on different Gaussian kernels using a KA Particle Swarm Optimization (PSO) approach along with an EI as infill criterion on a variety of benchmark problems with dimensions 20 and 30, finding that the squared exponential kernel becomes less suited as size increases. Moreover, Chugh et al. [16] utilize a KA approach paired with a Reference Vector Guided (RVG) EA. In it, they used a squared exponential kernel as well as an uncertainty metric to manage the model. Recently, Song et al. [19] used a KA method with a Two-Archive (Two-Arch) EA for an expensive many-objective optimization. They utilize the squared exponential kernel and three management criteria (EI, PoI & uncertainty), selected during the evolution based on the exploration and exploitation necessities.

On the other hand, evidence shows the inclusion of KA models with DE. Zhang et al. [17] used DE paired with a KM for multiobjective optimization problems using a squared exponential kernel and EI infill criterion. Moreover, Liu et al. [18] proposed a set of empirical rules applicable within an evolutionary search framework, considering a collection of configurations of DE. Their study found that, for some solutions, DE/rand/1/bin was unsuitable due to widely-spread child solutions being created, causing stress in constructing the surrogate model. More recently, Zhan & Xing [10] used a KADE approach with a squared exponential kernel and multi-point EI criterion as a pre-screening method for a set of benchmarks.

A summary of the implementations of KM on EA mentioned above is presented in Table 1. Even though there are existing KADE proposals in the literature, none have been applied to optimize BSA parameters. Therefore, a novel configuration of KADE is proposed tailored for the mentioned task.

Table 1. Summary of the related work presented.

Author	Year	EA	Kernel	Management
Emmerich et al. [13]	2006	ES	Exponential	LCB
Zhou et al. [14]	2007	GA	Squared exponential	PoI
Zhang et al. [17]	2009	DE	Linear	EI
Liu et al. [18]	2014	DE	Squared exponential	LCB
Tian et al. [15]	2017	PSO	Many	EI
Chugh et al. [16]	2018	RVG	Squared exponential	Uncertainty
Zhan & Xing [10]	2021	DE	Squared exponential	EI
Song et al. [19]	2021	Two-Arch	Squared exponential	EI - PoI - Uncertainty

4   KADE

As mentioned before, KA techniques have been tested along with EA methods, proving their efficiency and adaptability. Furthermore, including different infill sampling criteria makes KM a promising SA method. This work employs a KADE approach with DE's popular variant DE/rand/1/bin. Moreover, the KM is added to the fitness evaluation process.

The implementation starts with a random Latin Hypercube Sampling (LHS) step to create individuals of the initial population, all of which are evaluated in the expensive function. Subsequently, a KM is built using the information from the newly created points and their corresponding actual fitness values. After that, the normal course of DE/rand/1/bin takes place, using the KM as the evaluator. As model management and selection of individuals for re-evaluation criterion, two configurations are being considered:

- **KADE-5%:** The model update process occurs in every generation by selecting the 5% of the most uncertain (Uncertainty criterion) individuals among the population.
- **KADE-2-5%-EI:** The model update takes place every two generations by selecting the 5% of individuals according to the EI criterion.

Thereby, the main difference between these two configurations resides in the model management frequency and the selection criterion for re-evaluation. The framework for both approaches can be summarized in Algorithm 1.

In Table 2, the parameters used in all approaches are presented. These values were selected by independent Irace [20] executions. The consideration of these

Table 2. Parameter values used for DE and KADE algorithm.

Configuration →	DE/rand/1/bin	KADE-5%	KADE-2-5%-EI
Update freq	–	1	2
Selection Criterion (%)	–	Uncertainty (5%)	EI (5%)
NP %	50	50	50
Cr	2.86	2.43	3.49
F	0.65	0.20	0.39
Alpha	–	1E-06	1E-05
$N_{restart}$	–	20	40
Generations	200	50	100

Algorithm 1. Framework of the KADE implementation applicable to both KADE-5% and KADE-2-5%-EI

Require: Gen_M = Max. number of generations, NP = Population Size, U = Frequency of model update, $S_\%$ = Percentage of selection for re-evaluation. S_C = Selection criterion (Uncertainty or EI).

1: $Gen \leftarrow 0$
2: Generate NP initial sample points (ISP) using LHS and evaluate them in the expensive function (EF).
3: Insert the ISP and their fitness value to the first population P and to an archive DB, that will store all individuals evaluated in the expensive function.
4: Build an initial KM with a squared exponential kernel using the DB.
5: **while** $Gen < Gen_M$ **do:**
6:     Generate the offspring using the operators of DE/rand/1/bin and evaluate them using KM. Selection will be carried out as usual.
7:     **if** Gen mod U is 0 **then**
8:         Choose the $S_\%$ of P accordingly to S_C. Re-evaluate this new sample points (NSP) in the EF.
9:         Update the fitness values of NSP in P.
10:         Insert the NSP and their respective fitness values in DB.
11:         Re-build the KM using DB.
12:     **end if**
13:     $Gen \leftarrow Gen + 1$
14: **end while**
15: Re-evaluate all individuals in P using the EF. Report the best individual.

---

parameters included *alpha* and $N_{restarts}$. The first one refers to a value added to the diagonal of the kernel matrix during the fitting process, and the latter defines the number of restarts of the optimizer for finding the optimal kernel's parameters. All implementations and experiments were written in Python 3.9 and executed in an Intel Core i5-10300H CPU and 8 GB RAM.

5   Experiments and Results

Two experiments were conducted in this work. The case study is a synthetic signal in the first, whereas the implementation focuses on an EEG signal in the second one. The parameters shown in Table 2 hold, and the SNR value was set as the objective function for both experiments.

5.1   Experiment 1: Synthetic Signal

This first experiment uses a synthetic signal to have a more manageable environment. Following [2], one signal was created (Fig. 1). This was produced by the sum of sine signals ranging from 2 to 30 HZ with random power and random phase lags. In addition, white noise was introduced with a strength of 3. Finally, the signal has a length of 1001 elements, sampled at 1000 Hz. To compare the performance, 30 independent executions per approach were performed.

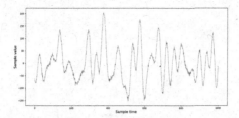

Fig. 1. Synthetic signal created with a length of 1001 elements at 1000 Hz.

Table 3. Statistical results of 30 independent executions. Values in boldface indicate the best value. Values in italics indicate the best value among KADE configurations. $H = 1$ means that a significant difference was found.

Statistic	DE/rand/1/bin	KADE-5%	KADE-2-5%-EI	Friedman Test	
				p-value	H
Best	**9.3117**	*9.2713*	9.2523	1.14E-11	1
Mean	**9.3049**	9.0448	*9.0808*		
Median	**9.3117**	9.0642	*9.0738*		
Worst	**9.2743**	8.7211	*8.9537*		
Std. Dev.	**0.0128**	0.1301	*0.0777*		
Time (mean)	1236.6165 s	*123.6597* s	190.0081 s		
	20.6103 min	*2.0610* min	3.1668 min		

In Table 3, the statistical analysis obtained in all executions and the Friedman test (95%-confidence) are presented. These results are upheld by a Bonferroni post-hoc test, displayed in Fig. 2. Details of the median run are shown in Table 4. Also, the convergence plots of the three implementations are presented in Fig. 3a. Finally, Fig. 3b includes the contrast between the original signal and the reconstructed signals.

From the results in Table 3, neither of the KADE configurations outperformed DE/rand/1/bin. However, KADE-2-5%-EI showed better capabilities than KADE-5% in all statistical values except for the best value and average time. Such observations are validated with the Friedman test and the Bonferroni post-hoc test (Fig. 2): The DE approach shows significant differences regarding the KADE implementations, while both KADE approaches do not exhibit this.

Moreover, it can be seen in Table 4a that the three approaches had a low AFR value, meaning that the spike trains produced are not saturated. On the other hand, Table 4b shows a similar threshold value for all algorithms, whereas the main differences are present in the filter size and cutoff frequency.

Regarding the convergence graph (Fig. 3a), the erratic behavior of the KADE configurations indicates that the KMs overstate the value of the best individual per generation, causing steep climbs. Nevertheless, the model update mechanism resolves some of these abnormalities. Finally, the visual comparison in Fig. 3b shows that the reconstructed signals do not exhibit major differences among the implementations despite the variations observed by the metric values.

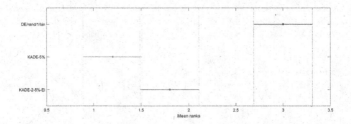

Fig. 2. Bonferroni post-hoc results of the approaches tested.

Table 4. Details of the median run by each algorithm. a) Metrics comparing the results of the algorithms. Values in boldface indicate a better result. b) Set of parameters found by each algorithm.

(a) Metrics achieved.

	DE/rand/1/bin	KADE-5%	KADE-2-5%-EI
SNR	**9.3117**	9.0690	9.0841
AFR	0.2547	**0.2488**	0.2617

(b) Parameters found.

	DE/rand/1/bin	KADE-5%	KADE-2-5%-EI
Fs	63	69	57
Cf	33.9450	34.9678	37.5677
Th	0.9429	0.9464	0.9407

5.2   Experiment 2: EEG Signal

This experiment deals with an EEG signal, a more chaotic and challenging instance. The database used was the CHB-MIT [21], which collects Scalp EEG signals of 24 pediatric patients suffering seizures. All recordings were sampled at 256 Hz. For this study, only the first 10000 elements in channel 'FP1-F7' of the patient 'CHB01-09' were used (Fig. 4). Similarly as before, 30 independent executions, per approach, were performed.

The statistical analysis obtained in all executions and the Friedman test (95%-confidence) are presented in Table 5 and those results are backed up by a Bonferroni post-hoc test, displayed in Fig. 5. Table 6 summarizes the values of the median run. Finally, Fig. 6 shows the convergence graphics and the contrast between the original and reconstructed signals of the three compared algorithms.

Similarly as before, both of the KADE configurations fall behind DE/rand/1/bin. Nevertheless, KADE-2-5%-EI showed better capabilities except for the average time. Moreover, the Friedman and Bonferroni post-hoc tests (Fig. 5) showed similar evidence as before, except for the KADE configurations; all tested methods showed significant differences. In this case, KADE-2-5%-EI presented better performance than KADE-5%. Unlike the last experiment, the three algorithms found, in the median run, similar values in all parameters (Table 6b), producing not-saturated spike train signals (Table 6a).

The same behavior in the convergence plots and the visual comparison (Fig. 6) holds in this experiment; an unstable convergence can be observed in both KADE configurations and, finally, no significant visual differences are observed.

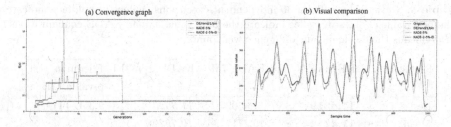

Fig. 3. Graphical results obtained. a) Convergence graph of the median execution by each algorithm. b) Visual comparison between the original signal and the reconstructed signals using the parameters found at the best execution.

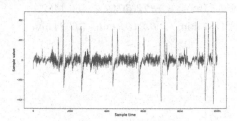

Fig. 4. EEG signal used with 10000 elements sampled at 256 Hz.

6 Discussion

Firstly, similar behavior of both KADE configurations against the DE approach in experiment 1 and experiment 2 was observable, showing that DE/rand/1/bin exhibited better performance, with significant differences, over the KADE variants. The performance of the proposed methods did not present significant differences in the first experiment, but it did in the second. Since both signals used differ in complexity, this could mean that as the difficulty increases, the performance of both KADE approaches diverges. Furthermore, both Bonferroni port-hoc tests imply that KADE-2-5%-EI is closer to the desired outcome.

Moreover, the quantitative differences were not greatly reflected during the visual evaluation. This also holds for the AFR metrics achieved and the parameters found in the median run. These observations suggest that the proposed KADE algorithm can produce not-saturated spike trains with similar reconstructed signals. Nevertheless, improving the method to achieve competitive SNR values is essential. A significant reduction of the execution time was obtained in the KADE models for both cases tested. These results indicate that the sought objective of decreasing the computational time was achieved.

Finally, the convergence behavior shows a notable difference between DE and KADE configurations. While the DE approach exhibits a typical, never-decreasing graph, the convergence of both KADE implementations is volatile. This issue arises from ranking predicted values from the KM as the best individual so far. Nevertheless, the fact that, in most cases, when a steep climb occurs, a

Table 5. Statistical results of 30 independent executions. Values in boldface indicate the best value. Values in italics indicate the best value among KADE configurations. *H=1* means that a significant difference was found.

Statistic	DE/rand/1/bin	KADE-5%	KADE-2-5%-EI	Friedman Test	
				p-value	H
Best	**11.4319**	11.3502	*11.3920*	9.35E-14	1
Mean	**11.4254**	10.2717	*10.79328*		
Median	**11.4249**	10.3062	*10.9719*		
Worst	**11.4018**	8.9670	*9.2674*		
Std. Dev	**0.0072**	0.6853	*0.5227*		
Time (mean)	14905.1196 s	*372.6049* s	456.1462 s		
	248.4187 min	*6.2101* min	7.6024 min		

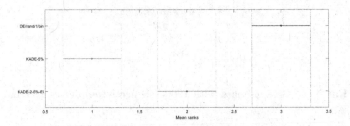

Fig. 5. Bonferroni post-hoc results of the approaches tested.

Table 6. Details of the median run by each algorithm. a) Metrics comparing the results of the algorithms. Values in boldface indicate a better result. b) Set of parameters found by each algorithm.

(a) Metrics achieved.

	DE/rand/1/bin	KADE-5%	KADE-2-5%-EI
SNR	**11.4249**	10.3062	10.9719
AFR	**0.2668**	0.2888	0.2788

(b) Parameters found.

	DE/rand/1/bin	KADE-5%	KADE-2-5%-EI
Fs	79	63	77
Cf	20.7163	20.7846	20.3622
Th	0.9513	0.9456	0.9529

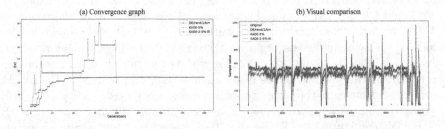

Fig. 6. Graphical results obtained. a) Convergence graph of the median execution by each algorithm. b) Visual comparison between the original signal and the reconstructed signals using the parameters found at the best execution.

sharp descent follows it indicates that the model management mechanism selects the 'false' best individual for re-evaluation, correcting the plot behavior.

7   Conclusions and Future Work

The main difference between SNNs and their predecessors is how information is handled. Given that SNNs use spike train signals, it is crucial to have proper methods for transforming analog signals to an impulse-based representation while being as efficient as possible. In previous studies, an exhaustive search [2] and EA techniques [12] were explored as optimizers for BSA parameters. Nevertheless, those approaches require significant computational time.

In this paper, the possibility of employing a KADE for the optimization of BSA parameters was explored. We tested two configurations, KADE-5% and KADE-2-5%-EI, against DE/rand/1/bin. It was found that, while the DE implementation outperformed both KADE algorithms, KADE-2-5%-EI showed better results than KADE-5% in both experiments. Also, despite the numerical variations, a visual comparison of the reconstructed signals showed minor differences between all methods. Moreover, both experiments demonstrated that the proposed algorithm was available to lower the computational time while finding competitive parameter values. Therefore, our findings showed that KADE could be considered an efficient and rapid optimizer of BSA parameters. The importance of lowering the computational time comes from the burden of the signal's length and quantity. In the second experiment, DE took over 4 h, even though only 10000 samples of the EEG signal from one channel was used. In fullness, CHB-MIT contains over 9000 h of signals from 24 patients, each with at least 18 channels. Finally, future work directions may include different venues: exploring the integration of a SA approach on different, more specialized, bio-inspired algorithms, evaluating the performance of other surrogate models (i.e., RBF), and assessing the possibility of using the classification performance of a SNN as the objective function.

Acknowledgments. The first author acknowledges support from the Mexican National Council for Science and Technology (CONACyT) through scholarship No. 1075919 to pursue graduate studies at the University of Veracruz.

References

1. Maass, W.: Networks of spiking neurons: the third generation of neural network models. Neural Netw. **10**(9), 1659–1671 (1997)
2. Petro, B., Kasabov, N., Kiss, R.M.: Selection and optimization of temporal spike encoding methods for spiking neural networks. IEEE Trans. Neural Netw. Learn. Syst. **31**(2), 358–370 (2020)
3. Schrauwen, B., Van Campenhout, I.: BSA, a fast and accurate spike train encoding scheme. In: Proceedings of the International Joint Conference on Neural Networks, 2003, vol. 4, pp. 2825–2830 (2003)

4. Jin, Y.: Surrogate-assisted evolutionary computation: recent advances and future challenges. Swarm Evol. Comput. **1**(2), 61–70 (2011)

5. Miranda-Varela, M., Mezura-Montes, E.: Constraint-handling techniques in surrogate-assisted evolutionary optimization. An empirical study. Appl. Soft Comput. **73**, 215–229 (2018)

6. Jin, Y.: A comprehensive survey of fitness approximation in evolutionary computation. Soft Comput. **9**(1), 3–12 (2005)

7. Diaz-Manriquez, A., Toscano-Pulido, G., Gomez-Flores, W.: On the selection of surrogate models in evolutionary optimization algorithms. In: IEEE Congress of Evolutionary Computation, vol. 788, pp. 2155–2162 (2011)

8. Matheron, G.: Principles of geostatistics. Econ. Geol. **58**(8), 1246–1266 (1963)

9. Shi, L., Rasheed, K.: A Survey of Fitness Approximation Methods Applied in Evolutionary Algorithms, pp. 3–28 (2010)

10. Zhan, D., Xing, H.: A population prescreening strategy for kriging-assisted evolutionary computation. In: Congress on Evolutionary Computation (2021)

11. Deng, Z., Rotaru, M.D., Sykulski, J.K.: Kriging assisted surrogate evolutionary computation to solve optimal power flow problems. IEEE Trans. Power Syst. **35**(2), 831–839 (2020)

12. López-Herrera, C., Acosta-Mesa, A., Mezura-Montes, E.: Spiking neural networks codification using bio-inspired computation. In: 3rd Workshop on New Trends in Computational Intelligence and Applications (CIAPP) (In press)

13. Emmerich, M.T.M., Giannakoglou, K.C., Naujoks, B.: Single and multiobjective evolutionary optimization assisted by Gaussian random field metamodels. IEEE Trans. Evol. Comput. **10**(4), 421–439 (2006)

14. Zhou, Z., Ong, Y.S., Nair, P.B., Keane, A.J., Lum, K.Y.: Combining global and local surrogate models to accelerate evolutionary optimization. IEEE Trans. Syst. Man Cybern. Part C **37**(1), 66–76 (2007)

15. Tian, J., Sun, C., Zeng, J., Yu, H., Tan, Y., Jin, Y.: Comparisons of different kernels in Kriging-assisted evolutionary expensive optimization. In: IEEE Symposium Series on Computational Intelligence, pp. 1–8 (2017)

16. Chugh, T., Jin, Y., Miettinen, K., Hakanen, J., Sindhya, K.: A surrogate-assisted reference vector guided evolutionary algorithm for computationally expensive many-objective optimization. IEEE Trans. Evol. Comput. **22**(1), 129–142 (2018)

17. Zhang, Q., Liu, W., Tsang, E., Virginas, B.: Expensive multiobjective optimization by MOEA/D with Gaussian process model. IEEE Trans. Evol. Comput. **14**(3), 456–474 (2009)

18. Liu, B., Chen, Q., Zhang, Q., Gielen, G., Grout, V.: Behavioral study of the surrogate model-aware evolutionary search framework. In: IEEE Congress on Evolutionary Computation, pp. 715–722 (2014)

19. Song, Z., Wang, H., He, C., Jin, Y.: A kriging-assisted two-archive evolutionary algorithm for expensive many-objective optimization. In: IEEE Transactions on Evolutionary Computation (2021)

20. López-Ibáñez, M., Dubois-Lacoste, J., Cáceres, L.P., Birattari, M., Stützle, T.: The irace package: iterated racing for automatic algorithm configuration. Oper. Res. Perspect. **3**, 43–58 (2016)

21. Shoeb, A.H.: Application of machine learning to epileptic seizure onset detection and treatment (2009)

Auto Machine Learning Based on Genetic Programming for Medical Image Classification

David Herrera-Sánchez(✉) [iD], Héctor-Gabriel Acosta-Mesa[iD],
and Efrén Mezura-Montes[iD]

Artificial Intelligence Research Institute, University of Veracruz, 91097 Xalapa,
Veracruz, Mexico
hersan19@hotmail.es, {heacosta,emezura}@uv.mx

Abstract. Medical imaging classification is an area that has taken relevance in recent years due to the capability to support the medical specialist at the time of diagnosis. However, there are different instruments to obtain images from the body, and each body organ is captured differently due to its chemical composition. In this way, there are some difficulties in working with different imaging modalities. Firstly, using different functions or methods to extract features from the images is necessary. Secondly, the classification performance depends on the relevant features extracted from the images, and thirdly, it is necessary to find the classifier that performs with the minimum error. Following the concept of Auto-Machine Learning (AutoML), where the feature engineering and the hyperparameter tuning of the classifier are done automatically, this work proposes an automated approach for feature extraction and image classification based on Genetic Programming. The approach modifies the functions and their parameters and the hyperparameters for the classifier. The results show that the approach can deal with different imaging modalities, demonstrating that feature extraction is necessary to increase the classification performance. For X-ray images, it achieves a classification accuracy of 0.99, and for computerized tomography, it achieves an accuracy of 0.96. On the other hand, the solutions given by the approach are easily reproducible and easy to interpret.

Keywords: Medical Image Classification · AutoML · Genetic Programming

1 Introduction

Medical image classification plays an important role in clinical care and treatment. Also, it can improve the diagnosis of professional radiologists, and it can support medical staff without wide experience. However, medical image classification is challenging due to different factors. The first is due to the image variations, such as noise or blur that come from the medical instruments at the time to capture the images. It is common to find Gaussian, salt and pepper, and speckle noise. Also, the blurred noise is present in some images principally caused by the movement of the patients at the time of capturing the image. The second one is the imaging modalities such as magnetic resonance imaging, ultrasound, radiography, and computer tomography, which complicate the classification due to the different features that contain each modality. In this way, it is important to make an image processing to extract important features to perform the classification with as few mistakes as possible.

The problem arises in finding the correct classifier and the appropriate image-processing techniques to complete the task. Due to this problem, it is proposed to use Automated Machine Learning (AutoML) defined as machine learning automation, automatizing the feature engineering, model selection, and the search of their parameters [16]. Specifically, the approach optimizes the set of functions for image classification. This work proposes an automated approach for image classification based on the Evolutionary Computation (EC) paradigm of Genetic Programming (GP). EC is a good technique for solving optimization problems. The representation of GP is seized due to the flexibility to include computer programs or functions. These functions make image processing, and the classifier and its parameters are selected. On the other hand, the set of functions makes a feature extraction and construction, and the objective is to find the most suitable functions with their parameters and the hyperparameters for the classifier.

Unlike the literature, this approach focuses on extracting and classifying two medical imaging modalities: x-ray images and computerized tomography. Also, the functions are supported in tensors to make the processing using GPU, speeding up the process and allowing the use of more than two thousand images. Consequently, the main contribution of this work lies in demonstrating that there is a classification improvement making a preprocessing to the medical images. Also, the capability of the automated approach to deal with two modalities of medical images that come from different instruments and present different patterns within the image.

2   Methodology

Based on Darwin's theory of evolution, EC works with a population that represents the potential solutions to the problem. Crossover and mutation operators create new individuals. The objective is to generate new potential solutions containing their parents' features. In this way, the new individuals are expected to solve the problem better than their parents [11]. The evolution process is performed until it reaches the stopping condition. As mentioned above, GP is the paradigm used to perform the AutoML in this work.

The highlight of this paradigm is that the solutions to the problem are represented by trees, which contain a set of functions to make a process within the images to extract important features and construct new ones. The improved version of the original GP is Strongly Type Genetic Programming [12]. It is used to restrict the data type of each function, which means if one function only works with integers, the data input for this function will be an integer.

GP has demonstrated the capability to handle tasks within computer vision. In image classification, a multi-tree representation was proposed in [2] to enhance the classification of skin images to detect melanoma. Fan *et al.* proposes a structured tree to perform image filtering, feature extraction, feature construction, and finally, classification. Also, propose a mutation operator based on the population's fitness to adjust the growth of the tree size of offspring. In [3], an ensemble approach is presented. The performance is outstanding. However, the computational cost is high due to the classifiers used within the solutions. In [4], an interesting approach is presented, where a dual-tree representation is used to extract features of the image. Then, concatenate the features into a vector. In this way, more information is obtained for each image.

In [15] presents a method for image segmentation. It introduces a convolution to construct multiple features from different images and then combine them to classify each pixel.

In [6], an image enhancement method is proposed where the node functions are functions and parameters of the software used to enhance the images. It proposes different pipelines for the procedure and uses different metrics to evaluate the performance of the images. The method performs well regarding the Generative Adversarial Networks. Other GP methods for computer vision tasks such as rendering, and object detection can be found in [9].

As part of the approach, the automated includes an image processing step, extracting features, and selecting the parameter automatically for functions included. Part of the improvement is that the functions are based on tensor structured data to reduce the computational time for the processing.

The functions used for image processing are divided into image enhancement, filter, and morphological operation. In the image enhancement, histogram equalization and brightness were chosen to improve the contrast of the images. The filter functions are the most used to deal with high frequencies in the images. Also, the morphological operators are the most common for preprocessing in medical images [13], which allows change in the structure of the mask used for the operations. Those operations allow for a decrease or increase in the size of some structures within the images.

The set of functions and the set of terminals for image processing are listed in Tables 1 and 2:

The parameters used for the evolutionary algorithm are taken from [8]. The population is set to 100 individuals, with a crossover rate of 0.8 and a mutation rate of 0.19. The maximum number of generations is 50. The minimum and maximum depth is 4 and 12, respectively. The selection method is by tournament with a size of 5. Finally, the elitism rate is 0.1 to keep the best individual during evolution. This ensures that the best individual survives during the evolution process.

In the input, the algorithm receives the set of images for training and their labels. The output is the best individual from the evolution process. The algorithm starts with generation zero, creating the population with the Ramped-Half-And-Half method as suggested in [11]. It produces different shapes and sizes of trees in the initial population, creating a major diversity initially. Then, the population is evaluated to start the evolution process. The parents are chosen

Table 1. Set of functions of GP

Function	Description (Input Data)
Image Enhacement	
Histogram equalization	Histogram equalization of the image (Tensor)
Brightness	Brightness adjustment of the image (Tensor, Float_1)
Filter	
Gaussian	Gaussian filter (Tensor, Integer_2)
Box Blur	Blur (Tensor, Integer_2)
Sharpen	Adjust the sharpness in the image (Tensor, Integer_1, Float_2)
Motion Blur	Movement Blur(Tensor, Integer_2, Integer_3)
Morphologic Operation	
Erosion	Apply the erosion in the image (Tensor, Integer_1)
Dilatation	Apply the dilation in the image (Tensor, Integer_1)
Opening	Apply the opening in the image (Tensor, Integer_1)
Closing	Apply the closing in the image (Tensor, Integer_1)
Gradient	Apply the gradient in the image (Tensor, Integer_1)
Classifier	
SVM	Support Vector Machine (Tensor, Kernel, Labels, Flag control)
RVM	Relevant Vector Machine (Tensor, Kernel, Labels, Flag control)

Table 2. Set of terminals of GP

Terminal	Description (Data type)
Data	Set of images (Tensor)
Integer_1	Kernel size. Value between 3 and 9 (Integer)
Integer_2	Kernel size. Odd value between 3 and 9 (Integer)
Integer 3	Value between 0 and 90 to adjust the angle (Integer)
Float_1	Value between -0.5 and 0.5 to adjust the brightness (Float)
Float_2	Value between 0.1 and 0.9 to adjust standard deviation for sharpen (Float)
Kernel	Kernel used for the classifier Sigmoid, Linear, Polynomial, RBF (String)

with the tournament selection method as part of the evolution process. Then the parents create new individuals with the operators. The new individuals replace the whole population and keep the ten best individuals directly into the next generation. The algorithm stops until it reaches the maximum number of generations. After that, the best individual is evaluated with the test set.

3 Experiments and Results

There were two experiments to compare the performance of two classification algorithms. In this case, the first experiment was tuned only with the Support Vector Machine [7], for the second one was adjusted by the Relevance Vector Machine [14].

In both experiments, the algorithms were run ten times. The algorithm was tested in two different sets of medical images. The first one contains two classes of tumor chest [1]. It contains 1601 computerized tomography images in grayscale. The second dataset includes three classes of images, COVID-19, pneumonia, and normal [5]. It includes 2754 X-ray images.

For both datasets, the image size is 256×256 pixels. Also, they were divided into 70% for training and 30% for tests. All the experiments were run on a computer with Windows 10, Intel i7-11700K processor with 24GB RAM, and GPU RTX 3060. It was developed in Python using the well-known package DEAP [10].

Table 3 shows the best values from the ten runs. In bold are marked the results that outperformed the other methods. There were executed tests for SVM and RVM without image processing, using the raw data from the images (first and second row). On the other hand, the third and fourth rows show the results of the approach with each classifier (SVM and RVM).

Table 3. Results of the best solution of ten runs

	Tumor		COVID-19	
Method	F1-Score	Accuracy	F1-Score	Accuracy
SVM	0.9287	0.9286	0.9294	0.9301
RVM	0.9124	0.9113	0.9021	0.9102
Approach + SVM	**0.9906**	**0.9991**	**0.9551**	**0.9622**
Approach + RVM	**0.9849**	**0.987**	**0.9507**	**0.9589**

The results for the Tumor dataset shown in Table 3 demonstrate that the search of the approach achieves good results. The T-student test was performed at a significance level of 0.05 to compare our approach with the classifier without preprocessing. For both datasets, the test demonstrated that there exists a significant difference. Regarding the best solutions, it performs a classification accuracy of 0.9991 and 0.987 for SVM and RVM, respectively. On the other hand, the performance is lower using the raw data without image preprocessing, achieving only 0.9286 and 0.9113 of accuracy for SVM and RVM.

The same situation is for the COVID dataset. The approach achieves 0.9622 and 0.9589 accuracy for SVM and RVM, respectively. The difference between SVM and RVM is only 0.01 units in accuracy. On the other hand, the SVM and RVM with raw data are 0.02 units of difference in accuracy with the best values. To compare our approach using the SVM and RVM classifier, the T-student test with a significance level of 0.05 was performed.

The results demonstrated that there is not a statistically significant difference among them. Both classifiers work statistically similarly for both datasets. It demonstrates that our approach can improve the performance of both classifiers.

The best solutions found by the approach for both databases using the RVM and the SVM classifier are presented in Figs. 1 – 4, respectively.

As we can see, the solutions are easily interpretable for reproducing the results only using the functions and their parameters that appear in the individual. Also, each tree shows the hyperparameter for the classifier.

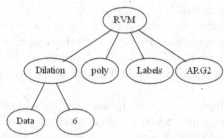

(a) Solution represented by functions. The original data passes through one preprocessing function, dilation with a kernel size of 6.

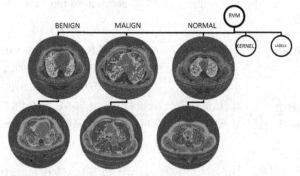

(b) Solution represented by images. Blue color represents low values, and red represents high values.

Fig. 1. Best solutions using Relevance Vector Machine Classifier for Tumor dataset.

The solutions represented with the functions are in 1(a) and 2(a). On the other hand, in 1(b) and 2(b) are the solutions with the images, where each node image represents the terminal/function. The images presented were selected randomly to make an explanation. In both solutions of the datasets, the same images are used.

For example, in Fig. 3(a), the leaf node which contains the data, in the 3(b) is shown the raw image that the solution receives as input. Then, following bottom-up, the operators that appear in 3(a) are applied in the images. The next node from the raw data node is the function histogram equalization. Then, the resulting images from the histogram equalization are the input for the next operator, the Gaussian Blur filter with a kernel size of 5. After that, the morphological operator of dilation with a kernel size of 4. Finally, the sharpen operator is applied with a kernel size of 5 and a standard deviation of 0.2. The images that appear at the top node before the classifier are images with the features most representative and utterly different from the original ones. This image-processing step allows extracting meaningful features to improve the classifier.

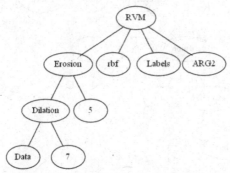

(a) Solution represented by functions. The original data is processed by two functions, dilation and erosion with kernel sizes of 7 and 5, respectively.

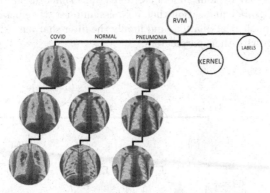

(b) Solution represented by images. Blue color represents low values, and red represents high values.

Fig. 2. Best solutions using Relevance Vector Machine Classifier for COVID-19 datset.

In the images, a colormap is used as part of the visualization to see changes that the image suffers during the feature engineering process. As we can see, the original data is processed, and the functions are applied to the images to extract regions of interest to facilitate the classifier's training and improve the performance.

The feature engineering functions for RVM are less than the SVM in both datasets. Consequently, the images using the RVM classifier suffer fewer changes than those using SVM, and qualitatively, the differences are very clear. On the other hand, the approach deals efficiently with the two types of imaging modalities. The results demonstrate that the classification is improved significantly, and the approach could find adequate functions for each dataset.

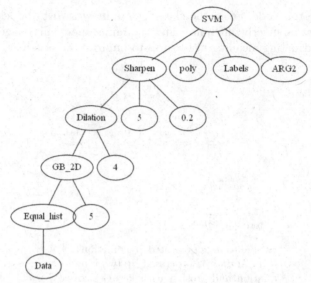

(a) Solution represented by functions. The original data is processed by four functions, firstly an enhancement with histogram equalization then a Gaussian filter, the operator morphological dilation, and finally the filter sharpen. All of them with their parameters.

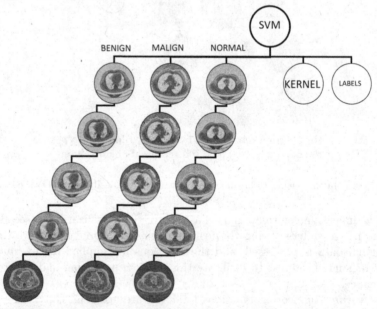

(b) Solution represented by images. Blue color represents low values, and red represents high values.

Fig. 3. Best solutions using Support Vector Machine Classifier for Tumor dataset.

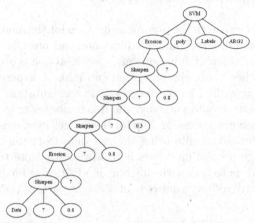

(a) Solution represented by functions. The solution only contains two functions, erosion and sharpen. However, their parameters are changed at each node, making six steps of image preprocessing.

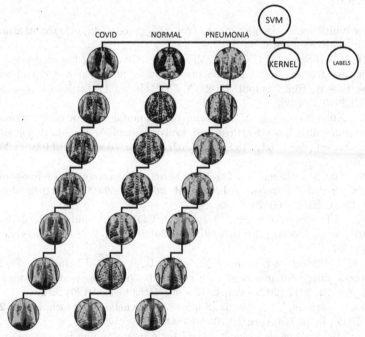

(b) Solution represented by images. Blue color represents low values, and red represents high values.

Fig. 4. Best solutions using Support Vector Machine Classifier for COVID-19 dataset.

4 Conclusions and Future Work

The approach deals well with the preprocessing step for the images. The solutions given for the algorithm show different functions and are the most suitable for each dataset. Also, different functions and parameters are observed with each dataset. On the other hand, the algorithm can make a hyperparameter tuning for the classifiers according to the images. The configuration for the functions and the classifier is done automatically, and it is unnecessary to make it by hand.

Finally, making a preprocessing step for the images increases the performance of the classifiers instead of only using the raw data from the images. The data type used for image processing allows it to work with more than 1000 images. Part of future work is to test with different modalities of images such as ultrasound. Also, test with different datasets of different parts of the body or different health issues.

Acknowledgments. The first author acknowledges support from the Mexican Council for Humanities, Science, and Technology (CONAHCYT) through a scholarship to pursue graduate studies at the University of Veracruz.

References

1. Tumor multiclass — kaggle. https://www.kaggle.com/datasets/amitkarmakar41/tumour-multiclass-dataset
2. Ain, Q.U., Al-Sahaf, H., Xue, B., Zhang, M.: Generating knowledge-guided discriminative features using genetic programming for melanoma detection. IEEE Trans. Emerg. Top. Comput. Intell. **5**, 554–569 (2021). https://doi.org/10.1109/TETCI.2020.2983426
3. Bi, Y., Xue, B., Zhang, M.: Genetic programming with a new representation to automatically learn features and evolve ensembles for image classification. IEEE Trans. Cybern. **51**, 1769–1783 (2021). https://doi.org/10.1109/TCYB.2020.2964566
4. Bi, Y., Xue, B., Zhang, M.: Dual-tree genetic programming for few-shot image classification. IEEE Trans. Evol. Comput. **26**, 555–569 (2022). https://doi.org/10.1109/TEVC.2021.3100576
5. Cohen, J.P., Morrison, P., Dao, L.: COVID-19 image data collection (2020). https://doi.org/10.1016/s0140-6736(20)30211-7, https://arxiv.org/abs/2003.11597v1
6. Correia, J., Rodriguez-Fernandez, N., Vieira, L., Romero, J., Machado, P.: Towards automatic image enhancement with genetic programming and machine learning. Appl. Sci. **12**, 2212 (2022). https://doi.org/10.3390/APP12042212
7. Cortes, C., Vapnik, V., Saitta, L.: Support-vector networks. Mach. Learn. **20**, 273–297 (1995). https://doi.org/10.1007/BF00994018
8. Fan, Q., Bi, Y., Xue, B., Zhang, M.: Genetic programming for feature extraction and construction in image classification. Appl. Soft Comput. **118**, 108509 (2022). https://doi.org/10.1016/j.asoc.2022.108509
9. Khan, A., Qureshi, A.S., Wahab, N., Hussain, M., Hamza, M.Y.: A recent survey on the applications of genetic programming in image processing. Comput. Intell. **37**, 1745–1778 (2021). https://doi.org/10.1111/COIN.12459

10. Kim, J., Yoo, S.: Software review: Deap (distributed evolutionary algorithm in python) library. Genet. Program. Evolvable Mach. **20**, 139–142 (2019). https://doi.org/10.1007/s10710-018-9341-4
11. Koza, J.R.: Genetic programming on the programming of computers by means of natural selection (1992). https://doi.org/10.1148/radiology.158.1.3940388
12. Montana, D.J.: Strongly typed genetic programming. Evolut. Comput. **3**, 199–230 (1995). https://doi.org/10.1162/EVCO.1995.3.2.199
13. Muthu, R., Rani, C., Saritha, S., Pearl Mary, S.: Morphological operations in medical image pre-processing. In: International Conference on Advanced Computing and Communication Systems, pp. 2065–2070 (2017)
14. Tipping, M.E.: The relevance vector machine. In: Advances in Neural Information Processing Systems, vol. 12 (1999)
15. Liang, Y., Zhang, M., Browne, W.N.: Wrapper feature construction for figure-ground image segmentation using genetic programming. In: Wagner, M., Li, X., Hendtlass, T. (eds.) ACALCI 2017. LNCS (LNAI), vol. 10142, pp. 111–123. Springer, Cham (2017). https://doi.org/10.1007/978-3-319-51691-2_10
16. Yao, Q., et al.: Taking human out of learning applications: a survey on automated machine learning (2018). https://arxiv.org/abs/1810.13306v4

Use of a Surrogate Model for Symbolic Discretization of Temporal Data Sets Through eMODiTS and a Training Set with Varying-Sized Instances

Aldo Márquez-Grajales[1]([✉]) [ID], Efrén Mezura-Montes[1] [ID],
Héctor-Gabriel Acosta-Mesa[1] [ID], and Fernando Salas-Martínez[2] [ID]

[1] Artificial Intelligence Research Institute, University of Veracruz, Campus Sur,
Calle Paseo Lote II, Sección Segunda No 112, Nuevo Xalapa,
91097 Xalapa-Enríquez, Veracruz, Mexico
li.aldomg@gmail.com, {emezura,heacosta}@uv.mx
https://www.uv.mx/iiia
[2] El Colegio de Veracruz, Carrillo Puerto 26, Zona Centro, Centro,
91000 Xalapa-Enríquez, Veracruz, Mexico
fersamtz@gmail.com
https://colver.com.mx/

Abstract. Time series classification is a supervised task in the field of temporal data mining. Time series naturally tend to be highly dimensional, requiring the use of reduction techniques such as discretization. eMODiTS is a data-driven method for symbolically discretizing time series, which determines the best scheme by modifying the number of time (word segments) and values (alphabet) cuts, generating a unique alphabet set for every word segment. However, due to the high computational cost required, a surrogate model is incorporated to minimize this cost, using the K-Nearest Neighbors approach for regression and Dynamic Time Warping (DTW) as the similarity measure. Results suggest that the surrogate model effectively estimates the objective functions' values similarly to the original ones, leading to similar classification rates. It is validated with the statistical test where there is no significant statistical difference between the surrogate and original models. The surrogate model produces modified acceptance index (d_j) values regarding predicting ability, indicating that the predictive performance is on average. On the other hand, the Mean Squared Error (MSE) consistently stays below 0.15, demonstrating that even when surrogate models cannot estimate the same values as the original model, the similarity of the values remains clear.

Keywords: Surrogate modelling · time series classification · symbolic discretization

1 Introduction

Time series classification is a supervised task in temporal data mining, whereby a classifier is trained to recognize a set of observations based on their class

H. Calvo et al. (Eds.): MICAI 2023 Workshops, LNAI 14502, pp. 360–372, 2024.
https://doi.org/10.1007/978-3-031-51940-6_27

label [14]. However, because a time series encompasses many ordered values collected through various sensors at irregular time intervals, an extensive amount of data is stored [4]. A preprocessing method known as discretization of time series has been developed to address this issue. Discretizing a time series entails converting continuously varying data points into a set of discrete values [14].

Several methods based on the well-known Symbolic Aggregate Approximation (SAX) algorithm have been proposed for the discretization of time series [1–3,9,17]. SAX partitions the temporal component of the time series using the Piecewise Aggregate Approximation (PAA) algorithm. PAA subdivides the temporal axis into equidistant cuts, known as word segments, and calculates the time series' mean values that are contained in each word segment. After generating the average values (PAA coefficients), SAX uses a Gaussian distribution to create breakpoints (alphabets) on the values axis and assigns a symbol to each one. The PAA coefficients are mapped to each alphabet and substituted with the corresponding symbol, resulting in a string that represents the final symbolic discretization produced by SAX.

One of these SAX-based approaches is the enhanced multi-objective symbolic discretization for time series (eMODiTS) [15]. eMODiTS stands out for extending the search for discretization schemes through the implementation of various cuts in value space (alphabet) for each specific time interval (word segment). Moreover, eMODiTS uses the widely recognized multi-objective evolutionary algorithm, NSGA-II, to search for appropriate discretization schemes for classifying temporal databases. The search is guided by the entropy, model complexity, and information loss functions, which are computed once the dataset discretization is performed. Nevertheless, discretizing and computing the functions in eMODiTS involves very high computational costs. Therefore, surrogate techniques have to be implemented to reduce the computational cost of eMODiTS.

Surrogate models have reduced the computational expense of the evolutionary optimization process in various applications [8,13]. To the best of our knowledge, we have not yet encountered techniques that minimize the computational cost of the time series discretization process.

The main contribution of this work is to propose surrogate models to reduce the computational cost of the eMODiTS discretization process during the objective function calculation by implementing surrogate models. To achieve this, the well-known k Nearest Neighbor classifier with DTW as the elastic similarity measure is used as a surrogate model for each function. Additionally, a methodology based on the Pareto Front is presented for updating each surrogate model. In summary, the most relevant highlights of this work can be described as follows.

1. To the best of our knowledge, this is the first approach to incorporating surrogate models in symbolic time series discretization.
2. Additionally, our methodology represents the initial endeavor of surrogate models in managing instances of diverse sizes.

The document's structure is described as follows. Section 2 provides a theoretical background for this research. Section 3 explains the integration of surrogate models in eMODiTS. Section 4 presents the experiment results and behavior of the proposal. Finally, Sect. 5 details the essential findings from the experiments.

2   Background

2.1   Surrogate-Based Optimization

Surrogate-based optimization (SBO) is the process of replacing computationally expensive parts of the optimization problem by another iterative and auxiliary process called *Surrogate models* [12]. Generally, this process is faster and more tractable than the original optimization process.

Constructing a surrogate model is the primary stage in SBO, entailing the creation of a training set to depict the problem's search space. A statistical approach is utilized for estimating a representative training set in a process referred to as *design of experiments*. Following this, the selection of an appropriate approximate method is imperative. Several classical techniques can be surrogate models, including Radial Basis Function, Support Vector Regression, Gaussian Process, and Kriging [8]. Nonetheless, any supervised learning method can estimate the value of computationally expensive processes [7]. Once the surrogate model has been built, its prediction capacity for new observations must be validated. This assignment uses partitioning techniques to divide the training dataset into multiple subsets, concealing one to assess the surrogate model's performance and making it resilient to unforeseen instances [12]. Finally, the surrogate model is optimized and evaluated for predictive accuracy, halting if a stopping criterion is met. If not, the surrogate model is revalidated and updated before reinitializing the process.

SBO has specifically been employed to optimize evolutionary computation. Primarily, the purpose is to estimate the value of computationally expensive objective functions, which direct the search towards the area where a satisfactory solution to the problem is located. Therefore, having a high-quality surrogate model is crucial.

Implementing SBO as an evaluation mechanism in an Evolutionary Algorithm (EA) requires three essential elements: a training set, evolution control, and model quality evaluation. The training set consists of solutions evaluated in the original model (known as exact solutions), which guides the surrogate model towards a reliable estimation. On the other hand, the evolution control determines which solutions shall be included in the training set for updating the model throughout the evolutionary process. Two strategies are commonly employed: individual-based and generation-based. In the individual-based approach, the original model scores a subset of individuals (the best, random, and worst) and selects them to join the training set.

Conversely, the generation-based approach updates the surrogate model after some generations. Finally, the model quality can be evaluated to see if it loses accuracy compared to the original. This element helps to take corrective actions to update the surrogate model.

2.2   Multiobjective Optimization Problem (MOP)

The Multiobjective Optimization Problem (MOP) entails simultaneously optimizing functions by maximizing, minimizing, or both. In addition, every possible solution negatively affects every function [5]. Equation 1 defines the MOP problem, where $x = [x_1, x_1, \ldots, x_d]$ represents the decision variables, F denotes the set of functions to be optimised, and $g(x)$ and $h(x)$ represent the inequality and equality constraints, respectively.

$$Minimize/Maximize$$
$$F = \{f_1(x), f_2(x), \ldots, f_M(x)\}$$
$$Subject \ to: \tag{1}$$
$$g(x) \leq 0$$
$$h(x) = 0$$

Instead of identifying a single solution, MOP generates a set of solutions that balance the objective functions using the principles of Pareto optimality [6]. This set of solutions is known as the Pareto optimal set, where a solution is considered Pareto optimal if no other solution decreases any criterion without causing a simultaneous increase in at least one criterion [5].

To identify the Pareto optimal set, Pareto developed a process of domination to determine which solution is superior among a set of solutions called *Pareto Dominance* ($\prec$) [5]. A feasible solution x_1 dominates to x_2 according to Pareto dominance, if and only if, $f_i(x_1) \leq f_i(x_2) \ \forall \ i \in \{1, 2, \ldots, F\}$ and $f(x_1) < f(x_2)$ at least in one function.

Moreover, the objective functions' values of the Pareto optimal set are called *Pareto front* ($\mathfrak{F}$), from which the user selects a final solution based on their preference.

2.3   KNN Regressor

K-Nearest Neighbor (KNN) computes the k nearest instances to an observation i based on the distance between i and the other instances in the dataset. Then, a count is taken of the values or class labels present in the neighborhood, and the class label that occurs most frequently is assigned to the instance i. This counting procedure is applicable only when the class label is categorical. However, if the class label is continuous, the observation i is assigned the average value of the class labels discovered in its neighborhood. This technique is denoted as the KNN Regressor (K-Nearest Neighbour Regressor) [16].

2.4   Dynamic Time Warping (DTW)

DTW is categorized as an elastic distance measure as it aligns two temporal sequences [18]. This technique of improving time series matching involves adjusting one time series by stretching or compressing it until it aligns with another.

DTW involves determining the path that minimizes the distance between the points of two distinct sequences [10]. Therefore, creating a distance matrix with all potential point combinations and applying dynamic programming to determine the most efficient path is necessary. However, this process can be computationally expensive for larger datasets. Consequently, various modifications and variations (including constraints and lower distances) have been suggested.

2.5 eMODiTS

eMODiTS is a discretization technique inspired by the Symbolic Aggregate Approximation (SAX) and the Piecewise Aggregate Approximation (PAA), symbolizing a time series. Unlike SAX, the eMODiTS approach explores a competitive discretization process by modifying the cut numbers in both the time (word segments) and value (breakpoints) axes. Furthermore, eMODiTS does not limit the number of breakpoints for all word segments. Alternatively, it assigns individual breakpoints for each segment, resulting in more adaptable discretization schemes than those offered by SAX.

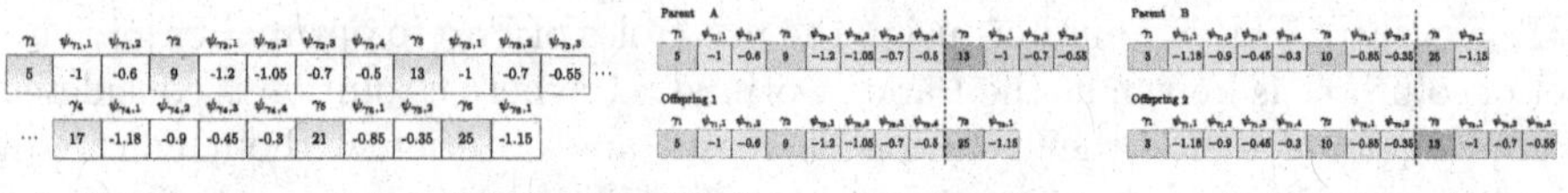

(a) Codification of the eMODiTS discretization scheme as an individual.

(b) Crossover operator based on the one-point approach. The dashed line represents the cuts performed by each parent

Fig. 1. eMODiTS elements

Regarding the discretization scheme search engine, eMODiTS utilizes the widely recognized multi-objective evolutionary algorithm NSGA-II, adapted to the eMODiTS schemes. The main modification is done on the crossover operator. Since eMODiTS discovers a flexible discretization scheme, each individual in the evolutionary search is represented by a real vector where every word segment is included, followed by its corresponding breakpoints. This codification is further illustrated in Fig. 1a. Subsequently, the possible crossover operators to use in this context are limited. Therefore, eMODiTS includes one of the most straightforward recombination operators, the one-point crossover. This operator selects a random vector cut point, and an exchange with the subparts is carried out. However, to avoid inconsistent discretization schemes, eMODiTS only defines random cuts at the segment word level. Figure 1b graphically shows the behavior of the crossover operator.

On the other hand, the eMODiTS objective functions aim to lead the search towards competitive schemes in classification. These functions consist of entropy estimation, model complexity, and information loss. The entropy estimation is expressed in Eq. 2, where $P((\mathbb{S}_i, c))$ represents the probability of assigning class c to a discretized time series $\mathbb{S}_i$, cm is a matrix containing these probabilities, and M represents the number of unique discretized time series. The Entropy function analyzes the disorder of the database after discretization with a scheme represented by an individual.

$$Entropy = \sum_{i=1}^{M} - \sum_{c=1}^{C} cm(P(\mathbb{S}_i, c)) \cdot \log_2 cm(P(\mathbb{S}_i, c)) \tag{2}$$

The second function, model complexity, is calculated using Eq. 3. In this equation, N represents the number of time series in the original dataset, and C represents the total number of classes in the problem. The function measures whether an individual stretches or shrinks a database, allowing it to have exact or extreme dimensions.

$$Complexity = \begin{cases} \frac{N-M-C}{N+(C-1)}, & M - C < 0 \\ \frac{M-C}{N+(C-1)}, & otherwise \end{cases} \tag{3}$$

Finally, the third function is the objective *Information Loss* (Infoloss), which is expressed in Eq. 4, where $\mathbb{R}$ represents a reconstructed time series, $|R|$ is the number of reconstructed time series, and S is the original temporal dataset. Since the discretization process involves dimensionality reduction, Information Loss measures the amount of loss incurred by an individual after the database is discretized. In all objective functions, values that are close to zero are preferred.

$$Infoloss = \frac{1}{|R|} \sum_{i=1}^{|R|} MSE(\mathbb{R}_i, S_i)$$

$$MSE(\mathbb{R}, S) = \frac{\sum(r_i - s_j)^2}{n - 1}, r_i \in \mathbb{R}, s_j \in S \tag{4}$$

3   Proposal

One drawback of eMODiTS is the significant computational expense needed to evaluate a single individual, as the database must be discretized for each objective function calculation. This cost escalates according to the size of the database. Thus, this paper proposes integrating surrogate models to estimate each objective function value for each individual. An overview of this integration is presented in Algorithm 1.

The creation of the training set utilized a uniform distribution due to variations in individual sizes within the population. As a result, most Design of Experiments methods are unsuitable for generating representative points. Each vector in Fig. 1a corresponds to an instance in the training set, with its respective objective function values as the response variable (class label). Furthermore, each surrogate model was developed utilizing the KNN Regressor with DTW as the similarity metric.

On the other hand, a criterion based on penalization was established for controlling evolution and evaluating surrogate models. This criterion utilizes a factor of penalization (PF) to regulate the number of updates of the surrogate models via a generation-based strategy. The effect of PF is to either accelerate or slow down this process depending on the surrogate model performance; if the performance is inadequate, PF will accelerate the update process; otherwise, PF will slow it down.

Algorithm 1. Algorithm for eMODiTS surrogate version

Require: PS: Population size, GN: Generation numbers, GUN: Number of times the surrogate models will be updated (generation-based strategy).

1: Generate an initial population of random size. PS.
2: Create a training set comprising twice the initial population size (2*PS).
3: Evaluate the initial population and training set utilizing the original objective functions.
4: Create the surrogate models.
5: **for** $g = 1$ to GN **do**
6:     Select parent set based on NSGA-II.
7:     Apply crossover operator as shown in Fig. 1b to generate the offspring set.
8:     Apply the mutation operator to the offspring set by randomly selecting a gene value from an individual and changing it based on a uniform distribution within its domain.
9:     Evaluate the offspring set using each surrogate model.
10:    Compute $\mathfrak{F}$ by the fast non-domination sorting and the crowding distance as NSGA-II.
11:    Replace the population using $\mathfrak{F}$.
12:    **if** g mod $GUN == 0$ **then** ▷ Start the process for updating surrogate models.
13:        $PF = NG/GUN$ ▷ PF is a penalization factor.
14:        Evaluate each surrogate model using $\mathfrak{F}(0)$ and (5). Record the results in *error*.
15:        **if** $error \leq 0.9$ **then**
16:            Update and retrain the surrogate models using $\mathfrak{F}(0)$ evaluated in the original functions.   ▷ (Individual-based strategy)
17:            $GUN = GUN + round(PF * error)$
18:            **if** $GUN == 0$ **then**   ▷ If GUN reaches zero, it will be updated and changed to one.
19:        **else**
20:            $PF = PF * 2$ ▷ As surrogate models provide high accuracy, the updating process is delayed.
    return $\mathfrak{F}(0)$.

The surrogate model updating process utilizes the individual from the first Pareto front to include them in the training set. It is crucial to note that each vector included in the training set must be evaluated in the original functions beforehand. Afterward, each surrogate model undergoes retraining for performance updates. Our proposal utilizes two evolution control strategies, namely generation-based and individual-based.

Finally, we evaluate each surrogate model using the modified index of agreement (d_j), a standardized measure of prediction ability in regression models [19]. In contrast to R^2, d_j does not allow negative values and ranges from 0 to 1, with 0 indicating complete disagreement between observed and predicted values and 1 indicating complete agreement. d_j is defined in Eq. 5, where Obs represents the observed values, $Pred$ represents the predicted values, and $\overline{Obs}$ denotes the average of the observed values. By default, j is typically set at one.

$$d_j = 1 - \frac{\sum_{i=1}^{|Obs|} |Obs_i - Pred_i|^j}{\sum_{i=1}^{|Obs|} |Pred_i - \overline{Obs}| + |Obs_i - \overline{Obs}|^j} \quad (5)$$

4 Experiments and Results

This section presents experiments and results to demonstrate the prediction capabilities of eMODiTS with the help of surrogate models. Our approach was executed with varied values of the KNN Regressor k parameter ($k = \{1, 3, 5, 7, 9\}$) while maintaining the remaining parameters under the suggestions made by eMODiTS [15] to ensure a fair comparison. Final solution selection was carried out by the Decision Tree classifier utilizing the training set. Solutions with the minimum misclassification rate were selected as the resulting solution.

Additionally, 25 datasets from the UCR archive [11] were used to evaluate proposal behavior. These datasets were selected based on their representative features, and their descriptions are included in Table 1.

Table 1. Description of the datasets used for testing the proposal performance.

Dataset	Train size	Test size	Length	No. Classes	Domain
ArrowHead	36	175	251	3	IMAGE
CBF	30	900	128	3	SIMULATED
Coffee	28	28	286	2	SPECTRO
DistalPhalanxOutlineAgeGroup	400	139	80	3	IMAGE
DistalPhalanxOutlineCorrect	600	276	80	2	IMAGE
DistalPhalanxTW	400	139	80	6	IMAGE
ECG200	100	100	96	2	ECG
ECG5000	500	4500	140	5	ECG
ECGFiveDays	23	861	136	2	ECG
FaceAll	560	1690	131	14	IMAGE
FacesUCR	200	2050	131	14	IMAGE
GunPoint	50	150	150	2	MOTION
ItalyPowerDemand	67	1029	24	2	EPG
MedicalImages	381	760	99	10	HAR
MiddlePhalanxOutlineAgeGroup	400	154	80	3	SENSOR
MiddlePhalanxTW	399	154	80	6	OTHER
MoteStrain	20	1252	84	2	SIMULATED
Plane	105	105	144	7	IMAGE
ProximalPhalanxOutlineAgeGroup	400	205	80	3	AUDIO
ProximalPhalanxTW	400	205	80	6	SENSOR
SonyAIBORobotSurface1	20	601	70	2	IMAGE
SonyAIBORobotSurface2	27	953	65	2	MISC
SyntheticControl	300	300	60	6	SENSOR
TwoLeadECG	23	1139	82	2	DEVICE
TwoPatterns	1000	4000	128	4	SENSOR

Firstly, Sect. 4.1 compares eMODiTS assisted by surrogate models with varying values of k and the original eMODiTS to determine which surrogate model exhibits competitive behavior concerning classification compared to the original model. The F1-score was computed using the Decision Tree classifier to compare each approach, as the datasets are unbalanced. Moreover, statistical tests used included Friedman's test and Nemenyi post-hoc at a 95% confidence level since the data does not exhibit a normal distribution. Subsequently, this study analyzes the predictive capabilities of the surrogate model with the highest performance. Prediction power was assessed using MSE and d_j.

4.1   Comparison Between Surrogate Versions and Original eMODiTS

Table 2 displays the results of comparing eMODiTS and each surrogate model. Importantly, eMODiTS performs better in identifying the best-performing databases than other models. This conclusion has been substantiated through statistical analysis. The surrogate model with $k = 7$ and $k = 9$ achieved the second-best outcomes.

Table 2. F1-score results obtained by each compared method. Bold numbers represent the best value regarding this metric.

Dataset	eMODiTS	k = 1	k = 3	k = 5	k = 7	k = 9
ItalyPowerDemand	0.937950006	0.958938222	0.904056256	0.896373107	**0.968977482**	0.963496842
SyntheticControl	**0.922979875**	0.85775603	0.851377827	0.838127401	0.824434072	0.856798927
ECG200	**0.832582465**	0.830898125	0.801170909	0.779655132	0.735541652	0.781720159
CBF	0.822231533	**0.895181149**	0.701055414	0.857572203	0.760879606	0.848818955
SonyAIBORobotSurface1	0.807560919	0.844157748	**0.857169305**	0.823064608	0.820062511	0.584868851
SonyAIBORobotSurface2	0.810978817	0.825195435	0.803748873	0.826974403	0.807372598	**0.832812292**
MoteStrain	0.716909764	0.776231817	0.696414854	**0.792811764**	0.745528668	0.755035059
TwoLeadECG	0.738039867	0.605028508	0.646517266	0.835064023	0.695162107	**0.847198294**
GunPoint	0.87	0.894786944	**0.949994999**	0.684191452	0.779647436	0.819711538
DistalPhalanxOutlineAgeGroup	0.792118248	0.810177682	**0.824636481**	0.798115681	0.81971152	0.792185579
Plane	**0.95675155**	0.848395279	0.825726851	0.913935935	0.899434611	0.910356478
MedicalImages	0.64944291	0.63909455	0.600628961	0.519143709	**0.673690966**	0.577321881
MiddlePhalanxOutlineAgeGroup	0.737011958	0.739659633	0.730434537	0.741011868	**0.749455946**	0.732708008
ProximalPhalanxOutlineAgeGroup	**0.845673369**	0.842230456	0.832319842	0.840862465	0.835769832	0.791010836
DistalPhalanxOutlineCorrect	0.802048851	0.775679866	0.810283889	0.769820792	0.784942994	**0.81190069**
ECG5000	0.91187338	0.889858615	**0.91923623**	0.896495456	0.912448917	0.909663458
ECGFiveDays	0.772995855	0.571520896	0.534644562	0.78162253	0.758366731	**0.795827238**
TwoPatterns	**0.81772447**	0.664027397	0.573579313	0.685654202	0.569735095	0.706801528
FacesUCR	0.543512273	0.566484253	0.437240353	0.40873316	**0.579546463**	0.489056821
MiddlePhalanxTW	0.549865135	**0.576898442**	0.540625295	0.54848353	0.544194865	0.574450682
FaceAll	**0.689907933**	0.617774435	0.622192924	0.541217952	0.575898535	0.676962453
ProximalPhalanxTW	0.759250602	0.766830728	0.757970464	0.761989921	0.741431698	**0.780389337**
DistalPhalanxTW	0.712292869	0.699758173	0.711757783	0.673939584	**0.716203727**	0.701475707
ArrowHead	**0.723312744**	0.505802664	0.682898606	0.615385908	0.671118954	0.658695555
Coffee	**0.820970696**	0.637276786	0.47462406	0.676082949	0.569490587	0.405473098

Figure 2 presents the statistical results, indicating that eMODiTS has the highest rank. However, the test suggests no significant differences between the methods, meaning that they exhibit comparable behaviors and can be employed to resolve any issue. This outcome is expected when deploying surrogate models for approximation as they do not enhance the original model but endeavor to match it to reduce the discrepancy between them closely.

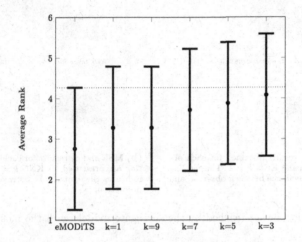

Fig. 2. The Friedman test and the Nemenyi post hoc test were performed at a 95% confidence level. Methods are ranked by their average rank in the test, with the lowest average rank indicating best performance.

4.2   Predictive Performance Analysis of Surrogate Models

Figure 3 illustrates the analysis of the predictive capabilities of the surrogate models with the most competitive classification rates observed in the previous section (with values of $k = 1$ and $k = 9$). The red line indicates the region where the estimated value matches the observed value.

The figure demonstrates that both models are highly similar. However, $k = 1$ displays a lower d_j value, denoting lower accuracy at more points (Fig. 3a) when contrasted with $k = 9$, which has a higher d_j value (Fig. 3b). Nevertheless, both surrogate models show lower MSE values in all functions, implying that, while precision may not suffice, observed and predicted values are closer, with a small number of points having a perfect match.

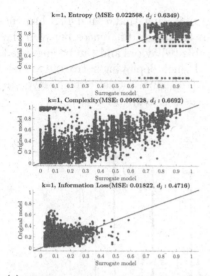
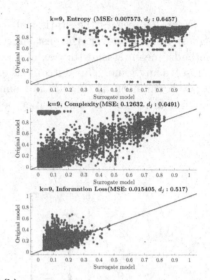

(a) MSE and d_j results obtained for each objective function using KNN $k = 1$. The red line indicates the equivalence between observed and prediction.

(b) MSE and d_j results for each objective function are presented for KNN $k = 9$. The red line indicates the equivalence between observed and prediction.

Fig. 3. Prediction power reached by the surrogate models with the most competitive performance

5   Conclusions and Future Work

This paper details the application of surrogate models in estimating the objective functions of the symbolic time series discretization method known as eMODiTS. This algorithm searches for adaptable schemes that tailor word segment boundaries in response to data and puts forward distinctive alphabet boundaries for each word segment. This results in different vectors of decision variables that diverge from each other, making a varying-size training set. Therefore, the KNN classifier was employed for regression tasks, using the DTW similarity measure to determine the proximity of a solution's neighbours.

The study's results show that the surrogate models implemented obtain classification errors comparable to those presented by eMODiTS. Nonetheless, the models do not achieve perfect accuracy in most estimated points compared to the actual model. This situation can be attributed to the individual-based approach to controlling evolution, which updates the training sets using the best individuals (Pareto front) found throughout the model run. This leads to a loss of representativeness and concentration in a small search space region.

As a result, it is proposed as a future work that further techniques for the management of the evolution be employed and investigated, together with alternative approaches to the representation of an eMODiTS discretization method within the training set.

Acknowledgments. The first author acknowledges to *Consejo Nacional de Humanidades, Ciencia y Tecnología (CONAHCYT)* for the postdoctoral research grant under the CVU No. 419862.

References

1. Acosta-Mesa, H.G., Rechy-Ramírez, F., Mezura-Montes, E., Cruz-Ramírez, N., Jiménez, R.H.: Application of time series discretization using evolutionary programming for classification of precancerous cervical lesions. J. Biomed. Inform. **49**, 73–83 (2014)
2. Ahmed, A.M., Bakar, A.A., Hamdan, A.R.: A harmony search algorithm with multi-pitch adjustment rate for symbolic time series data representation. Int. J. Mod. Educ. Comput. Sci. **6**(6), 58 (2014)
3. Bountrogiannis, K., Tzagkarakis, G., Tsakalides, P.: Distribution agnostic symbolic representations for time series dimensionality reduction and online anomaly detection. IEEE Trans. Knowl. Data Eng. (2022)
4. Chaudhari, P., Rana, D.P., Mehta, R.G., Mistry, N.J., Raghuwanshi, M.M.: Discretization of temporal data: a survey. CoRR abs/1402.4283 (2014). http://arxiv.org/abs/1402.4283
5. Coello, C.A.C., Lamont, G.B., Veldhuizen, D.A.V.: Evolutionary Algorithms for Solving Multi-objective Problems, vol. 5. Springer, Heidelberg (2007). https://doi.org/10.1007/978-0-387-36797-2
6. Deb, K., Deb, K.: Multi-objective optimization. In: Burke, E., Kendall, G. (eds.) Search Methodologies, pp. 403–449. Springer, Boston (2014). https://doi.org/10.1007/978-1-4614-6940-7_15
7. Forrester, A., Sobester, A., Keane, A.: Engineering Design via Surrogate Modelling: A Practical Guide. Wiley, Hoboken (2008)
8. Jiang, P., Zhou, Q., Shao, X.: Surrogate Model-based Engineering Design and Optimization. Springer, Singapore (2020). https://doi.org/10.1007/978-981-15-0731-1
9. Kegel, L., Hartmann, C., Thiele, M., Lehner, W.: Season-and trend-aware symbolic approximation for accurate and efficient time series matching. Datenbank-Spektr. **21**(3), 225–236 (2021)
10. Keogh, E., Ratanamahatana, C.A.: Exact indexing of dynamic time warping. Knowl. Inf. Syst. **7**, 358–386 (2005)
11. Keogh, E., et al.: The UCR time series classification archive (2018). https://www.cs.ucr.edu/eamonn/time_series_data_2018/
12. Koziel, S., Ciaurri, D.E., Leifsson, L.: Surrogate-based methods. In: Koziel, S., Yang, X.S. (eds.) Computational Optimization, Methods and Algorithms. Studies in Computational Intelligence, vol. 356, pp. 33–59. Springer, Heidelberg (2011). https://doi.org/10.1007/978-3-642-20859-1_3
13. Koziel, S., Pietrenko-Dabrowska, A.: Rapid multi-criterial antenna optimization by means of pareto front triangulation and interpolative design predictors. IEEE Access **9**, 35670–35680 (2021)
14. Lines, J., Bagnall, A.: Time series classification with ensembles of elastic distance measures. Data Min. Knowl. Disc. **29**, 565–592 (2015)
15. Márquez-Grajales, A., Acosta-Mesa, H.G., Mezura-Montes, E., Graff, M.: A multi-breakpoints approach for symbolic discretization of time series. Knowl. Inf. Syst. **62**(7), 2795–2834 (2020)

16. Miranda-Varela, M.-E., Mezura-Montes, E.: Surrogate-assisted differential evolution with an adaptive evolution control based on feasibility to solve constrained optimization problems. In: Pant, M., Deep, K., Bansal, J.C., Nagar, A., Das, K.N. (eds.) Proceedings of Fifth International Conference on Soft Computing for Problem Solving. AISC, vol. 436, pp. 809–822. Springer, Singapore (2016). https://doi.org/10.1007/978-981-10-0448-3_67

17. Muhammad Fuad, M.M.: Modifying the symbolic aggregate approximation method to capture segment trend information. In: Torra, V., Narukawa, Y., Nin, J., Agell, N. (eds.) MDAI 2020. LNCS (LNAI), vol. 12256, pp. 230–239. Springer, Cham (2020). https://doi.org/10.1007/978-3-030-57524-3_19

18. Tan, C.W., Petitjean, F., Webb, G.I.: FastEE: fast ensembles of elastic distances for time series classification. Data Min. Knowl. Disc. **34**(1), 231–272 (2020)

19. Willmott, C.J.: On the evaluation of model performance in physical geography. Spat. Stat. Models 443–460 (1984)

Estimation of Anthocyanins in Homogeneous Bean Landraces Using Neuroevolution

José-Luis Morales-Reyes[1]($\boxtimes$) iD, Elia-Nora Aquino-Bolaños[1] iD,
Héctor-Gabriel Acosta-Mesa[2] iD, and Aldo Márquez-Grajales[2] iD

[1] Centre for Food Research and Development, University of Veracruz, Xalapa, Veracruz, Mexico
jluismorey@hotmail.com
[2] Artificial Intelligence Research Institute, University of Veracruz, Xalapa, Veracruz, Mexico

Abstract. A pigment of great interest is the anthocyanins. It is due to the nutritional benefits discovered in various foods, such as common beans. In this work, we report the estimation of anthocyanins in homogeneous colored bean landraces using neuroevolution. Two neuroevolution techniques, NEAT and DeepGA, were implemented to find this task's suitable neural network structure. Both techniques were compared against a Convolutional Neural Network (CNN) experimentally developed called AnthEst-Net architecture, which found competitive results in anthocyanin estimation. The input data of the network architectures were two-color characterizations, two-dimensional histograms, and data vectors. The accuracies obtained on the test set in HSI color space were 85.38 ± 11.77 and 87.89 ± 9.67 for DeepGA and AnthEstNet architecture, respectively. Regarding CIE L*a*b* color space, DeepGA obtained an accuracy of 86.85 ± 11.08, while AnthEstNet got 87.08 ± 14.19. Results suggest that the architecture reported by DeepGA is suitable for anthocyanins estimation.

Keywords: Homogeneous · Neuroevolution · Anthocyanins · Color distribution · CNN

1 Introduction

Anthocyanins are water-soluble pigments that provide pigmentation in various shades, including red, purple, orange, blue, and brown in the plant kingdom. Their diverse properties have been widely investigated [1]. In plants, they attract pollinating insects and animals through their coloration and also offer tolerance to stress, low temperatures, salinity, drought, and ultraviolet radiation. Recent studies suggest that anthocyanins play an essential defense mechanism [2].

Food anthocyanins possess beneficial health properties, including anti-inflammatory, anticancer, and cardio-protective effects, and can prevent diabetes and other health conditions [1, 3]. Common beans are one of the primary crops in Mexico, with a vast array of domesticated varieties grown by farmers for personal use [4]. Understanding the concentration of anthocyanins enables us to determine their functional properties [5].

© The Author(s), under exclusive license to Springer Nature Switzerland AG 2024
H. Calvo et al. (Eds.): MICAI 2023 Workshops, LNAI 14502, pp. 373–384, 2024.
https://doi.org/10.1007/978-3-031-51940-6_28

The concentration of anthocyanins in common bean landraces can be determined using the pH Differential method, which involves invasive and expensive procedures performed in a laboratory [6].

On the other hand, various authors propose utilizing computer vision systems for estimating anthocyanin non-invasively and at a lower cost. These proposals are employed in diverse domains where the sample color is homogenous. Color reduction obtained from digital image processing is used to characterize the color of the samples [7–13]. The proposed color characterization, utilizing color distribution displayed as probability distributions, enables the examination of complete coloration in a group of seeds [14]. Various network architectures have been proposed for anthocyanin estimation in diverse domains [14–16].

Therefore, the primary challenge in architecture design is determining the required number of hidden or convolutional layers depending on the problem's complexity. In the case of convolutional neural networks, establishing the kernel size and number of filters is a critical question that arises during the architecture design for a specific task [17]. Therefore, it is crucial to implement methods for automatically discovering a fitting convolutional neural network structure to estimate anthocyanins in common beans, an area that requires further research.

Consequently, the work's primary accomplishment is producing a neural network architecture by neuroevolution for estimating anthocyanins in bean landraces of homogeneous color. Mainly, the research's objectives are twofold. First, to implement and evaluate the neural structure created by the NEAT (NeuroEvolution of Augmenting Topologies) is an evolutionary algorithm for our problem, and second, to assess the neural structure generated by the DeepGA algorithm.

The document's structure is as follows: Sect. 2 presents the methodology for estimating anthocyanin of homogeneous color via neuroevolution, and Sect. 3 displays the results of its implementation. Section 4 describes the discussions based on those results. Finally, Sect. 5 provides the conclusions.

2 Materials and Methods

2.1 Bean Landraces

This work used 40 different common bean landraces of homogeneous color (*Phaseolus vulgaris L.*). These landraces were collected in several municipalities of Oaxaca, Mexico. Each sample contains 60 grams of healthy and clean seeds. There are 18 landraces of black beans, nine of red beans, eight of yellow beans, four of white beans, and one of brown beans.

2.2 Quantification of Monomeric Anthocyanins

The bean seed coat quantified the anthocyanin levels using the differential pH method [18]. After a 12-h immersion in distilled water, the seeds underwent coat removal. Next, 25 ml of a 70:29.5:0.5 v/v/v acetone/water/acetic acid mixture was used to homogenize three grams of coat for 20 min (Wisetis homogenizer, HG-15-A, 110 V; DAIHAN brand,

Gang-won, Korea). After centrifugation at 4000 rpm for 20 min (Hettich centrifuge, Universal 32R, Tuttlingen, Germany), the supernatants were discarded, and the same procedure was repeated with the residues under identical conditions. Ultimately, the supernatant from each fraction was combined to determine the monomeric anthocyanin content.

Two extract dilutions were carried out: potassium chloride buffer at pH 1.0 and sodium acetate buffer at pH 4.5. The maximum absorbance was determined by obtaining the absorption spectrum within the 460–710 nm range (Spectrophotometer UV-1800, Shimadzu, Kyoto, Japan). The concentration of monomeric anthocyanins was calculated using the equation described by Giusti and Wrolstad [19]. Results were expressed in mg cyanidin-3-glucoside per gram of dry sample (mg C3G g^{-1}).

2.3 Acquisition System and Image Segmentation

The brightness is a distinctive feature of particular bean landraces. A controlled lighting environment was proposed to maintain uniform illumination and reduce specular brightness. The illumination setup contains an aluminum box with an aperture in the upper section, where the camera lens is installed. Moreover, the environment contains eight fluorescent light sources with a diffuser to mitigate glare [20].

A seed sample from each landrace was positioned on an individual sliding platform to capture images. For correct seed identification, it was necessary to contrast the seed color from the background. Consequently, blue was selected for this purpose.

On the other hand, a color-based region-growing segmentation algorithm was employed in this study, which groups adjacent pixels based on predetermined similarity measures [21, 22]. In this case, the similarity criterion employed was the background color because of its homogeneous coloration. Moreover, the color channels of the background were used as seed for the algorithm initialization.

2.4 Neuroevolution

Neuroevolution is a machine learning technique that uses evolutionary algorithms to evolve neural networks [23, 24]. In neuroevolution, a population of genetic neural network encodings is evolved to discover a network capable of solving a given task. The neuroevolution process begins by initializing a group of genomes. These genomes are simplified genetic representations that can be decoded into a neural network. Moreover, the genomes are then applied to the problem environment, after which each genome is assigned a fitness score based on the neural network's ability to complete the task effectively. Genetic operators create the next generation once all population members have been evaluated. The higher fitness codings mutate and interbreed with each other, and the resulting offspring replace the lower fitness genotypes in the population. Thus, the process constitutes an intelligent parallel search for better genotypes until a sufficiently high-fitness network is found [24, 25].

NEAT
NeuroEvolution of Augmenting Topologies (NEAT) is a tool for evolving neural networks (NN). The NEAT algorithm applies the idea of starting the evolution with small and

straightforward networks and letting them become increasingly complex over the generations. Therefore, this continuous evolution process allows finding very sophisticated and complex neuronal [26].

The genetic coding scheme of NEAT to represent each individual's genome is linear representation, including node genes that provide a list of inputs, hidden, and output nodes that can be connected. It also includes a list of linkage genes. Each linkage gene refers to the input node gene and output node gene, the linkage weight, the active linkage is expressed by an enable bit, and finally, an innovation number is used so individuals can be easily aligned during crossing.

Deep GA

Deep GA, a Convolutional Neural Network (CNN) structure optimization tool, uses a genetic algorithm to evolve CNNs. The result is a competitive CNN with fewer convolutional layers. Deep GA defines a hybrid coding to represent CNNs using two levels. In the first level, block coding is used, where each block represents a convolutional layer, batch normalization, and ReLU activation function. Additionally, it includes hyperparameters that define the filter size, the stride, and the type of reduction layer. The second level uses binary coding to describe the connection between layers, with connections as one and non-connections as zero [27].

Subsequently, both techniques use evolutionary algorithms with the difference that NEAT evolves NNs and DeepGA evolves CNNs. Using NEAT, network topology begins with zero hidden nodes, with inputs directly connected to the output. During the mutation process, network complexity intensifies, engendering novel architectures. Structures become more intricate as they become optimal by integrating nodes and connections within hidden layers.

In contrast, DeepGA evolves convolutional neural networks suitable to receive two-dimensional histograms. DeepGA regression allows evolving convolutional neural networks, and this algorithm will seek an efficient architecture for estimating anthocyanins. Its advantage is focused on optimizing the CNN architecture by reducing the number of parameters.

2.5 Data Splitting

Due to the similarity in coloration among the bean landrace seeds, it was possible to divide each landrace into four subgroups. Each subgroup was divided into three groups: the first with 70% of the seeds for training data, the second with 15% for validation, and the remaining 15% for testing. The above data partitioning procedure made creating three datasets of 160 data each possible.

2.6 Color Characterization Using a Probability Mass Function

Given two discrete random variables X and Y, a joint probability distribution or probability mass function (PMF) is defined as $f(X, Y) = P(X = x, Y = y)$, where $f(X, Y)$ represents the occurrence probability of x and y values subjects to following conditions.

1. $f(x, y) \geq 0$ for all (x, y),

2. $\sum_x \sum_y f(x, y) = 1$
3. $P(X = x, Y = y) = f(x, y)$

Based on the information presented above, it is possible to generate histograms utilizing the chromaticity channel data from the CIE L*a*b* and HSI color spaces. The 8-bit images consist of 256 unique grayscale levels, allowing the histogram creation of a resolution of 256 × 256. To account for the dominance of the chromaticity channels a* and b* [−128, 127] in CIE L*a*b*, the shift was computed by summing the values of each pixel along with the absolute value of the lower limit. For H and S, values were scaled to their corresponding values within the interval [0, 255].

The color representation of the bean landraces was 2D histograms of the chromaticity channels. Since NNs receive as input a feature vector, each histogram with dimension **256 × 256** was converted to a vector with 65,536 elements (see Fig. 1). Therefore, two-dimensional histograms can be the input for CNNs evolved with DeepGA.

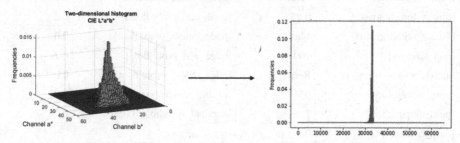

Fig. 1. Conversion of a two-dimensional histogram (**256 × 256**) to a vector (**1 × 65, 536**).

2.7 Metric Performance

Several metrics have been reported in the literature for the performance evaluation of regression models. In this work, the MAPE metric was employed because it allows the calculation of the error in percentage terms (Eq. 1). A lower error indicates that the precision value is close to the reference value.

$$MAPE = \frac{\sum_{i=1}^{n} \left| \frac{y_i - \hat{y}_i}{y_i} \right| * 100}{n} \tag{1}$$

Equation 2 was included to know the percentage of approximation of the estimated value to the value reported by the differential pH method, where a high value represents a good anthocyanin estimation.

$$Precision = 100 - MAPE \tag{2}$$

3 Experiments and Results

3.1 Experimental Design

The experiments presented in this work are designed to achieve the objectives defined in Sect. 1. The first experiment performed was the evolution of NNs using NEAT, where the values of the hyperparameters are shown in Table 1. Furthermore, the color characterization was done by converting the two-dimensional histograms into vectors.

Table 1. Hyperparameter values for the NEAT algorithm

Hiperparameter	Value	Hiperparameter	Value
[generations: int]	200	[weight_mutate_rate: float]	0.8
Objective function	Precision	[bias_mutate_rate: float]	0.7
[fitness_threshold: int]	0.95	[node_add_prob: float]	0.2
[fitness_criterion: str]	Max	[node_delete_prob: float]	0.0
[pop_size: int]	100	[edge_add_prob: float]	0.2
[activation_default: str]	Relu	[edge_delete_prob: float]	0.0
[num_inputs: int]	65536	[enable_mutate_rate: float]	0.01
[num_outputs: int]	1	[survival_threshold: float]	0.2
[num_hidden: int]	0		

On the other hand, the second experiment was the neuroevolution of CNNs using Deep GA. In order to know the proposed architectures, two trials were performed with Deep GA due to the information of each color space (HSI and CIE L*a*b*). Table 2 shows the values assigned to the hyperparameters of the DeepGA configuration. For comparison, the AnthEstNet architecture [14] was considered for data evaluation and result generation.

Table 2. Hyperparameters values for Deep GA in the regression task.

Hiperparámetros			
Evolutionary algorithm	Value	CNN	Value
Population size	20	Epochs number	30
Generations number	50	Learning rate	0.001
Objective function	MAPE	Optimization method	ADAM
		Loss function	MSE

3.2 Anthocyanin Estimation in Common Bean Landraces Using NN Generated by NEAT Algorithm

Table 3 shows the results of the networks generated by the NEAT algorithm with color characterization represented by a vector.

Table 3. Performance comparison of network models created using the NEAT algorithm. Bold number represents the best values according to the used metric.

Color space		HSI	CIE L*a*b*
Models		NEAT ANN	NEAT ANN
Train	Precision	72.70 ± 26.33	**88.99 ± 15.96**
	RMSE	2.1	**0.956**
	R2	0.0	**0.79**
Validation	Precision	72.49 ± 26.82	**88.55 ± 16.17**
	RMSE	2.10	**0.98**
	R2	0.0	**0.77**
Test	Precision	72.70 ± 26.26	**88.55 ± 16.17**
	RMSE	2.09	**0.98**
	R2	0.0	**0.77**

3.3 Anthocyanin Estimation in Common Bean Landraces Using CNN Generated by DeepGA Algorithm

DeepGA was run twice using two different color spaces. As a result, the same CNN architecture was generated for both color spaces. Figure 2 shows the resulting structure, where three convolutional layers with batch normalization and ReLU were obtained. The first layer contains 16 filters of 7×7, and the second layer contains 32 filters of 2×2. Both layers have padding and a stride of size 2 in max pooling. Finally, the third includes 46 filters of 6×6, a max pooling layer with padding of 4, and a stride of the size of 2. Finally, two fully connected and a regression layer were found to obtain the estimation value.

Moreover, the CNNs were trained for 200 epochs with a learning rate of 0.001 and Adaptive Moment Estimation. The performance evaluation results of the convolutional networks are shown in Table 4. Regarding the pH Differential method comparison, Figs. 3 and 4 show the performance of NEAT, DeepGA, and AnthEstNet architecture in each bean landrace.

4 Discussion

This work aimed to find an optimized architecture for anthocyanin estimation in homogeneous bean landraces. For this purpose, NEAT and DeepGA methods were implemented to find a suitable architecture for our problem. The CNN architecture returned

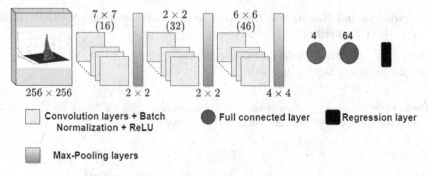

Fig. 2. CNN structure found by DeepGA algorithm.

Table 4. Anthocyanin estimation accuracy results with Deep GA and AnthEstNet architecture. Bold number represents the best values according to the used metric.

Color space		HSI		CIE L*a*b*	
Models		Deep GA	AntEstNet	Deep GA	AntEstNet
Train	Precision	93.02 ± 6.46	**93.73 ± 5.75**	92.76 ± 6.87	**93.04 ± 9.01**
	RMSE	0.38	**0.34**	**0.41**	0.43
	R2	0.96	**0.97**	**0.96**	0.95
Validation	Precision	85.57 ± 12.43	**88.05 ± 10.48**	86.94 ± 10.82	**87.19 ± 10.97**
	RMSE	0.59	**0.52**	**0.52**	**0.52**
	R2	0.92	**0.93**	0.92	**0.93**
Test	Precision	85.38 ± 11.77	**87.89 ± 9.67**	86.85 ± 11.08	**87.08 ± 14.19**
	RMSE	0.59	0.50	0.52	0.55
	R2	0.91	0.94	0.91	0.92

by DeepGA has 512,569 parameters, while the architectures returned by NEAT initially have 65,536 parameters. If the hidden layer has 10 neurons, it will increase the number of parameters, making the architecture returned by NEAT more computationally expensive.

The NEAT algorithm obtained an accuracy of 72.7 ± 26.26 on the HSI color space test set, while CIE L*a*b* got 88.55 ± 16.17.64. Based on the variance of the accuracies, the estimates are not as close to their reference values. Regarding the neural network structure, NEAT generates large and complex networks, which implies a high computational cost.

Consequently, we sought to apply neuroevolution methods to find a convolution-based neural architecture. For this purpose, DeepGA was used to find an optimal structure of convolutional neural networks for anthocyanin estimation. The results obtained for the architecture found by DeepGA on the test set were 85.38 ± 11.77 and 86.85 ± 11.08 for the HSI and CIE L*a*b* color spaces, respectively. These results were compared

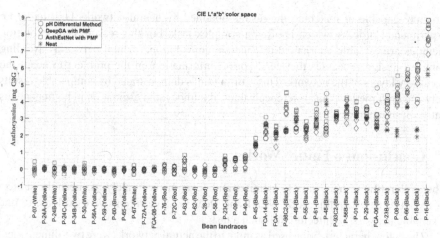

Fig. 3. Comparison of DeepGA, AnthEstNet, and NEAT estimated results using PMF and pH Differential method in CIE L*a*b color space.

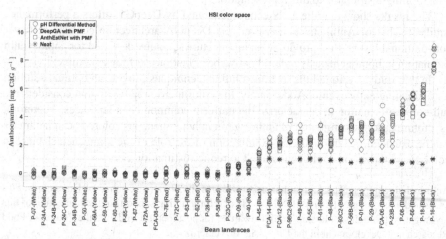

Fig. 4. Comparison of DeepGA, AnthEstNet, and NEAT estimated results using PMF and pH Differential method in HSI color space.

against a convolutional neural network applied to homogeneous bean landraces called AntEstNet [14]. The AntEstNet results on the test set were 87.89 ± 9.67 for HSI color space and 87.08 ± 14.19 for CIE L*a*b*, superior to those found by the DeepGA architecture.

Concerning the reference values and the estimates of the traditional method, Figs. 3 and 4 show a closeness between the results obtained by each CNN and the reference values. It is essential to mention that the AnthEstNet network was built experimentally and consists of six convolutional layers. In contrast, the network generated by Deep GA contains fewer convolutional layers, representing a less complex architecture and, therefore, a lower computational cost. In addition, DeepGA automatically finds a network structure capable of matching the results obtained by neural networks built under an experimental process, which is often a complex task. For this reason, neuroevolution methods are a viable alternative to generate convolutional neural networks, avoiding complex design processes that need apriori information on the problem to adapt the necessary layers of the network. This affirmation is demonstrated by comparable results between both methods and reference values of anthocyanin estimation in homogeneous native bean landraces.

5   Conclusions a Future Work

The exploration of new ways to estimate anthocyanins accurately has been the subject of recent research. In this work, NEAT and Deep GA algorithms were explored to find a suitable network architecture for such a task.

The color characterization used as input to the neural networks were two-dimensional histograms created from the chromaticity channels of the HSI and CIE L*a*b* color spaces. Additionally, color characterization represented by vectors was considered by transforming histograms to this representation.

The results show that the architecture generated by DeepGA offers a performance similar to that of AnthEstNet. However, the DeepGA architecture consists of fewer convolutional layers, thus providing an architecture specially designed for anthocyanin estimation in homogeneously colored native bean landraces of lower complexity.

In this study, we found that neuroevolution techniques provide an optimal solution for estimating anthocyanin. Additionally, the colorimetric representation considers the full-color information to characterize the uniform common bean landraces. Therefore, as a future task, the estimation of anthocyanin in heterogeneous colored bean landraces can be extended. Additionally, an examination of the role of the luminance channel in color characterization could be included to reduce estimation errors.

Acknowledgements. The first author acknowledges the National Council of Humanities, Sciences and Technologies (CONAHCyT) of Mexico for granting support for the realization of this investigation through scholarship 712056 awarded for postdoctoral studies at the Centre for Food Research and Development in the University of Veracruz.

References

1. Santos-Buelga, C., Mateus, N., De Freitas, V.: Anthocyanins. Plant pigments and beyond. J. Agric. Food Chem. **62**(29), 6879–6884 (2014). https://doi.org/10.1021/jf501950s
2. Kaur, S., et al.: Protective and defensive role of anthocyanins under plant abiotic and biotic stresses: an emerging application in sustainable agriculture. J. Biotechnol. **361**, 12–29 (2023). https://doi.org/10.1016/j.jbiotec.2022.11.009
3. Li, D., Wang, P., Luo, Y., Zhao, M., Chen, F.: Health benefits of anthocyanins and molecular mechanisms: update from recent decade. Crit. Rev. Food Sci. Nutr. **57**(8), 1729–1741 (2017). https://doi.org/10.1080/10408398.2015.1030064
4. Gepts, P., Papa, R.: Evolution during domestication. e LS (2001)

5. Chávez-Servia, J.L., et al.: Diversity of common bean (Phaseolus vulgaris L.) landraces and the nutritional value of their grains. In: Grain Legumes: IntechOpen (2016)
6. Wrolstad, R.E.: Color and pigment analyses in fruit products (1993)
7. Zhang, C., Wu, W., Zhou, L., Cheng, H., Ye, X., He, Y.: Developing deep learning based regression approaches for determination of chemical compositions in dry black goji berries (Lycium ruthenicum Murr.) using near-infrared hyperspectral imaging. Food Chem. **319**, 126536 (2020). https://doi.org/10.1016/j.foodchem.2020.126536
8. Chen, Y., Zheng, L., Wang, M., Wu, M., Gao, W.: Prediction of chlorophyll and anthocyanin contents in purple lettuce based on image processing. Presented at the 2020 ASABE Annual International Virtual Meeting, St. Joseph, MI (2020)
9. del Valle, J.C., Gallardo-López, A., Buide, M.L., Whittall, J.B., Narbona, E.: Digital photography provides a fast, reliable, and noninvasive method to estimate anthocyanin pigment concentration in reproductive and vegetative plant tissues. Ecol. Evol. **8**(6), 3064–3076 (2018). https://doi.org/10.1002/ece3.3804
10. Fernandes, A.M., Franco, C., Mendes-Ferreira, A., Mendes-Faia, A., da Costa, P.L., Melo-Pinto, P.: Brix, pH and anthocyanin content determination in whole Port wine grape berries by hyperspectral imaging and neural networks. Comput. Electron. Agric. **115**, 88–96 (2015). https://doi.org/10.1016/j.compag.2015.05.013
11. Chen, S., Zhang, F., Ning, J., Liu, X., Zhang, Z., Yang, S.: Predicting the anthocyanin content of wine grapes by NIR hyperspectral imaging. Food Chem. **172**, 788–793 (2015). https://doi.org/10.1016/j.foodchem.2014.09.119
12. Taghadomi-Saberi, S., Omid, M., Emam-Djomeh, Z., Ahmadi, H.: Evaluating the potential of artificial neural network and neuro-fuzzy techniques for estimating antioxidant activity and anthocyanin content of sweet cherry during ripening by using image processing. J. Sci. Food Agric. **94**(1), 95–101 (2014). https://doi.org/10.1002/jsfa.6202
13. Yoshioka, Y., Nakayama, M., Noguchi, Y., Horie, H.: Use of image analysis to estimate anthocyanin and UV-excited fluorescent phenolic compound levels in strawberry fruit. Breed. Sci. **63**(2), 211–217 (2013). https://doi.org/10.1270/jsbbs.63.211. (in English)
14. Morales-Reyes, J.L., Acosta-Mesa, H.-G., Aquino-Bolaños, E.-N., Herrera Meza, S., Márquez Grajales, A.: Anthocyanins estimation in homogeneous bean landrace (Phaseolus vulgaris L.) using probabilistic representation and convolutional neural networks. J. Agric. Eng. **54**(2) (2023). https://doi.org/10.4081/jae.2023.1421
15. Prilianti, K.R., Setiyono, E., Kelana, O.H., Brotosudarmo, T.H.P.: Deep chemometrics for nondestructive photosynthetic pigments prediction using leaf reflectance spectra. Inf. Process. Agric. **8**(1), 194–204 (2021). https://doi.org/10.1016/j.inpa.2020.02.001
16. Concepcion, R.S., II., Dadios, E.P., Cuello, J.: Non-destructive in situ measurement of aquaponic lettuce leaf photosynthetic pigments and nutrient concentration using hybrid genetic programming. AGRIVITA J. Agric. Sci. **43**(3), 589–610 (2021). https://doi.org/10.17503/agrivita.v43i3.2961
17. Zhou, X., Qin, A.K., Gong, M., Tan, K.C.: A survey on evolutionary construction of deep neural networks. IEEE Trans. Evol. Comput. **25**(5), 894–912 (2021). https://doi.org/10.1109/TEVC.2021.3079985
18. Xu, B.J., Yuan, S.H., Chang, S.K.C.: Comparative analyses of phenolic composition, antioxidant capacity, and color of cool season legumes and other selected food legumes. J. Food Sci. **72**(2), S167–S177 (2007). https://doi.org/10.1111/j.1750-3841.2006.00261.x
19. Giusti, M.M., Wrolstad, R.E.: Characterization and measurement of anthocyanins by UV-visible spectroscopy. Curr. Protocols Food Anal. Chem. (1), F1.2.1–F1.2.13 (2001). https://doi.org/10.1002/0471142913.faf0102s00
20. Morales-Reyes, J.L., Acosta-Mesa, H.G., Aquino-Bolaños, E.N., Herrera-Meza, S., Cruz-Ramírez, N., Chávez-Servia, J.L.: Classification of bean (Phaseolus vulgaris L.) landraces

with heterogeneous seed color using a probabilistic representation. Presented at the 2021 IEEE International Autumn Meeting on Power, Electronics and Computing (ROPEC) (2021)

21. Tang, J.: A color image segmentation algorithm based on region growing. In: 2010 2nd International Conference on Computer Engineering and Technology, vol. 6, pp. V6-634–V6-637. IEEE (2010). https://doi.org/10.1109/ICCET.2010.5486012

22. Gonzalez, R.C., Woods, R.E.: Digital Image Processing. Prentice Hall, Upper Saddle River (2002)

23. Chandra, R., Tiwari, A.: Distributed Bayesian optimisation framework for deep neuroevolution. Neurocomputing **470**, 51–65 (2022). https://doi.org/10.1016/j.neucom.2021.10.045

24. Lehman, J., Miikkulainen, R.: Neuroevolution. Scholarpedia **8**(6), 30977 (2013)

25. Stanley, K.O., Clune, J., Lehman, J., Miikkulainen, R.: Designing neural networks through neuroevolution. Nat. Mach. Intell. **1**(1), 24–35 (2019). https://doi.org/10.1038/s42256-018-0006-z

26. Stanley, K.O., Miikkulainen, R.: Evolving neural networks through augmenting topologies. Evol. Comput. **10**(2), 99–127 (2002). https://doi.org/10.1162/106365602320169811

27. Vargas-Hákim, G.-A., Mezura-Montes, E., Acosta-Mesa, H.-G.: Hybrid encodings for neuroevolution of convolutional neural networks: a case study. Presented at the Proceedings of the Genetic and Evolutionary Computation Conference Companion, Lille, France (2021). https://doi.org/10.1145/3449726.3463133

Representation of Expert Knowledge on Product Design Problems Using Fuzzy Cognitive Maps

Hector-Heriberto Rodriguez-Martinez[(⊠)], Jesus-Adolfo Mejia-de Dios[ⓘ], and Irma-Delia García-Calvillo[ⓘ]

Centro de Investigación en Matemáticas Aplicadas, Universidad Autonoma de Coahuila, 25280 Saltillo, Coahuila, Mexico
{hector.rodriguez,adolfomejia,irma.garcia}@uadec.edu.mx
https://www.cima.uadec.mx/

Abstract. An increasing number of companies are seeking to integrate design as a strategic capability to address contemporary business and societal challenges. However, effectively integrating design into an organization poses challenges due to the limited understanding of how to manage this process and assess its impact. This study explores a model based on fuzzy cognitive maps for measuring the impact of design within projects at an organizational level. This model establishes causal relationships among four key layers: design decisions, design metrics, business metrics, and product lifecycle. These layers encompass underlying concepts that can be represented using fuzzy variables, enabling the modeling of a complex system as a Fuzzy Cognitive Map (FCM) in order to visually represent expert knowledge. This approach facilitates the visualization and examination of potential scenarios within the organizational context. The proposed model provides an alternative or complement to existing methods for measuring the impact of design on projects within an organization. To illustrate this, a real-world problem is presented by describing the application of an FCM to evaluate the workflow of a specific operation known as "Digital CVV on-off" feature.

Keywords: Fuzzy Logic · Fuzzy Cognitive Maps · Product Design

1 Introduction

Product design is a critical discipline in today's industry, playing an essential role in creating and enhancing products to meet consumer needs, maintain market competitiveness, and ultimately drive organizational success. In this context, the effective representation of expert knowledge in product design is a crucial component for informed decision-making and innovation in product development [8]. Legarda et al. [4], emphasize the significance of comprehending the influence of design on organizations and its potential contribution to achieving business success. Their work offers valuable insights into the strategic management of

H. Calvo et al. (Eds.): MICAI 2023 Workshops, LNAI 14502, pp. 385–396, 2024.
https://doi.org/10.1007/978-3-031-51940-6_29

design as an asset and underscores the necessity for an effective assessment and measurement of this impact.

Despite the growing awareness of the importance of design in organizations, there is a dichotomy in the literature concerning the representation of expert knowledge in product design [4]. This gap emphasizes the need for research that addresses how expert knowledge in product design can be captured and optimally utilized to drive the product design process. This is precisely the central issue that this article aims to address.

Taking into consideration the model proposed in [4], it is possible to obtain causal models that represent the various interrelationships among organizational elements. In this context, it is hard to determine these relationships based on statistical information, nor can they be expressed probabilistically; therefore, approaches like Bayesian networks [7] are ruled out, especially when incorporating design decision modeling and organizational goals. An alternative approach for causal modeling in such circumstances is the use of Fuzzy Cognitive Maps (FCMs) [3,6]. FCMs are successfully used for representing causal reasoning through fuzzy graphs structures, where nodes represent concepts, arcs (with weights) are representing connections and their strength among nodes.

This research focuses on the use of FCMs as a tool for representing expert knowledge in product design (e.g. design of mobile applications, web platforms, among others). FCMs enable the modeling and understanding of complex and often ambiguous relationships among factors influencing product design, providing a solid foundation for decision-making and continuous improvement in this critical field.

This study seeks to contribute to the existing literature by providing an innovative and effective approach to representing expert knowledge in product design, which may lead to significant advances in the efficiency and quality of the design process. Furthermore, the results of this research are expected to have practical applications in industry, enabling organizations to use expert knowledge more effectively in product design decision-making and ultimately enhance their market competitiveness.

For the purpose of this study, we will illustrate our approach with a case study involving an organization providing financial services through mobile applications, web platforms, and physical branches. This organization boasts a dedicated design and user experience department where various roles, including design managers as experts in customer experience and stakeholders as experts in business models, collaborate to define the product and service design strategy which is currently impacting at least 17 millions users in Mexico. This real-world application serves as a practical illustration of how our proposed model can be applied in a multifaceted setting, encompassing both digital and physical channels, highlighting the relevance of effective design management in modern organizations.

To achieve the stated objective, this article is structured as follows: Sect. 2 presents the theoretical framework related to the model for measuring and managing the impact of design on organizations and FCMs; subsequently, the theory for creating the FCM from expert knowledge is overviewed in Sect. 3. Section 4 details the proposed methodology on using FCMs to represent expert knowledge.

The practical application of the model to a case study is described in Sect. 5. Finally, Sect. 6 presents the conclusions and future paths of research.

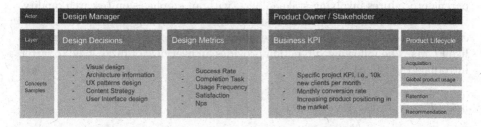

Fig. 1. Layers of organization, concepts and actors involved.

2 Measuring and Managing the Impact of Design on Organizations

Design Impact Measurement and Management (DIMM) [4] is a model designed to measure and manage the impact of design within organizations, with a focus on helping managers enhance their organization's design capabilities. The model establishes connections between various ways design affects an organization, including changes in organizational perception, processes, and culture, and links them to strategic goals and business outcomes. It guides organizations in reflecting on how design creates value and assists them in creating action plans to develop design capabilities based on their maturity level and specific needs. Considering the first two primary levels:

- **Level 1:** Outcome Identification focuses on defining design-related objectives and employs traditional business metrics like turnover, market share, and project profitability to evaluate their achievement.
- **Level 2:** Design Perception takes a dual approach by considering both external stakeholder perspectives, using metrics like Net Promoter Score (NPS) and brand value, and internal perceptions gathered from employee satisfaction surveys. This level ensures a comprehensive evaluation, even accommodating subjective assessments when standard metrics are unavailable.

We propose to reorganize these levels in layers that fit into our organization structure in Fig. 1 so that a cause-and-effect relationship can be established. This approach allows for the establishment of design-associated perception concepts to be linked with business metrics effectively.

3 Fuzzy Cognitive Maps for Representing Expert Knowledge

The increasing complexity of decision-making models, incorporating expert opinions and diverse stakeholder perspectives, can lead to confusion in choosing the best analytical and adaptive methodology for managing and assessing design impact on organizations. Therefore, we advocate an approach rooted in experts' definitions of concepts. Additionally, our model employs systematic and transparent procedures involving stakeholders, with an emphasis on research quality. Effective decision-making in such systems relies on analyzing intricate cause-and-effect relationships, emphasizing the need for system conceptualization graphs like FCMs. The following subsection provides an overview of fuzzy logic to latter introduce the FCMs.

3.1 Fuzzy Logic Overview

Fuzzy logic is an extension of conventional logic that deals with partial truth values, allowing for smooth transitions between completely true and completely false [9]. Unlike traditional logic with binary true/false values, fuzzy logic permits elements to partially belong to multiple sets, representing an intermediate state. For instance, a glass of water can be described as lukewarm, falling between cold and hot.

As mentioned previously, fuzzy sets encompass members to varying degrees. Let F represent a fuzzy set with a universe of discourse $X = x$, defined as the mapping $\mu F(x) : X \to [0, \alpha]$. The universe of discourse spans all possible real scalar values associated with a measurement or information items we intend to make fuzzy. This mapping assigns a value in the range $[0, \alpha]$ to each x. When $\alpha = 1$, the set is referred to as normal. A fuzzy set incorporates a distribution, also known as a membership function. If a distribution has zero width, the membership function condenses into a singularity, akin to the conventional case of a crisp set. When these singularities can assume only one of two possibilities, they perform binary logic. μF is denoted as the membership grade or truth degree of x. Fuzzy sets, while often resembling triangular or Gaussian distributions, can take on various forms, with no single shape proven to be superior.

Assume a variable named "usability" which is associated with how efficient and useful is a product or service in the design context. Employing fuzzification, a variable or concept can be classified into one or several fuzzy sets. In this particular case, there are three fuzzy sets to which "usability" can be classified into, denoted "LOW" "MEDIUM," and "HIGH". The membership functions of these fuzzy sets are Gaussian, and the universe of discourse X contains usability grades between 0 and 100 percent. Figure 2 shows the fuzzification of "usability".

3.2 Fuzzy Cognitive Maps

An FCM is employed as a graphical representation of a system to illustrate how different concepts relate to each other, allowing us to describe its behavior in a simple and symbolic manner [2,3]. In an FCM, concepts are depicted as nodes, while the connections between them are represented as edges, reflecting the perceived relationships among these concepts. These relationships in an FCM are logically established by connecting concepts through directed links that carry semantic or other meaningful significance, displaying causality between them.

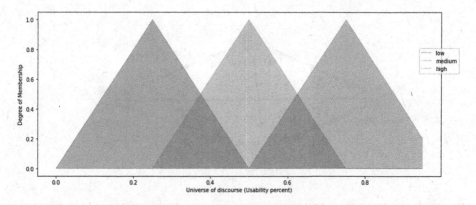

Fig. 2. Fuzzification of the concept "usability" with triangular function. The crisp value of X belongs to both the "Medium" and "High" sets, with different degrees of membership.

Another perspective on FCMs is from the standpoint of Neural Networks and Fuzzy Logic. Therefore, techniques and learning algorithms can be borrowed and applied to train FCMs and adjust the weights of concept interconnections.

To ensure the effective operation of the system, FCMs incorporate knowledge and experience accumulated from experts who understand how the system operates under various conditions. Before extracting knowledge from experts, feedback is collected from stakeholders to determine which concepts should be integrated. Especially in the formulation of decision-making policies, this knowledge extraction involves transforming all linguistic variables into numerical values through a defuzzification process. This results in a set of concepts denoted as $C_i (i = 1, 2, ..., n)$ (graph nodes) with their interrelations denoted as w_i (graph directed edges). After defuzzification, concepts are assigned a value within the range $[0, 1]$, and weights are assigned values in the range $[-1, 1]$ to capture negative and positive causality. A positive value of the weight w_{ij} indicates that an increase (decrease) in the value of concept C_i results in an increment (decrement) in the value of concept C_j. Similarly, a negative weight w_{ij} indicates that an increase (decrease) in the value of concept C_i results in a decrement (increment) in the value of concept C_j, while a weight of zero denotes the absence of a relationship between concepts C_i and C_j, respectively. Figure 3 illustrates an FCM.

Considering the connections among the concepts within an FCM, it becomes straightforward to create the corresponding adjacency matrix. Each concept C_i on the graph is assigned a value A_i, representing the magnitude of its associated physical value obtained through the previously described defuzzification process. During each simulation step, the value A_i of C_i is calculated, essentially signifying the impact of all other concepts C_j on C_i (inference). The three most widely used inference rules are: Kosko's inference (1), Modified Kosko's inference (2),

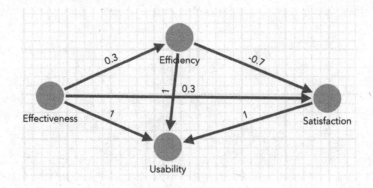

Fig. 3. Simple "Usability" FCM and its adjacency matrix, modeled in a custom web app to generate and modeling FCMs

and Rescale inference (3), which are illustrated by the following three activation functions, respectively.

$$A_i(k+1) = f\left(\sum_{j=1,j\neq i}^{N} w_{ij}A_j(k)\right) \tag{1}$$

$$A_i(k+1) = f\left(A_i(k) + \sum_{j=1,j\neq i}^{N} w_{ij}A_j(k)\right) \tag{2}$$

$$A_i(k+1) = f\left((2A_i(k)-1) + \sum_{j=1,j\neq i}^{N} w_{ij}(2A_j(k)-1)\right) \tag{3}$$

Moreover, f represents the threshold (transformation) function, which can take one of the following forms: (a) bivalent, (b) trivalent, (c) sigmoid, or (d) hyperbolic tangent, as demonstrated in the subsequent four equations, respectively.

$$f_a(x) = \begin{cases} 1 \; if \; x > 0 \\ 0 \; if \; x \leq 0 \end{cases} \qquad f_b(x) = \begin{cases} 1 & if \; x > 0 \\ 0 & if \; x = 0 \\ -1 & if \; x < 0 \end{cases}$$

$$f_c(x) = \frac{1}{1+e^{-\lambda x}} \qquad f_d(x) = \tanh(\lambda x)$$

Here, λ is a positive real number ($\lambda > 0$) that governs the steepness of the continuous function f, and x represents the value $A_i(k)$ at the equilibrium point [5]. It is important to highlight that the sigmoid threshold function guarantees that the computed value for each concept falls within the range $[0,1]$. In cases where concept values can be negative and fall within the interval $[-1, 1]$, the hyperbolic tangent function may be employed as an alternative.

Experts leverage their knowledge in the relevant field to construct an FCM through a structured process. Initially, they identify the primary concepts involved and establish causal relationships among them. The next step involves quantifying the strength of these causal connections, either using precise numeric values within the $[-1, 1]$ range or linguistic variables that are later transformed into numeric values through defuzzification.

Additionally, experts can enhance an existing FCM by collectively analyzing the essential system characteristics and reassessing the graph's structure and connections using fuzzy conditional statements or fuzzy rules. The algorithm utilized for developing an FCM is outlined below, [1].

4   Proposed Procedure

This part describes the methodology used to represent expert knowledge from data acquisition to constructing the corresponding fuzzy cognitive map. The main idea of the proposal is to run a workshop with design managers and stakeholders, where we actively engage them in knowledge elicitation sessions to tap into their extensive expertise.

- **Step 1:** Experts select the concepts C_i that constitute the FCM graph.
- **Step 2:** Each expert defines the causal relationship between any two concepts, specifying whether it is positive, negative, or neutral.
- **Step 3:** Experts meticulously determine the strength of the relationship between the two concepts.
- **Step 4:** Initially, experts describe the causal influence using linguistic variables such as "low", "medium", "high", and so on. The sign of each weight (+ or −) signifies the nature of the influence between concepts. There are three types of interconnections between two concepts, C_i and C_j:
  - $w_{ij} > 0$ indicates that a change in concept C_i leads to the same directional change in concept C_j.
  - $w_{ij} < 0$ implies that a change in concept C_i results in the opposite directional change in concept C_j.
  - $w_{ij} = 0$ signifies that there is no discernible relationship between concepts C_i and C_j.

The extent of influence between these two concepts is indicated by the absolute value of w_{ij}. During the simulation, the value of each concept is calculated based on the Kosko's modified rule showed in Eq. 2 and tanh function as threshold function as described in Sect. 3.2.

Table 1. Concepts list defined by Design Managers

Concept	Concept Name	Organization Layer
C_1	Visual Design	Design decision
C_2	User Interface	Design decision
C_3	Architecture information	Design decision
C_4	Completion Task Time	Design metric
C_5	Success Rate	Design metric
C_6	Usage Frequency	Design metric
C_7	Intuitive Experience	Design metric
C_8	Satisfaction	Design metric
C_9	Monthly deferred purchases with no interest	Business KPI
C_{10}	Global Invoiced Balance	Business KPI
C_{11}	Global Application Usage Percent	Product lifecycle
C_{12}	Recommendation	Product lifecycle
C_{13}	Retention	Product lifecycle

5   Proof of Concept and Results

This proof of concept (POC) is focused on evaluating the workflow of a specific operation known as "Digital CVV on-off" feature, which is a digital button that allows you to view an automatically generated and short-lived CVV for online purchases. The construction of this FCM allowed for a detailed analysis and representation of the cause-and-effect relationships involved in this critical process for the company. Through this proof of concept, the aim was to demonstrate the effectiveness and utility of fuzzy logic and FCM in assessing and enhancing the Digital CVV on-off within the financial sector.

In order to conduct this POC, our team defined specific scenarios relevant to our study. These scenarios were carefully selected to represent a range of potential situations that could influence the behavior of the system under investigation. By varying key parameters and inputs within the FCM model, we aimed to explore how the system responds to different conditions and perturbations.

For the first step, design managers (experts) defined the following concepts C_i that constitute the FCM graph, as showed in Table 1.

For the causal relationship between concepts, they specified whether it is positive, negative, or neutral and determine the strength of the relationship between the concepts, describing the causal influence using linguistic variables such as "Negative very high", "Negative high", "Negative medium", "Negative low", "Negative very low", "Non Relationship", "Very low", "Low", "Medium", "High", "Very high". Combining concepts and their relationships, the FCM appears as shown in Fig. 5[1].

---

[1] Data used to generate the presented FCM can be provided upon request.

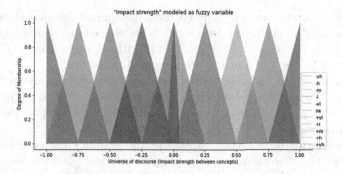

Fig. 4. Impact Strength modeled as fuzzy variable to construct FCM

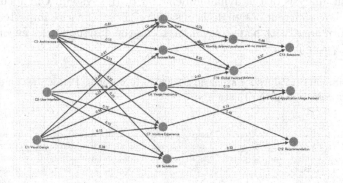

Fig. 5. The FCM taking only a few concepts for each organization layer

Once the problem has been represented using FCMs, it opens the door to conducting scenario analysis. This involves simulating changes or modifications in the concepts involved in the FCM, such as increasing or decreasing the values of some of the concepts C_i. This approach to scenario analysis allows us to explore how different adjustments or perturbations in the concepts can impact the overall behavior of the modeled system, which is crucial for understanding its dynamics and making informed decisions (Fig. 4).

5.1 Experimental Setup

To carry out the simulations, we employed Kosko's modified inference rule, a well-established method in fuzzy logic and $f(x) = \tanh(\lambda x)$ as activation function with $\lambda = 1$. This choice of inference rule allowed us to analyze and interpret the outcomes of the scenarios and gain insights into the dynamics of the FCM over this first approach. The combination of expert-driven scenario definition and the utilization of Kosko's modified inference rule provided valuable results that contribute significantly to our understanding of the system's behavior and its potential implications for real-world applications. For the simulations in this POC, two primary scenarios were formulated:

$S1$: Defined as the case where designers make optimal design decisions, often referred to as "Perfect Design Decisions". This scenario is represented by the state vector $V_{s_1} = \{C_1 : 1, C_2 : 1, C_3 : 1\}$, with all other concepts set to 0. Simulation values for this scenario are presented in Table 2.

$S2$: Defined as the case where designers make non-optimal design decisions, often termed "Worst Design Decisions". This scenario is represented by the state vector $V_{s_2} = \{C_1 : -1, C_2 : -1, C_3 : -1\}$, with all other concepts also set to 0. Simulation values for this scenario are presented in Table 3.

These scenarios were chosen to explore the system's response to extreme design decisions, allowing us to gain insights into the potential consequences and dynamics of such choices within the context of our study.

During the simulation and evolution of scenarios $S1$ and $S2$, we conducted an in-depth analysis of the dynamic behavior of the system. Our focus was on examining how different design decisions within these scenarios impacted the system's performance, with a specific emphasis on the design metrics encompassing concepts C_4 to C_8.

Table 2. Simulation values for scenario $S1$ which represents "Perfect design decisions", FCM values converge in the 15 state

iteration	C_1	C_2	C_3	C_4	C_5	C_6	C_7	C_8	C_9	C_{10}	C_{11}	C_{12}	C_{13}
0	1.0	1.0	1.000000	0.000000	0.000000	0.000000	0.000000	0.000000	0.000000	0.000000	0.0	0.000000	0.000000
1	1.0	1.0	0.883171	−0.978879	0.641077	0.000000	0.000000	0.000000	0.000000	0.000000	0.0	0.641077	0.485381
2	1.0	1.0	0.849859	−0.996365	0.879647	0.757467	0.631524	0.000000	0.145402	0.477138	0.0	0.879647	0.756726
3	1.0	1.0	0.838867	−0.996296	0.922435	0.952961	0.882898	0.525128	0.343566	0.830471	0.0	0.922435	0.848898
4	1.0	1.0	0.835082	−0.996229	0.928220	0.969077	0.927484	0.862113	0.518173	0.918233	0.0	0.928220	0.872113
5	1.0	1.0	0.833760	−0.996205	0.928897	0.970186	0.933458	0.931003	0.636469	0.931564	0.0	0.928897	0.877359
6	1.0	1.0	0.833295	−0.996197	0.928948	0.970267	0.934220	0.939998	0.702025	0.933375	0.0	0.928948	0.878492
7	1.0	1.0	0.833132	−0.996194	0.928941	0.970273	0.934316	0.941081	0.733837	0.933613	0.0	0.928941	0.878726
8	1.0	1.0	0.833075	−0.996193	0.928935	0.970273	0.934328	0.941211	0.748192	0.933643	0.0	0.928935	0.878771
9	1.0	1.0	0.833055	−0.996192	0.928932	0.970273	0.934330	0.941226	0.754445	0.933646	0.0	0.928932	0.878778
10	1.0	1.0	0.833047	−0.996192	0.928931	0.970273	0.934330	0.941228	0.757127	0.933646	0.0	0.928931	0.878779
11	1.0	1.0	0.833045	−0.996192	0.928931	0.970273	0.934330	0.941228	0.758269	0.933646	0.0	0.928931	0.878779
12	1.0	1.0	0.833044	−0.996192	0.928930	0.970273	0.934330	0.941228	0.758754	0.933646	0.0	0.928930	0.878779
13	1.0	1.0	0.833044	−0.996192	0.928930	0.970273	0.934330	0.941228	0.758959	0.933646	0.0	0.928930	0.878778
14	1.0	1.0	0.833044	−0.996192	0.928930	0.970273	0.934330	0.941228	0.759046	0.933646	0.0	0.928930	0.878778

Table 3. Simulation values for scenario $S2$ which represents "Worst design decisions", FCM values converge in the 15 state

iteration	C_1	C_2	C_3	C_4	C_5	C_6	C_7	C_8	C_9	C_{10}	C_{11}	C_{12}	C_{13}
0	−1.0	−1.0	−1.000000	0.000000	0.000000	0.000000	0.000000	0.000000	0.000000	0.000000	0.0	0.000000	0.000000
1	−1.0	−1.0	−0.883171	0.978879	−0.641077	0.000000	0.000000	0.000000	0.000000	0.000000	0.0	−0.641077	−0.485381
2	−1.0	−1.0	−0.849859	0.996365	−0.879647	−0.757467	−0.631524	0.000000	−0.145402	−0.477138	0.0	−0.879647	−0.756726
3	−1.0	−1.0	−0.838867	0.996296	−0.922435	−0.952961	−0.882898	−0.525128	−0.343566	−0.830471	0.0	−0.922435	−0.848898
4	−1.0	−1.0	−0.835082	0.996229	−0.928220	−0.969077	−0.927484	−0.862113	−0.518173	−0.918233	0.0	−0.928220	−0.872113
5	−1.0	−1.0	−0.833760	0.996205	−0.928897	−0.970186	−0.933458	−0.931003	−0.636469	−0.931564	0.0	−0.928897	−0.877359
6	−1.0	−1.0	−0.833295	0.996197	−0.928948	−0.970267	−0.934220	−0.939998	−0.702025	−0.933375	0.0	−0.928948	−0.878492
7	−1.0	−1.0	−0.833132	0.996194	−0.928941	−0.970273	−0.934316	−0.941081	−0.733837	−0.933613	0.0	−0.928941	−0.878726
8	−1.0	−1.0	−0.833075	0.996193	−0.928935	−0.970273	−0.934328	−0.941211	−0.748192	−0.933643	0.0	−0.928935	−0.878771
9	−1.0	−1.0	−0.833055	0.996192	−0.928932	−0.970273	−0.934330	−0.941226	−0.754445	−0.933646	0.0	−0.928932	−0.878778
10	−1.0	−1.0	−0.833047	0.996192	−0.928931	−0.970273	−0.934330	−0.941228	−0.757127	−0.933646	0.0	−0.928931	−0.878779
11	−1.0	−1.0	−0.833045	0.996192	−0.928931	−0.970273	−0.934330	−0.941228	−0.758269	−0.933646	0.0	−0.928931	−0.878779
12	−1.0	−1.0	−0.833044	0.996192	−0.928930	−0.970273	−0.934330	−0.941228	−0.758754	−0.933646	0.0	−0.928930	−0.878779
13	−1.0	−1.0	−0.833044	0.996192	−0.928930	−0.970273	−0.934330	−0.941228	−0.758959	−0.933646	0.0	−0.928930	−0.878778
14	−1.0	−1.0	−0.833044	0.996192	−0.928930	−0.970273	−0.934330	−0.941228	−0.759046	−0.933646	0.0	−0.928930	−0.878778

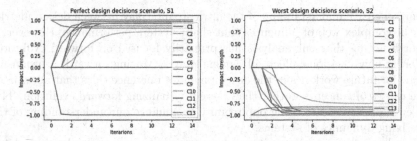

Fig. 6. Evolution of concepts for the FCM on $S1$ and $S2$, perfect and worst design decisions.

5.2 Discussion

Considering the particular FCM in Fig. 5 one of the key findings from our analysis pertains to the concept "Completion Task Time" C_4, which exhibited a particularly intriguing behavior showed in Fig. 6. In scenario S1, characterized by optimal design decisions, we observed a rapid and substantial decrease in the value of C_4 over time. This decline in task completeness time may be attributed to designers prioritizing other aspects of the design process over task completion, leading to a decrease in the overall completeness of tasks.

Conversely, in scenario S2, characterized by suboptimal design decisions, we observed a contrasting trend (see Fig. 6). Here, the value of C_4 displayed a steady increase, indicating that poor design choices may inadvertently lead to an emphasis on task completion within the system. This finding suggests that under suboptimal design conditions, the system may inherently prioritize task completeness as a compensatory mechanism for potential design deficiencies.

These insights provide valuable implications for decision-makers in the field of design, highlighting the intricate interplay between design decisions and system behavior. It underscores the importance of a holistic approach to design that considers not only individual design metrics but also the dynamic relationships between them.

6 Conclusions and Future Work

This study has explored the application of Fuzzy Cognitive Maps (FCMs) as a powerful tool for modeling and analyzing complex systems, with a specific focus on the impact of design decisions within the context of a financial services company. Our research has demonstrated the effectiveness of FCMs in capturing the cause-and-effect relationships among various design-related concepts, shedding light on how different decisions can influence system behavior.

One of the key takeaways from this study is the importance of considering the dynamic nature of these relationships. By simulating extreme scenarios (S1 and S2), we have uncovered nuanced insights into how design decisions can lead to unexpected shifts in system behavior. These findings underscore the need for a comprehensive understanding of the intricate interplay between design metrics and the complex web of influences that shape system performance. However, it is worth noting that our analysis has primarily focused on forward evolution, exploring how decisions affect subsequent states. Moving forward, a promising avenue for future work lies in the development of inference rules that allow us to trace causes of concepts, rather than merely visualizing forward evolution. This challenge presents an exciting opportunity to enhance our understanding of the underlying dynamics of systems.

To effectively trace causes, it may be necessary to explore the development of backward inference rules in future work. These rules would enable us to identify the factors or decisions that led to a particular state or outcome, offering a more comprehensive perspective on the system's behavior.

Acknowledgements. We extend our gratitude to three exceptional coworkers whose invaluable contributions greatly enhanced our research project. Ilse Rosas's creativity, Nora Tejeda's expertise, and Marlen Osorios's commitment significantly enriched our work.

References

1. Borrero-Domínguez, C., Escobar-Rodríguez, T.: Decision support systems in crowd-funding: a fuzzy cognitive maps (FCM) approach. Decis. Support Syst. **173**, 114000 (2023). https://doi.org/10.1016/j.dss.2023.114000, https://www.sciencedirect.com/science/article/pii/S0167923623000751
2. Dickerson, J., Kosko, B.: Virtual worlds as fuzzy cognitive maps. Presence **3**, 173–189 (1994). https://doi.org/10.1162/pres.1994.3.2.173
3. Kosko, B.: Fuzzy cognitive maps. Int. J. Man-Mach. Stud. **24**, 65–75 (1986)
4. Legarda, I., Iriarte, I., Hoveskog, M., Justel-Lozano, D.: A model for measuring and managing the impact of design on the organization: insights from four companies. Sustainability **13**(22) (2021). https://doi.org/10.3390/su132212580, https://www.mdpi.com/2071-1050/13/22/12580
5. Nápoles, G., Grau, I., Concepción, L., Koutsoviti Koumeri, L., Papa, J.P.: Modeling implicit bias with fuzzy cognitive maps. Neurocomputing **481**, 33–45 (2022). https://doi.org/10.1016/j.neucom.2022.01.070, https://www.sciencedirect.com/science/article/pii/S092523122200090X
6. Osoba, O., Kosko, B.: Beyond DAGs: modeling causal feedback with fuzzy cognitive maps (2019)
7. Scanagatta, M., Salmerón, A., Stella, F.: A survey on Bayesian network structure learning from data. Progr. Artif. Intell. **8**(4), 425–439 (2019). https://doi.org/10.1007/s13748-019-00194-y
8. Silva, A., Wood, K., Venkatesh, S.: Measuring design impact across disciplines, industries and scale. Ph.D. thesis, Singapore University of Technology and Design (2021). https://doi.org/10.13140/RG.2.2.19484.41609
9. Zadeh, L.: Fuzzy sets. Inf. Control (1965)

Neural Architecture Search for Placenta Segmentation in 2D Ultrasound Images

José Antonio Fuentes-Tomás[1]([✉]), Héctor Gabriel Acosta-Mesa[1],
Efrén Mezura-Montes[1], and Rodolfo Hernandez Jiménez[2]

[1] Artificial Intelligence Research Institute, University of Veracruz,
91097 Xalapa, Veracruz, Mexico
josefuentes.at@gmail.com
[2] Private Obstetrician and Gynecologist, Xalapa, Mexico

Abstract. Monitoring the placenta during pregnancy can lead to early diagnosis of anomalies by observing their characteristics, such as size, shape, and location. Ultrasound is a popular medical imaging technique used in placenta monitoring, whose advantages include the non-invasive feature, price, and accessibility. However, images from this domain are characterized by their noise. A segmentation system is required to recognize placenta features. U-Net architecture is a convolutional neural network that has become popular in the literature for medical image segmentation tasks. However, this type is a general-purpose network that requires great expertise to design and may only be applicable in some domains. The evolutionary computation overcomes this limitation, leading to the automatic design of convolutional neural networks. This work proposes a U-Net-based neural architecture search algorithm to construct convolutional neural networks applied in the placenta segmentation on 2D ultrasound images. The results show that the proposed algorithm allows a decrease in the number of parameters of U-Net, ranging from 80 to 98%. Moreover, the segmentation performance achieves a competitive level compared to U-Net, with a difference of 0.012 units in the Dice index.

Keywords: Neural Architecture Search · Genetic Programming · U-Net · Placenta Segmentation

1 Introduction

Monitoring the state of the placenta during pregnancy is of utmost importance for the early detection of pathologies that may endanger the fetus's or the mother's health. Characteristics such as size, shape, location, and degree of calcification may indicate abnormalities in the placenta. Granumm et al. [7] propose a maturity scale of four grades, from 0 to 3, based on the morphology and texture of the placenta. A premature calcification and risk situation is attributed if a grade 3 placenta is present before 32–36 weeks of gestation [16], which can lead to complications such as preeclampsia, low fetal weight, intrauterine growth retardation.

H. Calvo et al. (Eds.): MICAI 2023 Workshops, LNAI 14502, pp. 397–408, 2024.
https://doi.org/10.1007/978-3-031-51940-6_30

One method to evaluate the placenta's state is ultrasound, whose accessibility, cost, and non-invasive characteristics stand out compared to other medical imaging domains such as computed tomography or X-rays. However, its interpretation depends on the expert where the images present different sound artifacts.

Convolutional neural networks (CNNs) have been applied in medical imaging, obtaining a good performance in visual recognition tasks and allowing specialists to assess different abnormalities for better intervention. One of the limitations of the application of CNNs is the network topology design. This task requires extensive knowledge whose impact on the model performance is more significant than optimizing synaptic weights in isolation [21]. U-Net [18] is a CNN that has shown good performance for medical image segmentation. U-Net is an example of a manually designed neural network with a fixed, general-purpose architecture. To overcome designing limitations, techniques based on evolutionary computation have emerged for the design and/or optimization of neural networks, called Neuroevolution (NE). A continuous cost function is usually employed to optimize network synaptic weights. On the other hand, network architecture search (NAS) is a complicated problem-dependent task. Genetic programming (GP) [14] is an evolutionary computation technique that uses a tree representation of functions applied to a set of terminals. The GP's representation encourages diversity in the search space, expressiveness, and easy decoding of the solutions.

This paper proposes a search algorithm to build U-Net-type networks whose architecture is exclusively for the application domain, particularly 2D ultrasound images. For this purpose, the evolutionary paradigm of GP is employed to automatically design a network architecture capable of segmenting placentas in 2D ultrasound images, with a competitive level in terms of Dice similarity coefficient and compact complexity, to obtain a lower computational cost and a shorter prediction time.

The rest of the text is as follows. The next section provides related work about NAS and placenta segmentation. Section three describes the materials and methods employed, including the dataset, the proposed algorithm, and the evaluation metrics. Section four shows the experiments and results. Finally, in section five, we provide some conclusions.

2   Related Work

Recently U-Net has been the trend in medical image segmentation tasks, specifically in placenta segmentation [8,12,22]. U-Net consists of a contracting path and an expanding symmetric path with skip connections. Hu et al. [12] use it for placenta segmentation on 2D ultrasound images with the help of a preprocessing algorithm to capture acoustic shadows. To reduce the trainable parameters of U-Net, Han et al. [8] proposed the use of depth-wise separable convolution [10] and a reduction in the kernels' size. Under the ultrasound images domain, Zimmer et al. [22] use a multitask approach to obtain a more extensive training dataset considering the variability of the data. We remark that U-Net commonly is employed as the backbone NAS methods [5,15] have been used.

3   Materials and Methods

The steps involved in developing our research include a constant collaboration with a medical expert, gynecologist, and obstetrician, who performs data acquisition, knowledge acquisition, and evaluation of results. The materials and methods are described in this section.

3.1   Data and Knowledge Acquisition

A dataset of one hundred images was collected gradually throughout the project, corresponding to 28 pregnant women from March 4 to June 17, 2020. The ages of the patients ranged from 18 to 42 years old. The images were acquired with a Samsung Accuvix 2016 machine. The size of the images is 768×1024 pixels and resized to 288×480 for experimental purposes. The dataset has been divided into 70% for training, 15% for testing, and 15% for validation. Software has been implemented in MATLAB for the expert to segment the placenta in each image manually.

3.2   Neural Architecture Search Algorithm

The proposed algorithm has been named **NASGP-Net**, which refers to the neuroevolutionary approach (NAS) using Genetic Programming (GP) as the evolutionary computation paradigm and based on the U-Net (Net) architecture backbone. NASGP-Net uses a cell-based coding that evolves a fragment of the network so that it is possible to build the complete network following the symmetric shape of U-Net by doubling the number of channels of each layer convolution as the profound increases in the decoder path and replacing the pooling operation by transposed convolution in the decoder path, where the number of channels is halved and the spatial dimensions doubled. Therefore, a crucial previous step in the fitness evaluation is the model construction based on the evolved cell, as is shown in Fig. 1.

Inspired by [2,3] and the properties of Strongly Type Genetic Programming (STGP) to get type consistency in the function set, our proposal employs a multi-layer operation structure that includes feature extraction, connectivity, recalibration, feature construction, and pooling operations, as shown in Fig. 2. The rectangles with dotted lines in the left subfigure indicate flexible layers, indicating that operations from these layers may not evolve.

The NASGP-Net algorithm is shown in Algorithm 1. A initial population of size T with a bounded height determined by the parameters h_{min} and h_{max} is evolved. Similarly, the h_{off} is a bloat control static limit randomly selecting an individual's parent if the offspring exceeds the limit. Subsequently, the model corresponding to each tree is built in the evaluation stage, and fitness is computed. Then, the population is updated using the crossover and mutation operators. Then, the new population is evaluated. The process continues iteratively until a maximum number of generations, G_{max}. Finally, the best individual is trained during epochs and returned.

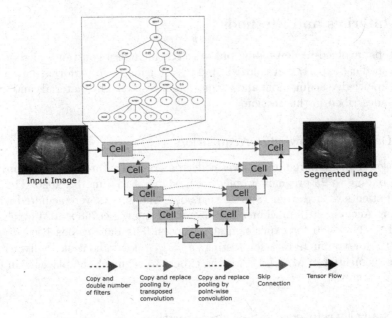

Fig. 1. The NASGP-Net employs the U-Net backbone to build the model from the evolved syntax tree. Dotted lines represent the construction heuristics and continued lines show the flow of the feature maps.

The fitness function depends on the number of trainable parameters of the generated model and their segmentation performance based on the Dice similarity coefficient (DSC). Let $\hat{Y}(\theta)$ and Y corresponding to the predicted segmentation mask and the ground-truth mask, respectively, DSC is calculated by Eq. 1, where θ represents the trainable model parameters and $|.|$ denotes the set cardinality.

$$DSC = \frac{2|Y \cap \hat{Y}(\theta)|}{|Y| + |\hat{Y}(\theta)|} \tag{1}$$

To measure the models' complexity, we use a complexity fact (CF) expressed Eq. 2 described as the difference between the trainable parameters' number θ and the maximum allowed parameter's number, θ_{max}, which is a user-defined parameter.

$$CF = \frac{|\theta|_{max} - |\theta|}{|\theta|_{max}} \tag{2}$$

Therefore, the fitness function is a linear combination between the average DSC on the test set, DSC_{mean} and CF, as expressed in Eq. 3 where w indicates the measure's contribution.

$$fitness = (1 - w) \cdot DSC_{mean} + w \cdot CF \tag{3}$$

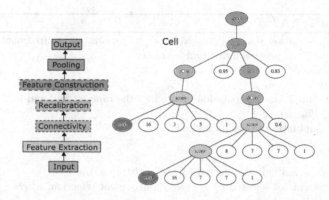

Fig. 2. The NASGP-Net structure and an example of a tree generated

A crucial step in designing a GP system is the choice of the function set and terminal set. Based on *state-of-art*, we have selected operations to construct a convolutional neural network, described in Table 1.

We use the *depth-wise-separable* convolution [10] (*sconv*) that can reduce the trainable parameters. The size of the applied convolution filters is not necessarily square. A group normalization (GN) layer is added for each convolution layer, followed by a ReLU activation layer. Also, residual [9] (*rCon*) and dense blocks [13] (*dCon*) can be constructed by applying the pattern connections in a sequence of convolution layers. The residual and dense type connections are considered since they have performed better than their sequential versions. The squeeze-and-excitation operation (*se*) [11] is an attention mechanism that can suppress or enhance the channel information. Since the *se* operation requires an increase in the trainable parameters of the model, it is limited to be applied only once for each sequence of convolution layers, dense block or residual block. Similarly to [19], padding channels and interpolation functions are used to allow the element-wise addition, subtraction, and concatenation operations with *add*, *sub*, and *cat* functions, respectively.

Table 2 describes the terminal set and their value range for the considered functions. The terminal *mod* is a container that concatenates the convolution layers created with the operations in the feature extraction tier. The inputs can be considered as tensors or feature maps once the model has been constructed. The parameters n_1 and n_2 are random numbers that weigh the functions in the feature construction (*add* and *sub*). The parameter θ corresponds to the compression factor used in the *dCon* function, which reduces the number of channels of the dense block output. The terminals of the extraction operations (*conv* and *sconv*) include the number of filters, ($n_{filters}$), kernel size, (k_{size}), and dilation rate (d_{rate}). The number of filters has been limited to 32 to ensure fewer filters per convolution layer than U-Net.

The implementation of the algorithm is based on Pytorch [17], Albumentations [4], and DEAP [6] libraries. Experiments were performed using four Graphic Processing Units card Volta V100. Moreover, for faster and safer execution, we

---

Algorithm 1: NASGP-Net

Data: T: population size, G_{max}:maximum generations, t_{sz}: tournament size,
n_{elit}:elitist process individuals' number

Result: ind_{best} :best individual

Cache $\leftarrow \emptyset$

$P_0 \leftarrow$ a random T-sized population created by the ramped half-and-half method
using $h \in [h_{min}, h_{max}]$

$ind_{best} \leftarrow$ population's best individual

$g \leftarrow 0$

repeat

 $S_e \leftarrow n_{elit}$ individuals selected by the elitist operator

 $S_t \leftarrow$ parent set selected by t_{sz}-sized tournament selection, where
 $|S_t| = T - |S_e|$

 $Q_{g+1} \leftarrow$ offspring generated and $h \leq h_{off}$

 Evaluate individuals from Q_{g+1} using Eq. 3

 $P_{g+1} \leftarrow Q_{g+1} \cup S_e$

 Update ind_{best} with the population's best individual.

 $g \leftarrow g + 1$

until G_{max}

Train ind_{best} for 100 epochs

return ind_{best}

---

use a checkpoint mechanism, multiprocessing evaluations, and a cache memory to avoid individual evaluations founded in previous generations.

3.3 Parameters Setting

The parameter setting is shown in Table 3. We are using a population size, $T = 100$. The election of $|\theta|_{max}$ is based on the number of parameters of U-Net. The number of epochs to train each model is limited to ten to save computational cost, $n_{epochs} = 10$. The combo loss function [20] is used to train each model, which consists of a trade-off between the Dice index and the cross-entropy controlled by a parameter, α, and false positives and false negatives trade-off controlled by a parameter β. The parameter values are t$\alpha = 0.5$ and $\beta = 0.4$ to penalize more the false positives than false negatives. Based on gradual experiments, the weight that controls the contribution of the complexity models in the fitness function is set to minimal values, $w = 0.001$, allowing considerable size models but penalizing models with a major complexity than U-Net. The probabilities for variation operators (p_{cross}, p_{mut}) and the number of generations G_{max} are based on the commonly used in GP systems [1–3].

3.4 Evaluation Metrics

In addition to using the Dice index to guide the search (see Eq. 1), we have used it to measure our algorithm's segmentation performance. Moreover, the

Table 1. The function set

Layer	Function	Input	Output	Description
Pooling	mpool	$Module$	$Module$	Peforms average pooling with a $k_{size} = 2$ and $stride = 2$
	apool	$Module$	$Module$	Performs average pooling with a $k_{size} = 2$ and $stride = 2$
Feature construction	add	$Module_1$, $module_2$, ω_1, ω_2	$Module$	Performs a weighted addition to the output of two models
	sub	$Module_1$, $module_2$, ω_1, ω_2	$Module$	Performs a weighted subtraction to the output of two models
	cat	$Module_1$, $module_2$	$Module$	Concatenates the output of two modules in the channel dimension
Recalibration	se	$Module$	$Module$	Apply *squeeze-and-excitation* operation with a factor reduction, $r = 8$
Connectivity	dCon	$Module$	$Module$	Apply a *dense connection* to a sequence of convolutional layers. The number of channels of output is compressed by a factor $\theta \in [0, 1)$
	rCon	$Module$	$Module$	Apply a *residual connection* to a sequence of convolutional layers every two layers. The first layer is not connected if the sequence contains an odd number of layers. If the sequence contains a unique layer, the residual connection is applied
Feature extraction	conv	$Module$, $n_{filters}, k_{size}$, d_{rate}	$Module$	Generate a convolution-GN-ReLU operation
	sconv	$Module$, $n_{filters}, k_{size}$, d_{rate}	$Module$	Generate a depth-wise separable convolution-GN-ReLU

Intersection over the Union (IoU) and Hausdorff distance (HD) are considered for comparison purposes.

Considering each image as a set of pixels, where one corresponds to the mask segmented by an expert, Y, and the other to the mask predicted by the model, $\hat{Y} \cap Y(\theta)$. IoU is the intersection between both sets divided by their union. Its values lie in the interval $[0, 1]$, where a higher value indicates better performance, as expressed in Eq. 4.

$$IoU = \frac{|\hat{Y} \cap Y(\theta)|}{|Y \cup \hat{Y}(\theta)|}. \tag{4}$$

Table 2. The terminal set

Layer	Terminal	Range/Type	Description
Feature construction	n_1, n_2	$[0.00, 1.00)$	Random values for *add* and *sub* functions
Connectivity	θ	$[0.30, 0.80)$	Compression rate for *dCon* function
Feature extraction	$n_{filters}$	$\{8, 16, 32\}$	Number of filters applied to *conv* and *sconv*
	k_{size}	$\{3, 5, 7\}$	Height and width of convolutional kernel used in *conv* and *sconv* functions, represented as a 2-tuple
	d_{rate}	$\{1, 2\}$	Dilation rate that adds spaces between the elements of the convolutional kernel. For the dilation rate equal to 1, it is a normal convolution
Input	*mod*	List	A container that storage the convolution operations sequence used for the model construction

Table 3. The NASGP-Net parameters

Parameter	Value	Parameter	Value
T	100	h_{off}	10
G_{max}	20	n_{epochs}	10
t_{size}	3	p_{max}	31,038,000
n_{elit}	1	w	0.01
p_{cross}	0.8	α	0.5
p_{mut}	0.19	β	0.4
h_{min}, h_{max}	[2, 6]		

Unlike metrics such as DSC or IoU, based on image overlap, this shape comparison-based metric is a longitudinal measure of the discrepancy between two sets. A lower value represents better performance. It determines the distance of each mask point obtained by a model from the reference mask and vice versa. Formally:

$$HD = max(h(Y, \hat{Y}(\theta)), h(\hat{Y}(\theta))) \tag{5}$$

where $h(Y, \hat{Y}(\theta) = max_{y \in Y}\{min_{\hat{y}(\theta) \in \hat{Y}(\theta)}\{\|y - y(\theta)\|\}\}$ and $\| \cdot \|$ is the Euclidean distance.

The mean value of each metric with the test data set takes the performance. On the other hand, the complexity of the model is given by the number of trainable parameters of the model.

4   Experiments and Results

In this section, we provide the results of the experiments and the comparison with the U-Net method in our Placenta dataset. We executed ten runs using the NASGP-Net and U-Net algorithms to validate the competitiveness of NASGP-Net. Also, the number of parameters were compared.

A statistical analysis was performed using the non parametric Mann-Whitney U test with a 95% confidence level to establish significant differences in the DSC, IoU, and HD metrics by taking the average achieved in the validation image set. The statistics are shown in Table 4, with the best results in bold. According to the statistics and the p-value, the superiority of U-Net over NASGP-Net is evident. However, we can attribute a higher accuracy to NASGP-Net when comparing the standard deviation concerning DSC. The p-value obtained when comparing the HD metric shows us that there are no significant differences.

Regarding model complexity, NASGP-Net achieves a 92–98% reduction over U-Net's 31×10^6 parameters, denoted as $|\theta|$, where the NAsGP-Net models' parameters reached values between 0.37×10^6 to 2×10^6.

The solutions' quality improvement of NASGP-Net during generations is shown in the convergence graph in Fig. 3 corresponding to the median DSC

Table 4. The statistic regarding ten runs on the placentas dataset. The Mann-Whitney U test is performed, considering the Dice, IoU, and HD metrics. The best results between the two methods are bolded

	Segmentation metrics					
	Dice		IoU		HD	
	U-Net	NASGP-Net	U-Net	NASGP-Net	U-Net	NASGP-Net
Best	**0.943**	0.929	**0.891**	0.871	**8.353**	8.354
Worst	**0.927**	0.917	**0.866**	0.852	**8.876**	9.007
Mean	**0.934**	0.922	**0.879**	0.861	**8.656**	8.727
Median	**0.934**	0.921	**0.880**	0.859	**8.691**	8.731
Std.	0.0046	**0.0037**	**0.0066**	0.0068	**0.1792**	0.1985
p-value (5%)	3.10E−04		4.25E−04		0.4727	

execution. Figure 4 shows the evolved cell from the whole model built. Using the *sconv* function indicates the bias to compact models, reducing the number of trainable parameters. Furthermore, the application of residual connection and the weighted subtraction operation shows the potential of the proposed structural primitive set to reach models with good segmentation performance and lower complexity than fixed architectures, such as U-Net.

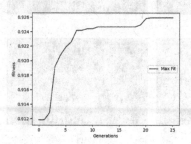

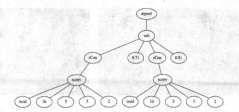

Fig. 3. Convergence Graph

Fig. 4. Tree representing the evolved cell to construct the model corresponding to the median according to the Table 4

Figures 5 and 6 show samples of the predicted segmentation mask obtained with U-Net and NASGP-Net in scenarios with good and poorly segmentation results based on the Dice Index.

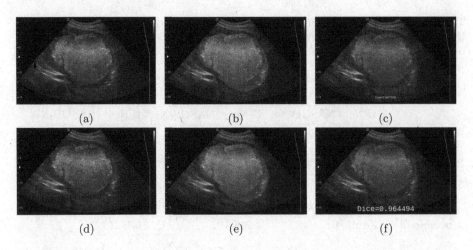

Fig. 5. Best segmentation images obtained with NASGP-Net (((a)–(c)) and U-Net ((d)–(f)). The first column indicates the input image, the middle column the segmentation mask, and the third column represents the predicted mask

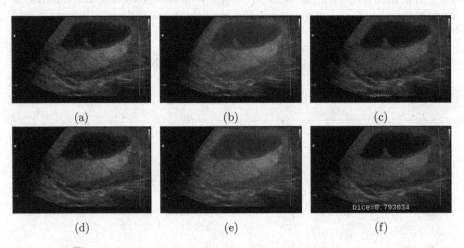

Fig. 6. Worst segmentation images obtained with NASGP-Net (((a)–(c)) and U-Net ((d)–(f)). The first column indicates the input image, the middle column the segmentation mask, and the third column represents the predicted mask

5 Conclusions

The results make clear the potential of tree-based PG to perform neuroevolution. The features offered by STGP allow us to design and organize different functions related to deep learning, such as pooling, convolution, and connectivity operations. In addition, it enables the integration of new functions simply.

Our approach takes *combo loss* as a cost function and uses layers of type *group normalization*, operations specifically designed for the image segmentation task.

On the other hand, *depth-wise-separable-convolution* reduces the computational cost of the convolution operation. In contrast, *dilated-convolution* operation considers a receptive field of a larger area without increasing the complexity of the model. In our proposal, such operations can be applied in a composite form, allowing the application of the *dept-wise-separable-convolution* operation in its dilated version.

The results show that NASGP-Net can obtain competitive segmentation with a more compact model than U-Net, with a difference of 0.012 units in the *DSC* metric. The present study can also establish the basis for future research in placental calcifications.

In future work, we plan to study the generality of the proposed algorithm when evaluated in more databases. Choosing the w value in the fitness function 3 is not trivial. A multi-objective optimization approach would contribute to avoiding trying different values of w, since the computational cost of the algorithm is considerable (between 18–24 h) despite the implemented strategies (*checkpoint, cache memory, multiprocessing*). Similarly, applying techniques such as *surrogated models* can contend against the high computational cost.

Acknowledgments. The first author acknowledges support from the Mexican National Council of Humanities, Science, and Technology (CONAHCyT) through a scholarship to pursue graduate studies at the University of Veracruz. The authors thankfully acknowledge computer resources, technical advice, and support provided by Laboratorio Nacional de Supercómputo del Sureste de México (LNS), a member of the CONAHCYT national laboratories, with project No. 202201016n.

References

1. Bi, Y., Xue, B., Zhang, M.: An automatic feature extraction approach to image classification using genetic programming. In: Sim, K., Kaufmann, P. (eds.) EvoApplications 2018. LNCS, vol. 10784, pp. 421–438. Springer, Cham (2018). https://doi.org/10.1007/978-3-319-77538-8_29
2. Bi, Y., Xue, B., Zhang, M.: An evolutionary deep learning approach using genetic programming with convolution operators for image classification. In: 2019 IEEE Congress on Evolutionary Computation (CEC), pp. 3197–3204. IEEE (2019)
3. Bi, Y., Zhang, M., Xue, B.: An automatic region detection and processing approach in genetic programming for binary image classification. In: 2017 International Conference on Image and Vision Computing New Zealand (IVCNZ), pp. 1–6 (2017). https://doi.org/10.1109/IVCNZ.2017.8402469
4. Buslaev, A., Iglovikov, V.I., Khvedchenya, E., Parinov, A., Druzhinin, M., Kalinin, A.A.: Albumentations: fast and flexible image augmentations. Information **11**, 125 (2020). https://doi.org/10.3390/INFO11020125, https://www.mdpi.com/2078-2489/11/2/125/htm
5. Fan, Z., Wei, J., Zhu, G., Mo, J., Li, W.: Evolutionary neural architecture search for retinal vessel segmentation. arXiv preprint arXiv:2001.06678 (2020)
6. Fortin, F.A., De Rainville, F.M., Gardner, M.A.G., Parizeau, M., Gagné, C.: DEAP: evolutionary algorithms made easy. J. Mach. Learn. Res. **13**(1), 2171–2175 (2012)

7. Grannum, P.A., Berkowitz, R.L., Hobbins, J.C.: The ultrasonic changes in the maturing placenta and their relation to fetal pulmonic maturity. Am. J. Obstet. Gynecol. **133**(8), 915–922 (1979)

8. Han, M., et al.: Automatic segmentation of human placenta images with U-net. IEEE Access **7**, 180083–180092 (2019). https://doi.org/10.1109/ACCESS.2019.2958133

9. He, K., Zhang, X., Ren, S., Sun, J.: Deep residual learning for image recognition (2015)

10. Howard, A.G., et al.: MobileNets: efficient convolutional neural networks for mobile vision applications. arXiv preprint arXiv:1704.04861 (2017)

11. Hu, J., Shen, L., Sun, G.: Squeeze-and-excitation networks. In: Proceedings of the IEEE Conference on Computer Vision and Pattern Recognition, pp. 7132–7141 (2018)

12. Hu, R., Singla, R., Yan, R., Mayer, C., Rohling, R.N.: Automated placenta segmentation with a convolutional neural network weighted by acoustic shadow detection. In: 2019 41st Annual International Conference of the IEEE Engineering in Medicine and Biology Society (EMBC), pp. 6718–6723. IEEE (2019). https://doi.org/10.1109/EMBC.2019.8857448

13. Huang, G., Liu, Z., van der Maaten, L., Weinberger, K.Q.: Densely connected convolutional networks (2018)

14. Koza, J.R., Koza, J.R.: Genetic Programming: On the Programming of Computers by Means of Natural Selection, vol. 1. MIT Press, Cambridge (1992)

15. Liu, H., Simonyan, K., Yang, Y.: DARTS: differentiable architecture search. arXiv preprint arXiv:1806.09055 (2018)

16. McKenna, D., Tharmaratnam, S., Mahsud, S., Dornan, J.: Ultrasonic evidence of placental calcification at 36 weeks' gestation: maternal and fetal outcomes. Acta Obstet. Gynecol. Scand. **84**(1), 7–10 (2005)

17. Paszke, A., et al.: PyTorch: an imperative style, high-performance deep learning library. In: Wallach, H., Larochelle, H., Beygelzimer, A., d' Alché-Buc, F., Fox, E., Garnett, R. (eds.) Advances in Neural Information Processing Systems, vol. 32, pp. 8024–8035. Curran Associates, Inc. (2019). http://papers.neurips.cc/paper/9015-pytorch-an-imperative-style-high-performance-deep-learning-library.pdf

18. Ronneberger, O., Fischer, P., Brox, T.: U-net: convolutional networks for biomedical image segmentation. In: Navab, N., Hornegger, J., Wells, W.M., Frangi, A.F. (eds.) MICCAI 2015. LNCS, vol. 9351, pp. 234–241. Springer, Cham (2015). https://doi.org/10.1007/978-3-319-24574-4_28

19. Suganuma, M., Shirakawa, S., Nagao, T.: A genetic programming approach to designing convolutional neural network architectures. In: Proceedings of the Genetic and Evolutionary Computation Conference, pp. 497–504 (2017)

20. Taghanaki, S.A., et al.: Combo loss: handling input and output imbalance in multi-organ segmentation. Comput. Med. Imaging Graph. **75**, 24–33 (2019). https://doi.org/10.1016/j.compmedimag.2019.04.005

21. Turner, A.J., Miller, J.F.: The importance of topology evolution in NeuroEvolution: a case study using cartesian genetic programming of artificial neural networks. In: Bramer, M., Petridis, M. (eds.) Research and Development in Intelligent Systems XXX, pp. 213–226. Springer, Cham (2013). https://doi.org/10.1007/978-3-319-02621-3_15

22. Zimmer, V.A., et al.: A multi-task approach using positional information for ultrasound placenta segmentation. In: Hu, Y., et al. (eds.) ASMUS/PIPPI -2020. LNCS, vol. 12437, pp. 264–273. Springer, Cham (2020). https://doi.org/10.1007/978-3-030-60334-2_26

Experimental Study of the Instance Sampling Effect on Feature Subset Selection Using Permutational-Based Differential Evolution

Jesús-Arnulfo Barradas-Palmeros[1]([✉]), Rafael Rivera-López[2],
Efrén Mezura-Montes[1], and Héctor-Gabriel Acosta-Mesa[1]

[1] Artificial Intelligence Research Institute, University of Veracruz, Xalapa, Mexico
`zs21000456@estudiantes.uv.mx`, {`emezura,heacosta`}`@uv.mx`
[2] Departamento de Sistemas y Computación, Instituto Tecnológico de Veracruz,
Veracruz, Mexico
`rafael.rl@veracruz.tecnm.mx`

Abstract. Wrapper approaches for feature subset selection are computationally intensive because they require training and evaluation of a machine learning algorithm to assess the goodness of a subset of features. This proposal combines the permutational-based differential evolution for feature selection (DE-FS^{PM}) algorithm as a wrapper approach with three instance sampling strategies: fixed, incremental, and evolving sampling fraction. These sampling schemes are applied to the search process to reduce the instance set used in an individual evaluation, resulting in overall computational time savings. In addition, the DE-FS^{PM} algorithm is modified to have adaptive parameter control using success history-based parameter adaptation for differential evolution (SHADE). The experimental results show that using a reduced number of instances permits a reduction in computational cost with no significant differences in performance. The algorithm's use of adaptive parameter control did not improve its capabilities.

Keywords: Feature Selection · Data Preprocessing · Differential Evolution

1 Introduction

Feature subset selection is a data preprocessing step in machine learning (ML) and data mining. The process aims to select a subset of the available features where redundant, irrelevant, and noisy features are discarded. The feature selection procedure offers data dimensionality reduction, improving algorithm performance, and reduction of computational resources demand [1]. Feature selection is a challenging problem, given its associated search space size. For a dataset containing n features, the search space contains 2^n possible solutions [13].

© The Author(s), under exclusive license to Springer Nature Switzerland AG 2024
H. Calvo et al. (Eds.): MICAI 2023 Workshops, LNAI 14502, pp. 409–421, 2024.
https://doi.org/10.1007/978-3-031-51940-6_31

As seen in [4], the three main approaches to solving the feature selection problem are filter, wrapper, and embedded. The wrapper approach uses an ML algorithm to evaluate different subsets of features. The classifier's accuracy or error rate is commonly used as the performance measure. Wrapper approaches present advantages such as a better accuracy performance of a classifier given its interaction during the search process. Nonetheless, wrappers are prone to overfitting and are computationally intensive since an ML model needs to be trained and evaluated in each performance assessment [1].

In [7], it is proposed that the feature selection process can be performed using only a subset of the instances during the search. The authors show minimal impact on the classification performance of an ML algorithm with the resulting selected features. The proposal in that paper defines an amount of instances, such as 100, 250, 500, 1000, 1500, and 2000, to be selected randomly from the dataset and perform the feature selection process. A significant computational cost reduction is observed since classifiers for the final evaluation are trained faster. Nonetheless, the authors only considered filter methods for feature selection and predefined that the total of selected features is 10.

In this work, the instance reduction scheme from [7] is extended using three different sampling strategies for instance reduction and applying them to the search process of the permutational-based Differential Evolution for feature selection algorithm (DE-FS^{PM}) proposed in [9]. The DE-FS^{PM} algorithm is a well-performing wrapper approach for feature selection that adapts the Differential Evolution algorithm [10] into a permutational search space. The sampling strategies used in the proposal include a fixed, incremental, and evolving sampling fraction. In addition, the success-history based parameter adaptation for differential evolution (SHADE) [11] is also adapted to the DE-FS^{PM} algorithm with the sampling strategies. To the best of the authors' knowledge, no other works in the state-of-the-art attempt to reduce the computational cost of the DE-FS^{PM} algorithm or adapt the SHADE algorithm to a permutational space.

The rest of this paper is organized into four additional sections. Section 2 presents the differential evolution algorithm, its adaptation to permutational search space, and how a proposal as SHADE [11] can be used to adapt the parameters of the algorithm. Section 3 details the proposed sampling strategies to reduce the number of instances used in the feature selection search process. Section 4 presents the experimentation details and the results. Finally, in Sect. 5, the conclusions of this work are presented, and some directions for future work are defined.

2   Differential Evolution Background

This section first presents the Differential Evolution (DE) algorithm in its basic version proposed in [10] for real-value optimization. Then, the adaptation of DE to a permutational search space is presented as suggested in [8] and used in [2,9] for feature selection. Finally, the details of a parameter adaptation proposal for DE as SHADE [11] are covered and adapted to the permutational search space.

2.1   The Differential Evolution Algorithm

DE is a population-based metaheuristic for optimization over continuous spaces proposed in [10]. The procedure requires four parameters to guide the search: population size (NP), scaling factor (F), crossover rate (CR), and the maximum number of generations. $DE/rand/1/bin$ is the basic version of DE, where $rand$ represents the aleatory selection of r_0, r_1, and r_2 for the mutation procedure that uses 1 vector difference and bin refers to the binary crossing procedure. More details are given below.

At the beginning of DE, an initial population of NP individuals is randomly generated. Then, a series of iterations called generations are conducted. Each generation consists of computing a trial vector u_i for each individual (target vector x_i) in the population and facing them in a binary tournament to select which one will be included in the next-generation population. To generate u_i, it is first necessary to compute a noise vector v_i following Eq. 1. As stated before, r_0, r_1, and r_2 are randomly chosen individuals.

$$v_i = r_0 + F(r_1 - r_2) \tag{1}$$

Once v_i is calculated, the crossing strategy takes place according to Eq. 2. The procedure consists of generating a random vector position called J_{rand}; this value will guarantee that the trial vector takes at least one value from u_i. After that, the process considers one vector position j at a time. If a random value is less or equal to CR or $j = J_{rand}$, the trial vector will take the value from v_i. Otherwise, it is taken from x_i.

$$u_{i,j} = \begin{cases} v_{i,j} & \text{if } (rand_j \leq CR) \text{ or } (j = J_{rand}); j = 1, ..., |x_i| \\ x_{i,j} & \text{otherwise} \end{cases} \tag{2}$$

2.2   Permutational-Based Differential Evolution

As proposed in [8], DE can be adapted to combinatorial problems using a permutational approach. The individuals are represented with a permutation instead of real-value vectors. The main variation in the method occurs in the calculation of v_i; instead of using Eq. 1, Eqs. 3 and 4 are applied. In Eq. 3, a permutation matrix $\mathbf{P}$ is calculated for r_1 and r_2. This matrix represents the difference between the selected individuals. Then, Eq. 4 is used to scale the permutation matrix previously obtained with F and apply it to r_0.

$$r_1 = \mathbf{P}r_2 \tag{3}$$

$$v_i = \mathbf{P}_F r_0 \tag{4}$$

The scaled permutation matrix $\mathbf{P}_F$ from Eq. 4 is calculated following the algorithm in [8]. For each column i in the matrix $\mathbf{P}$, if there is a 0 in the diagonal position $\mathbf{P}[i,i]$ and a random number $rand_i$ is greater than F, find the row j

where $\mathbf{P} = 1$ and swap rows i and j in $\mathbf{P}$. This way, the $\mathbf{P}_F$ allows only a fraction of the permutations from $\mathbf{P}$. If F is set as 0, the $\mathbf{P}_F$ will become the diagonal matrix; when applied to a permutation, the permutation remains unchanged. If F is set to 1, $\mathbf{P}_F = \mathbf{P}$.

In [9], the permutational-based DE is applied in the feature selection problem with the DE-FS^{PM} algorithm. The individuals are represented with a permutation with integer values from 0 to the number of features in the dataset. The decoding occurs as follows: each integer represents the index of a feature and the number 0 acts as a division between the selected and discarded features. This way, the indexes at the left of the zero are selected to reduce the dataset.

The DE crossing procedure occurs as explained previously using Eq. 2. Nonetheless, it is noticed that the resulting u_i vector could no longer represent a valid permutation since it could have repeated and missing values. A repair mechanism is applied in [9] where the repeated values in the u_i vector are eliminated. Then, the missing values are copied from the x_i vector to complete the permutation. By doing this, it is guaranteed that all individuals in the population are valid permutations.

2.3 Success-History Based Adaptive Differential Evolution

In [11], the Success-History-based Adaptive DE (SHADE) is proposed as a technique for parameter adaptation using a historical memory containing successful parameter configurations. At the beginning of the process, two memories are initialized using 0.5 as values; one is for F (M_F) and one for CR (M_{CR}). The memory size (H) is a user-defined parameter. In each generation, the algorithm calculates each individual's F_i, CR_i, and p_i. Based on a randomly selected position r_i in M_{CR} and M_F, Eqs. 5, 6, and 7 perform the aforementioned calculations:

$$CR_i = randn_i(M_{CR,r_i}, 0.1) \tag{5}$$

$$F_i = randc_i(M_{F,r_i}, 0.1) \tag{6}$$

$$p_i = rand[p_{min}, 0.2] \tag{7}$$

where $randn_i(\mu, \sigma^2)$ and $randc_i(\mu, \sigma^2)$ are values obtained from a normal and a Cauchy distribution, respectively. In both cases, μ and σ represent the mean and the variance from the distributions. In the calculation of CR_i, if the resulting value is lower than 0, it is replaced by 0. Similarly, if the CR_i value is larger than 1, it is replaced by 1. For F_i, the same criterion is applied for values larger than 1. However, in case of values lower than 0, the value generation using Eq. 6 is repeated until a valid value is calculated. p_{min} corresponds to a random value bounded by the fraction of the population represented by two individuals and the 20% of the population.

Since the SHADE procedure is based on a previous parameter adaptation proposal called JADE [14], it considers the usage of an optional external archive to maintain diversity. If the archive is enabled, a target vector x_i outperformed by its trial vector u_i is stored instead of discarded. The archive size is recommended to be NP. When the archive is complete and an individual is expected to be added, a random individual is deleted.

Then, the SHADE procedure uses Eq. 8 to compute v_i. r_0 is selected randomly from the population. r_1 is selected randomly from the population if the archive is disabled or from the union between the population and the archive if the archive is enabled. x_{pbest} is an individual selected at random from the top $NP *$ p_i individuals of the population. Once the v_i calculation is completed, u_i is computed following Eq. 2.

$$v_i = x_i + F_i(x_{pbest} - x_i) + F_i(r_0 - r_1) \tag{8}$$

In this work, Eq. 8 is similarly adapted to the permutational-based procedure that uses Eqs. 3 and 4. A series of steps is proposed for v_i calculation. First, the permutation matrix that maps x_{pbest} and x_i is computed as shown in Eq. 9. After that, the associated scaled-permutation matrix is calculated and applied to x_i, as shown in Eq. 10, and the result is an auxiliary permutation called a_i. In Eq. 11, the permutation matrix that maps r_0 and r_1 is computed along with the scaled permutation matrix that is applied to a_i in Eq. 12.

$$x_{pbest} = \mathbf{P}x_i \tag{9}$$

$$a_i = \mathbf{P}_F x_i \tag{10}$$

$$r_0 = \mathbf{P}r_1 \tag{11}$$

$$v_i = \mathbf{P}_F a_i \tag{12}$$

When a trial vector is generated and outperforms the target vector in each generation, the parameters used to generate the v_i are added to the lists S_F and S_{CR}. Given this, S_F and S_{CR} store the successful parameters in a generation, and their values are used to update the memory using Eqs. 13 and 14. In these Equations, the values are not updated if the lists of successful parameters S_F and S_{CR} are empty. The position in the memory to be updated is controlled by an index n. At the beginning of the procedure, n is initialized with 1. Each time the values in M_{CR} and M_F are updated, n increases in 1. When n is larger than the memory size H, the n value is reinitialized with 1.

$$M_{CR,n,G+1} = \begin{cases} mean_{WA}(S_{CR}) & \text{if } S_{CR} \neq \emptyset \\ M_{CR,n,G} & \text{otherwise} \end{cases} \tag{13}$$

$$M_{F,n,G+1} = \begin{cases} mean_{WL}(S_F) & \text{if } S_F \neq \emptyset \\ M_{F,n,G} & \text{otherwise} \end{cases} \tag{14}$$

In Eq. 13 the weigthed mean ($mean_{WA}$) is calculated following Eq. 15. In Eq. 14, the Lehmer mean ($mean_{WL}$) is computed with Eq. 16. Additional calculations are performed using Eqs. 17 and 18. In Eq. 18, Δf_j represents the amount of improvement of a successful trial over its associated target.

$$mean_{WA}(S_{CR}) = \sum_{j=1}^{|S_{CR}|} w_j \cdot S_{CR,j} \tag{15}$$

$$mean_{WL}(S_F) = \frac{\sum_{j=1}^{|S_F|} w_j \cdot S_{F,j}^2}{\sum_{j=1}^{|S_F|} w_j \cdot S_{F,j}} \tag{16}$$

$$w_j = \frac{\Delta f_j}{\sum_{j=1}^{|S_{CR}|} \Delta f_j} \tag{17}$$

$$\Delta f_j = |f(u_{j,G}) - f(x_{j,G})| \tag{18}$$

3   Proposal Details

In this proposal, the DE-FS^{PM} algorithm from [9] and its adaptation to the SHADE procedure from [11] are used with three different sampling strategies during the search process to save computational resources. The strategies proposed are using a fixed, incremental, and evolving sampling fraction. The main goal is to let some individual evaluations be computed with fewer dataset instances than required in the original process where all instances are used. The details of each sampling scheme are shown in Subsects. 3.1, 3.2, and 3.3.

The DE-FS^{PM} algorithm uses the k-nearest-neighbors (KNN) classifier. It comprises a series of steps: dataset preprocessing, finding the most appropriate k value for the classifier, evaluating the complete dataset, and the search process. In the adopted configuration from [5,9], the preprocessing consists of imputing missing values (using the mean for numerical features and the mode for categorical features), converting nominal to numerical values, and min-max normalization. In DE-FS^{PM}, calculating the most appropriate k value for the KNN classifier consists of evaluating the data with a 10-fold cross-validation (CV) using k values from 1 to 19 with a step of 2. A 5-fold stratified CV is used as the fitness function during the search process. The dataset with all the features and the dataset with only the selected features are evaluated with a 10-fold stratified CV. Figure 1 shows a scheme of the feature selection process followed by the DE-FS^{PM} algorithm along with the sampling proposals.

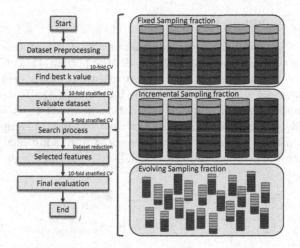

Fig. 1. Schematics of the feature selection procedure including the proposed variations in the search process.

3.1 Fixed Sampling Fraction

The first proposal uses a fixed sampling fraction to reduce the dataset instances at the beginning of the search process. The sampling fraction (sf) is a user-defined parameter for the algorithm. In all generations of the search process, the reduced dataset is used, and the individuals are evaluated with a 5-fold stratified CV. Ultimately, the best individual is evaluated using all the dataset instances with a 10-fold stratified CV.

3.2 Incremental Sampling Fraction

In this sampling strategy, the search process begins with an initial subset of instances determined by a sampling fraction sf, a user-defined parameter. Then, given a second user-defined parameter called the number of divisions, divide the generations of the search process into blocks in which more instances are added. In the last block, all the dataset instances will be used. This way, the procedure is expected to find promising individuals with a less costly evaluation in the initial generations of the search process.

Consider an initial sf of 0.6, a total number of generations of 200, and 5 divisions as an example of how the instances are used in the search process. The first block of generations (1 to 40) will use 60% of the dataset instances. Then, the second block will use 70% of the instances, and so on, until the last block (161 to 200) will use all the instances. This sampling strategy requires an additional set of evaluations for the individuals in the population each time more instances are included in the search process. Therefore, the population is reevaluated at the beginning of each division, adding computational cost to the process.

3.3 Evolving Sampling Fraction

The proposal of evolving the sampling fraction along the process requires an additional value coded in an individual. Besides the permutation of integer numbers the DE-FS^{PM} algorithm usually uses, a real-value component is added at the end as shown in Fig. 2. This value is the individual's sf. The individual sf is generated randomly for the initial population bounded to at least 0.1 and at most 1. In decoding an individual, sf is used to apply a sampling method to the dataset instances. This proposal does not require additional user-defined parameters since the evolving procedure handles finding an adequate sf value.

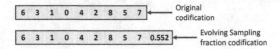

Fig. 2. Codification differences between the original DE-FS^{PM} algorithm and the evolving sampling fraction proposal.

The mutation procedure is the same for the permutational part of the individual, Eqs. 3 and 4 or 9, 10, 11, and 12 are used. Nonetheless, the real-value element of the individual is considered in an independent process where Eqs. 1 or 8 are employed. The crossing procedure is the classical with Eq. 2. The repair mechanism mentioned in Subsect. 2.2 is applied to the individual permutation part. A different mechanism is utilized for the real-value part if a value is lower than 0.1 or greater than 1. In that case, a new value is calculated as twice the exceeded limit minus the out-of-bounds value.

The evolving sampling fraction proposal does not require the definition of additional parameters. However, the values evolved inside the process could tend to the allowed limits, difficulting the process. For example, if the sf encoded in most individuals is close to 1, the number of avoided instances will be low, representing scarce resource savings. Conversely, a low sf could imply more savings since just a few instances are used, but overfitting the sample could become a problem.

4 Experiments and Results

Five datasets are selected from the UCI machine learning repository as shown in Table 1 to evaluate the performance of the three proposals defined in the previous section. Six different methods are compared with the DE-FS^{PM} algorithm original procedure. The three sampling proposals are applied to the DE-FS^{PM} algorithm and its adaptation under the SHADE scheme. Ten runs of each method are computed for each dataset. At the beginning of each method run, the dataset instances are shuffled. The DE-FS^{PM} algorithm early stopping mechanism that stops the procedure when the best solution has not changed in 100 generations was turned off for the tests.

Table 1. Selected datasets for experimentation.

Dataset	Instances	Features	Classes
Ionosphere	351	34	2
Sonar	208	60	2
SPECTF	267	44	2
Vehicle	846	18	4
Hill-Valley	1212	100	2

The parameters of the DE-FS^{PM} algorithm are proposed in [9]. NP is five times the number of features in the dataset, and this value is bounded to at least 200 and at most 450 individuals. F and CR are 0.1514 and 0.8552, respectively. The maximum number of generations is 200. On the other hand, in the methods that use the SHADE scheme, NP is 2.5 times the number of features bounded to at least 100 and 225 individuals at most. The maximum number of generations in this case is 400. Memory size H is set to 6, as proposed in [12] and [3]. The initial s_f is 0.6 for the methods with the fixed and incremental sampling fraction proposals. The s_f value was defined with preliminary experimentation where it was noticed that larger values represented less cost savings and smaller values implied worsened algorithm performance more substantially. The number of blocks for the incremental sampling fraction methods is ten.

Tables 2 and 3 present the tested methods' accuracy results and selected features. The mean value and standard deviation are computed from the ten runs of the procedure. The results for each dataset are ranked, and the best result is marked in bold. The methods are identified as DE-FS^{PM} for the algorithm from [9], Fixed, Incremental and Evolving for the DE-FS^{PM} algorithm with the sampling proposals and SHADE fix, inc, and evo for the SHADE scheme applied to the DE-FS^{PM} algorithm with the fixed, incremental and evolving sampling strategies. It is seen in the accuracy performance that the original DE-FS^{PM} method outperformed the other proposals in four of the five datasets. Statistical tests were computed with the results in Table 2. First, the Friedman test with a p-value of 0.0038 indicates significant differences among the means of the DE-FS^{PM} procedure and the proposals. Then, the Nemenyi post-hoc test indicated that there is no significant difference between the DE-FS^{PM} algorithm and the fixed, the incremental, the evolving, and the SHADE inc proposals given their respective p-values of 0.053, 0.900, 0.385, and 0.739. On the other hand, according to the p-values of 0.021 and 0.008 from the test, the SHADE fix and SHADE evo proposals present significant differences.

Table 4 presents the average sampling fraction coded in the best individuals in the ten runs of the evolving proposals. It is seen that the coded sampling fraction falls next to the bounds defined for the value. It selects a few number of instances, or it selects almost all the instances. The high sampling fraction in the hill-valley dataset is expected to be related to the good performance of these proposals in that specific data. A critical problem of selecting too few instances is

Table 2. Accuracy performance of the proposals.

Method	Ionosphere			Sonar			SPECTF			Vehicle			Hill-valley		
	Average	Std dev	Rank	Average	Std dev	Rank	Average	Std dev	Rank	Average	Std dev	Rank	Average	Std dev	Rank
DE-FS^{PM}	**94.64**	0.82	(1)	**94.00**	1.32	(1)	**83.67**	1.02	(1)	**74.31**	0.76	(1)	71.51	0.50	(2)
Fixed	91.51	1.62	(5)	88.02	2.64	(6)	81.54	1.83	(4)	72.34	1.42	(5)	68.62	1.29	(6)
Incremental	93.22	1.07	(3)	92.84	1.24	(2)	82.21	1.68	(3)	73.67	1.16	(3)	71.10	0.76	(3)
Evolving	89.77	3.42	(6)	90.69	7.26	(4)	77.28	1.52	(7)	74.15	1.18	(2)	**71.53**	0.78	(1)
SHADE fix	91.99	1.49	(4)	89.29	2.81	(5)	80.10	1.30	(5)	72.06	1.19	(7)	68.57	1.50	(7)
SHADE inc	94.48	0.57	(2)	91.60	1.36	(3)	83.00	1.69	(2)	73.50	0.87	(4)	68.99	0.87	(5)
SHADE evo	87.79	2.09	(7)	85.61	1.95	(7)	77.79	2.32	(6)	72.31	0.94	(6)	70.12	1.06	(4)

Table 3. Number of selected features from the different proposals.

Method	Ionosphere			Sonar			SPECTF			Vehicle			Hill-valley		
	Average	Std dev	Rank	Average	Std dev	Rank	Average	Std dev	Rank	Average	Std dev	Rank	Average	Std dev	Rank
DE-FS^{PM}	**6.30**	0.82	(1.5)	23.40	3.44	(4)	**4.20**	1.14	(1)	10.10	0.99	(7)	9.60	2.07	(2)
Fixed	6.40	1.35	(3.5)	**18.00**	4.92	(1.5)	8.20	3.65	(3)	8.80	1.14	(2.5)	12.50	5.99	(5)
Incremental	6.40	1.35	(3.5)	**18.00**	4.19	(1.5)	4.80	2.39	(2)	**8.40**	1.35	(1)	**9.50**	3.21	(1)
Evolving	13.10	10.73	(6)	20.80	6.53	(3)	29.50	15.97	(7)	9.50	1.08	(4)	10.10	2.23	(3)
SHADE fix	7.20	2.20	(5)	28.30	7.57	(5)	15.00	9.02	(5)	8.80	1.62	(2.5)	10.90	3.11	(4)
SHADE inc	**6.30**	0.67	(1.5)	28.50	5.80	(6)	8.30	6.17	(4)	9.90	1.79	(6)	19.70	10.48	(7)
SHADE evo	18.00	11.86	(7)	32.10	14.24	(7)	23.90	11.52	(6)	9.60	1.07	(5)	14.40	5.02	(6)

that the procedure overfits the feature selection to that small sample of the data, and it is difficult to generalize and perform well when considering the remaining instances.

Table 4. Average sampling fraction in the best individual found during the search process using the evolving sampling fraction proposals.

Method	Ionosphere	Sonar	SPECTF	Vehicle	Hill valley
Evolving	0.2761	0.8511	0.1101	0.8262	0.9833
SHADE evo	0.1288	0.1607	0.1068	0.5288	0.9784

The number of instances in the procedures used to evaluate individuals was stored for further comparisons as the one presented in Table 5, where the average percentage of avoided instances is shown with the reference of the DE-FS^{PM} original procedure. The most significant savings for each dataset are marked in bold. The value of instances used is calculated by the sum of the number of instances considered in each fitness evaluation of the individuals. Table 5 also includes the average reduction by each method, along with a rank. It is seen that the fixed fraction procedures represent more considerable savings in the number of instances used. The incremental procedures presented a smaller number of avoided instances. Finally, Table 6 presents the computational time and percentage of reduction of the different proposals with respect to the original DE-FS^{PM} algorithm. The methods are ranked and the most prominent savings for each dataset are marked in bold. The fixed sampling fraction proposals achieved a larger average time reduction. It is seen that the methods present a variant behavior in the different datasets tested.

Table 5. Percentage of avoided instances in the search process in the proposals.

Method	Ionosphere	Sonar	SPECTF	Vehicle	Hill valley	Average	Rank
Fixed	39.89%	39.90%	40.07%	39.95%	40.02%	39.97%	(3)
Incremental	16.51%	16.62%	16.46%	16.45%	16.43%	16.49%	(6)
Evolving	15.01%	9.87%	**86.06%**	10.92%	2.46%	24.87%	(4)
SHADE fix	40.04%	**40.25%**	40.49%	**40.10%**	**40.17%**	**40.21%**	(1)
SHADE inc	18.50%	18.87%	18.81%	18.44%	18.42%	18.61%	(5)
SHADE evo	**52.84%**	32.17%	80.96%	25.75%	8.48%	40.04%	(2)

Table 6. Computational time and time reduction percentage of the proposals.

Method	Ionosphere		Sonar		SPECTF		Vehicle		Hill valley		Average	Rank
	Time(s)	%Reduc	Time(s)	%Reduc	Time(s)	%Reduc	Time(s)	%Reduc	Time(s)	%Reduc		
DE-FS^{PM}	1581.06	–	2902.87	–	1983.83	–	4346.58	–	7062.38	–		
Fixed	1368.57	13.44%	2672.10	7.95%	1791.81	9.68%	3827.40	11.94%	5981.04	15.31%	11.66%	(1)
Incremental	1482.20	6.25%	2942.49	-1.36%	1854.10	6.54%	4236.24	2.54%	7034.53	0.39%	2.87%	(4)
Evolving	1691.74	-7.00%	2826.59	2.63%	1899.06	4.27%	3000.23	30.97%	6153.40	12.87%	8.75%	(3)
SHADE fix	1474.20	6.76%	2576.50	11.24%	1969.18	0.74%	2901.30	33.25%	7009.70	0.75%	10.55%	(2)
SHADE inc	1388.83	12.16%	2946.93	-1.52%	2156.45	-8.70%	5147.95	-18.44%	7654.61	-8.39%	-4.98%	(6)
SHADE evo	1584.57	-0.22%	2797.40	3.63%	1897.17	4.37%	4255.78	2.09%	7341.71	-3.96%	1.18%	(5)

5   Conclusions and Future Work

This work presented six proposals for using sampling methods to reduce the computational cost of a wrapper approach for feature selection. The results show that reducing the resource demand is possible, but the algorithm's performance is affected. Contrary to the findings in [7], where using fewer instances minimally impact filter approaches, wrapper approaches seem more sensitive. Nonetheless, some of the proposals presented no significant differences in performance from the original procedure. Overfitting is presented in [1] as a disadvantage for wrappers. Subsequently, using fewer instances could make the overfitting problem more severe. Using a parameter adaptation scheme such as SHADE did not improve the performance of the feature selection procedure; in some cases, it was diminished.

In future work, dealing with the overfitting problem could improve the feature selection procedure performance. Using sampling strategies during the search, as presented in this work, must be refined using additional mechanisms to improve the performance and increase resource savings. More recent DE parameter adaptation schemes, as proposed in [3] and [12], can be adapted to the problem expecting better results. The problem of evaluating repeated individuals shown in [6] can be considered for computational cost reduction of the feature selection process. Additionally, the proposed sampling strategies can be adapted to a multi-objective approach for feature selection to evaluate the impact on the performance of the procedure.

Acknowledgements. Mexico's National Council of Humanities, Science, and Technology (CONAHCYT) awarded a scholarship to the first author (CVU 1142850) for graduate studies at the University of Veracruz.

References

1. Abdulwahab, H.M., Ajitha, S., Saif, M.A.N.: Feature selection techniques in the context of big data: taxonomy and analysis. Appl. Intell. **52**(12), 13568–13613 (2022). https://doi.org/10.1007/s10489-021-03118-3

2. Barradas-Palmeros, J.A., Mezura-Montes, E., Acosta-Mesa, H.G., Rivera-López, R.: Fitness function comparison for unsupervised feature selection with permutational-based differential evolution. In: Rodríguez-González, A.Y., Pérez-Espinosa, H., Martínez-Trinidad, J.F., Carrasco-Ochoa, J.A., Olvera-López, J.A. (eds.) Pattern Recognition. MCPR 2023. LNCS, vol. 13902, pp. 58–68. Springer, Cham (2023). https://doi.org/10.1007/978-3-031-33783-3_6

3. Brest, J., Maučec, M.S., Bošković, B.: iL-shade: improved l-shade algorithm for single objective real-parameter optimization. In: 2016 IEEE Congress on Evolutionary Computation (CEC), pp. 1188–1195 (2016). https://doi.org/10.1109/CEC.2016.7743922

4. Dhal, P., Azad, C.: A comprehensive survey on feature selection in the various fields of machine learning. Appl. Intell. **52**(4), 4543–4581 (2022). https://doi.org/10.1007/s10489-021-02550-9

5. Engelbrecht, A.P., Grobler, J., Langeveld, J.: Set based particle swarm optimization for the feature selection problem. Eng. Appl. Artif. Intell. **85**, 324–336 (2019). https://doi.org/10.1016/j.engappai.2019.06.008

6. Kitamura, T., Fukunaga, A.: Duplicate individuals in differential evolution. In: 2022 IEEE Congress on Evolutionary Computation (CEC), pp. 1–8 (2022). https://doi.org/10.1109/CEC55065.2022.9870366

7. Malekipirbazari, M., Aksakalli, V., Shafqat, W., Eberhard, A.: Performance comparison of feature selection and extraction methods with random instance selection. Expert Syst. Appl. **179**, 115072 (2021). https://doi.org/10.1016/j.eswa.2021.115072

8. Price, K.V., Storn, R.M., Lampinen, J.A.: Differential Evolution: A Practical Approach to Global Optimization. Springer, Berlin, Heidelberg (2005). https://doi.org/10.1007/3-540-31306-0

9. Rivera-López, R., Mezura-Montes, E., Canul-Reich, J., Cruz-Chávez, M.A.: A permutational-based differential evolution algorithm for feature subset selection. Pattern Recognit. Lett. **133**, 86–93 (2020). https://doi.org/10.1016/j.patrec.2020.02.021

10. Storn, R., Price, K.: Differential evolution - a simple and efficient heuristic for global optimization over continuous spaces. J. Glob. Optim. **11**(4), 341–359 (1997). https://doi.org/10.1023/A:1008202821328

11. Tanabe, R., Fukunaga, A.: Success-history based parameter adaptation for differential evolution. In: 2013 IEEE Congress on Evolutionary Computation, pp. 71–78 (2013). https://doi.org/10.1109/CEC.2013.6557555

12. Tanabe, R., Fukunaga, A.S.: Improving the search performance of shade using linear population size reduction. In: 2014 IEEE Congress on Evolutionary Computation (CEC), pp. 1658–1665 (2014). https://doi.org/10.1109/CEC.2014.6900380

13. Xue, B., Zhang, M., Browne, W.N., Yao, X.: A survey on evolutionary computation approaches to feature selection. IEEE Trans. Evol. Comput. **20**(4), 606–626 (2016). https://doi.org/10.1109/TEVC.2015.2504420

14. Zhang, J., Sanderson, A.C.: JADE: adaptive differential evolution with optional external archive. IEEE Trans. Evol. Comput. **13**(5), 945–958 (2009). https://doi. org/10.1109/TEVC.2009.2014613

Computational Learning in Behavioral Neuropharmacology

Isidro Vargas-Moreno[1] (ID), Héctor Gabriel Acosta-Mesa[2] (ID),
Juan Francisco Rodríguez-Landa[3] (ID), Martha Lorena Avendaño-Garido[4] (ID),
and Socorro Herrera-Meza[5]([⊠]) (ID)

[1] Doctorado en Neuroetología, Instituto de Neuroetología, Universidad Veracruzana, Xalapa, Mexico

[2] Instituto de Investigaciones en Inteligencia Artificial, Universidad Veracruzana, Xalapa, Mexico
heacosta@uv.mx

[3] Instituto de Neuroetología, Universidad Veracruzana, Xalapa, Mexico
juarodriguez@uv.mx

[4] Facultad de Matemáticas, Universidad Veracruzana, Xalapa, Mexico
maravendano@uv.mx

[5] Instituto de Investigaciones Psicológicas, Universidad Veracruza, Xalapa, Mexico
soherrera@uv.mx

Abstract. Behavioral neuropharmacology is an area of neuroscience, which is responsible for the study of behavioral modifications through the administration of substances, treatments or experimental manipulations. Particularly to determine the effect on behavior, this area uses classic statistical techniques for comparing measures of central tendency; the analysis of variables is mostly carried out in a univariate manner, where the interpretation of the results obtained is often limited. There are other areas that also provide tools for data analysis, such as computational learning, through prediction models we can determine the characteristic behavioral patterns of each treatment administered. In the present study, computational learning data analysis techniques were used, specifically, supervised machine learning applied to a behavioral neuropharmacology experiment, where 3 doses of allopregnanolone (0.5, 1, and 2 mg) were evaluated in maze tests. Raised arms and motor activity test. We identified with classical statistical methods that the 2 mg dose of allopregananolone has an anxiolytic-type effect, similar to that exerted by the reference drug diazepam. Additionally, with computational learning methods, we can identify the characteristic patterns of each treatment based on the combination of the variables of both behavioral tests, likewise, we demonstrate with mathematical support the most important variables for the identification of anxioselective effects. In conclusion, computational learning methods promote enrichment in the results of neuropharmacology reflected in the characteristic patterns that are modified by the administration of different drugs, and provide foundations to support the importance of the most relevant variables of behavioral tests.

Keywords: Behavioral neuropharmacology · Computational Learning · Allopregnanolone

© The Author(s), under exclusive license to Springer Nature Switzerland AG 2024
H. Calvo et al. (Eds.): MICAI 2023 Workshops, LNAI 14502, pp. 422–431, 2024.
https://doi.org/10.1007/978-3-031-51940-6_32

1   Introduction

Behavioral neuropharmacology (BNP) is a branch that studies the effects of chemical substances, such as drugs, natural extracts, functional foods, or experimental manipulations, that can modify behavior through mechanisms that act at the level of the nervous system, as well as their effects on brain function [1]. In the preclinical phase, laboratory animals, such as rats or mice, are used in the research of psychoactive drugs, such as amphetamines, cannabinoids, and opiates, on behavior [2]. These studies are associated with anxiety [3], depression [4], the analysis of learning and memory [5], and even neuropsychiatric disorders [6].

In the field of BNP, data analysis is typically carried out using a univariate approach. In this approach, the dependent variable is defined, which is generally an indicator of various types of behavior, such as anxiety, depression, memory, and locomotion, among others. Similarly, the selection of groups is determined, representing the independent variable, particularly in this field, which is the treatment or experimental intervention. Behavioral variables are obtained through the exposure of animals to behavioral tests, and the data collected are subsequently analyzed [7].

In addition, in BNP research, various statistical techniques are employed to analyze and visualize the data obtained from experiments and studies in order to determine their effects on behavior. Among the most commonly used methods in BNP are the Student's t-test and the Mann-Whitney U test, as well as various-way ANOVAs and the Kruskal-Wallis test, all of which are used to compare measures of central tendency (mean or median) between groups [8]. Repeated measures analysis is also utilized when the goal is to observe any improvement with the treatments administered over time [9]. Similarly, correlation analyses are employed to determine the degree of influence of an independent variable on the variable of interest [10]. However, due to assumptions being met or the robustness of the statistical tests, it is often not possible to discern the treatment's effects.

Therefore, other data analysis tools can enhance BNP research. Computational learning can play a crucial role in enabling a more profound and sophisticated analysis of the data. It aids in identifying complex patterns and relationships, as well as optimizing pharmacological treatments to enhance the comprehension and treatment of neuropsychological disorders in rats, and ultimately, in humans.

With this concept in mind, the present study aims to utilize supervised machine learning techniques to identify significant features of experimental treatments administered to Wistar rats. Furthermore, it aims to compare the results obtained using classical statistical methods with those derived from computational learning analysis in the field of behavioral neuropharmacology.

2   Materials and Methods

2.1   Subjects

Forty adult male rats of the Wistar strain, weighing between 250 and 300 g at the start of the experiment, were housed in translucent acrylic boxes in a biotherium with a controlled temperature of 25 °C (± 1 °C) and a 12 × 12-h light/dark cycle (light was turned on at 7:00 am). Rats had ad libitum access to water and food. All experimental care and procedures were conducted in compliance with the national standard, NOM-ZOO-062 [11]. Additionally, at the international level, the research followed the guidelines outlined in the "Guide for the Care and Use of Laboratory Animals' by the Institute of Laboratory Animal Resources" [12].

2.2 Experimental Groups

The experimental subjects were divided into five groups, with eight rats in each group: 1) the vehicle group received sterile water (PiSA Laboratory, Mexico City, Mexico), 2) a 0.5 mg/kg dose of allopregnanolone, 3) a 1 mg/kg dose of allopregnanolone, 4) a 2 mg/kg dose of allopregnanolone (Sigma-Aldrich Co., St. Louis, MO, USA), and 5) a 2 mg/kg dose of diazepam (Roche Laboratory, Mexico City, Mexico) (Fig. 1). These treatments were administered intraperitoneally as a one-time, acute treatment, with a volume of 1 mL per kg of body weight. The effects were evaluated in behavioral tests 30 min after administration.

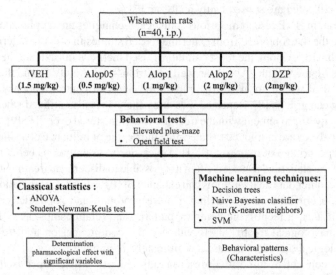

Fig. 1. Experimental design (VEH: vehicle; Alop05: allopregnanolone dose of 0.5 mg; Alop1: allopregnanolone dose of 1 mg; Alop2: allopregnanolone dose of 2 mg; DZP: diazepam dose of 2 mg)

2.3 Behavioral Tests

The sessions were videotaped for 5 min, and after the exposure of each subject, the apparatus was cleaned with a 15% ethanol solution to prevent volatile particles from interfering with the behavior. Subsequently, the videos were analyzed by two expert behavior analysts using UVehavior software to obtain variables of interest.

2.3.1 Elevated Plus-Maze Test

The Elevated Plus-Maze test is widely used to measure anxiety-related indicators by exposing rats to two natural fears: open spaces (agoraphobia) and heights (acrophobia) [13]. The test utilized an apparatus with two open arms (50 × 10 cm) and two closed arms (50 × 10 × 50 cm), arranged in the shape of a cross and elevated 50 cm above the floor. The animals were individually placed in the central box, with their eyes directed towards an open arm. The following variables were obtained: a) time spent in open arms (seconds), b) number of entries in open arms, c) number of entries in closed arms, d) total number of entries in arms (open + closed), e) percentage of entries in open arms, and f) the anxiety index, calculated using the formula proposed by Cohen and collaborators [14]:

$$AI = 1 - \left(\frac{\left(\frac{TOA}{TTT} \right) + \left(\frac{OAE}{TE} \right)}{2} \right) \tag{1}$$

(TOA: time spent in the open-arms; OAE: number of entries in open-arms; TTT: total test time; TE: total number of entries in arms).

2.3.2 Locomotor Activity Test (Open Field)

The Open Fields test is commonly used to assess the spontaneous activity of animals, measuring their exploratory behavior, motivation, and motor activity [15]. The test involved a sky-blue glass box (44 × 33 × 40 cm in length, width, and height, respectively) with a floor marked with 12 squares (11 × 11 cm each). The test lasted for 5 min, during which the animals were individually placed in the same corner and facing the same direction. The evaluated variables included: a) the number of squares crossed, b) the time spent rearing, and c) the time spent grooming.

2.4 Data Analysis

2.4.1 Classical Statistical Analysis

The data were analyzed using one-way ANOVA, with the necessary assumptions verified through the Shapiro-Wilks test for normality and the Bartlett test for homogeneity. Post-hoc testing was conducted using the Student-Newman-Keuls test to compare differences between groups. All statistical tests were assessed with a significance level of $\alpha = 0.05$ to evaluate each set of hypotheses. The analysis was performed using SigmaPlot software (Version 12.0).

2.4.2 Computational Learning Analysis

The data were analyzed using the J48 decision tree algorithm within Weka software (version 3.8.1) to examine all variables from both behavioral tests. Subsequently, we selected four variables based on their significance and existing literature in the behavioral neuropharmacology field and made the following model adjustments:

- Naive Bayes (with naivebayes library, laplace = 1).
- K-nearest neighbors (with the kknn library, using the Ker-nel = optimal, a K = 3 for classification)
- Support vector machines (with the e1071 library, using the linear Kernel = function, obtaining 3 vectors for classification).

For the implementation of these techniques, we used the R Studio program (Version 1.3.1093). For all the adjustments made, a training set was used, consisting of 70% of the base, the remaining set was the test set.

3   Results

3.1   Classical Statistics

Statistical analysis revealed significant differences in the frequency of open arm entries in the elevated arm maze. The groups administered 1 mg and 2 mg of allopregnanolone, as well as the diazepam group, showed increased frequency compared to the vehicle group ($p < 0.05$). The same differences were observed in the duration of this behavior, with the 2 mg allopregnanolone and diazepam groups showing longer durations ($p < 0.05$). All groups exhibited significant differences in the percentage of time spent in open arms compared to the vehicle group ($p < 0.05$). Regarding the anxiety index, all doses showed significant differences compared to the vehicle, with the 2 mg doses of allopregnanolone and diazepam being the most effective, leading to a notable reduction in this parameter ($p < 0.05$) (Table 1).

In the open field test, statistically significant differences were observed in the time spent in rearing and grooming behaviors. Specifically, the 0.5 mg and 1 mg doses of allopregnanolone differed significantly from the vehicle group in terms of time spent in rearing, with an increase ($p < 0.05$). Regarding grooming behavior, the only group that showed significant differences was the 1 mg dose of allopregnanolone, indicating a decrease in time dedicated to grooming ($p < 0.05$) (Table 1).

Table 1. Statistics of the behavioral variables through the 1-way ANOVA test.

| VAR | Groups | | | | | | | | | | | | | | | Statistics | |
| | VEH | | | Alop05 | | | Alop 1 | | | Alop 2 | | | DZP | | | | |
	MN	STE	STD	MN	STE	STD	MN	STE	STD	MN	STE	STD	MN	STE	STD	F	p-value
TOA	26.6	3.8	10.8	27.2	5.09	14.4	142	8.68	24.5	185	4.36	12.3	198	10.6	30	141.61	0.000
OAE	3	0	1	5	0	1	7	1	2	7	1	2	8	1	3	6.48	0.001
CAE	4	0	1	4	0	1	4	0	1	3	0	1	4	1	2	0.83	0.517
TE	7	1	2	9	1	2	10	1	3	9	1	3	12	2	5	2.80	0.410
POA	0.47	0.06	0.18	0.6	0.04	0.13	0.65	0.02	0.05	0.71	0.02	0.06	0.73	0.04	0.11	6.83	0.000
AIN	0.72	0.03	0.08	0.66	0.02	0.07	0.44	0.02	0.04	0.34	0.02	0.05	0.3	0.02	0.07	71.20	0.000
C	32	3	9	40	6	17	37	3	9	45	4	10	41	5	13	1.27	0.299
TR	10.7	1.74	4.92	21.7	2.52	7.14	23	3.47	9.82	17.1	0.94	2.66	18.6	2.13	6.03	4.31	0.006
TG	19.9	1.46	4.14	24.3	1.61	4.56	12	1.19	3.37	20.8	2.51	7.09	24.4	2.57	7.27	6.74	0.000

(VAR: variable; VEH: vehicle; Alop05: allopregnanolone dose of 0.5 mg; Alop1: allopregnanolone dose of 1 mg; Alop2: allopregnanolone dose of 2 mg; DZP: diazepam dose of 2 mg; TOA: time spent in open-arms; OAE: number of open-arms entries; CAE: number of closed-arms entries; TE: total entries; POA: percentage in open-arms; AIN: anxiety index; C: crossing; TR: time spent rearing; TG: time spent grooming; reporting MN: mean, STE: standard error and STD: standard deviation)

3.2 Computational Learning

On the other hand, computational learning using the decision tree revealed that the most significant variable is the time spent in open arms (TOA). Values less than or equal to 61.32 s classified the vehicle (VEH) and the 0.5 mg dose of allopregnanolone (Alop05) groups, with further subdivision based on the time spent in rearing behavior. VEH exhibited values less than or equal to 15.41 s, while the Alop05 group exceeded this cutoff point (Fig. 2).

To identify the 1 mg (Alop1) and 2 mg (Alop2) doses of allopregnanolone, we consider two characteristics. The first involves a range of values greater than 61.32 s but less than or equal to 171.31 s for TOA. The second characteristic pertains to grooming time, where values less than or equal to 19.19 s classify the Alop1 group, while values greater than this threshold classify the Alop2 group (Fig. 2).

Finally, to identify the group administered with diazepam, we consider two sets of characteristics. The first set includes having a duration greater than 171.31 s in open-arm duration and more than 9 open-arm entries (OAE). The second set is associated with three variables: having more than 171.31 s in TOA, less than or equal to 9 OAE, and less than or equal to 43 crossings. When a number is assigned to the last characteristic, we identify the Alop2 group (Fig. 2). It's important to note that this model has a 75% accuracy rate.

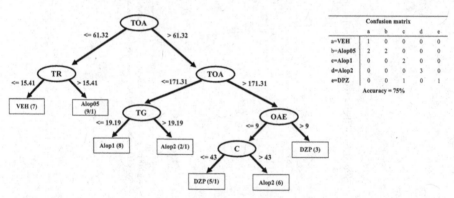

Fig. 2. Decision tree with the variables obtained in the two behavioral tests. (TOA: time spent in open arms; TR: time spent in rearing; TG: time spent grooming; OAE: number of open arms entries; C: crossing; VEH: vehicle; Alop05: allopregnanolone dose of 0.5 mg; Alop1: allopregnanolone dose of 1 mg; Alop2: allopregnanolone dose of 2 mg; DZP: diazepam dose of 2 mg)

Regarding the model fits with different classifiers, we found that the highest accuracy was achieved with the general model using the Support Vector Machine (SVM) classifier, reaching 75% accuracy. The Naive Bayesian (NB) and k-Nearest Neighbors (KNN) classifiers did not perform as well in the general model. However, when we selected specific variables for comparison, we observed that the accuracy improved for NB and KNN classifiers, while SVM accuracy decreased slightly (Table 2). It's important to note that the selected variables were time spent in open arms, crossings, time spent rearing, and grooming behavior.

Table 2. Accuracy percentages of the model fit for each classifier.

Accuracy comparison

	NB	KNN	SVM
GM	58.33	41.67	75
SELVM	66.67	66.67	58.33

(NB: naive bayes classifier: KNN; k-nearest neighbors; SVM: support vector machines; GM: general model; SELVM: selected variables model)

4 Discussion

The present work sought to identify the most relevant characteristics of behavioral variables through the application of computational learning methods in the area of behavioral neuropharmacology, in this case, the identification of the effect of allopregnanolone on anxiety and motor activity. As well as comparing the findings of classical statistical methods with respect to the application of computational learning methods.

4.1 Computational Learning Contribution

Initially, using the decision tree technique, we identified distinct behavioral patterns associated with the treatments administered to rats. A decision tree is a prediction model designed for inductive learning from observations and logical constructions [16]. These logical constructions are formed by combining relevant variables to enhance classification.

In this manner, we applied multivariate techniques to identify distinct behavioral patterns associated with each administered treatment. The decision tree method proved advantageous over traditional measures of central tendency comparison due to its ability to capture interactions between variables characterizing each treatment. Additionally, this predictive model helped us pinpoint the most crucial variables. In our study, the time spent in open arms (TOA) emerged as the most critical variable for treatment classification, consistent with the findings of Fernández-Demeneghi and colleagues [17]. They, too, employed a decision tree and found that TOA is a key variable for determining anxiolytic effects in rats, a result supported by our research.

In terms of model fitting with different classifiers, we found that better accuracy is achieved by selecting specific variables that are both significant in the literature and through statistical methods. Comparative studies in disease classification have shown that the Naive Bayesian (NB), k-Nearest Neighbor (KNN), and Support Vector Machines (SVM) classifiers demonstrate similar accuracy [18]. In our study, the selected variables were a combination of results from both behavioral tests, including TOA, crossings, time spent in upright behavior, and grooming. This underscores the importance of these variables in determining treatment effects under the experimental conditions established in our study.

4.2 Neuropharmacological Behavior

Through statistical methods, we identified six significant variables. In the elevated arm maze, the duration and frequency in open arms were significant. Allopregnanolone doses of 1 and 2 mg exhibited an anxiolytic effect similar to that of diazepam. These significant variables also showed a higher percentage of time spent in open arms and lower values in the anxiety index, indicating a lower degree of anxiety. This aligns with previous research in behavioral neuropharmacology that seeks to demonstrate anxiolytic effects with reference drugs [17, 19]. In the open field test, the time spent in vertical behavior increased for doses of 0.5 and 1 mg of allopregnanolone, with the latter also decreasing the time spent in grooming. It's important to note that, generally, the open field test does not reveal differences that would suggest treatment effects on locomotor activity [20]. Extreme increases or decreases in grooming are associated with high levels of stress [21]. However, in our study, even though the expression of this behavior decreased, we cannot conclude that the 1 mg dose is associated with stress levels based on the parameters observed in the elevated plus-maze.

5 Conclusions and Future Work

The data analysis methods offered by computational learning, specifically supervised machine learning, demonstrated that their application provides evidence of treatment-dependent behavioral patterns, as well as mathematical support to determine which variables are most relevant to the behavioral tests evaluated. With respect to the comparison between analysis methods, we can use both classical statistical and computational learning approaches, where the latter provide additional results that promote a more enriched explanation for conclusions of behavioral analysis effects. Finally, the 1 and 2 mg doses of allopregnanolone show parameters associated with an anxiolytic effect, the latter being the most effective, similar to the effect of the anxiolytic reference drug: diazepam. It's worth noting that the use of computational learning tools in the field of behavioral neuropharmacology shows significant promise for investigating the pharmacological effects of various substances in preclinical research. This extends to encompass not only variables derived from behavioral tests but also those associated with the expression of biomarkers, providing insights into characteristics influenced by the administered treatment.

Acknowledgements. The authors gratefully acknowledge CONAHCyT to grant scholarship 628503 to Isidro Vargas-Moreno for Postgraduate Studies in Neuroethology.

References

1. Branch, M.N.: Behavioral pharmacology. In: Techniques in the Behavioral and Neural Sciences, vol. 6, pp. 21–77. Elsevier (1991)
2. Sañudo-Peña, M.C., Walker, J.M.: Effects of intrapallidal cannabinoids on rotational behavior in rats: interactions with the dopaminergic system. Synapse **28**(1), 27–32 (1998)

3. Landgraf, R., Wigger, A.: High vs low anxiety-related behavior rats: an animal model of extremes in trait anxiety. Behav. Genet. **32**, 301–314 (2002)
4. Planchez, B., Surget, A., Belzung, C.: Animal models of major depression: drawbacks and challenges. J. Neural Transm. **126**, 1383–1408 (2019)
5. Blokland, A., Geraerts, E., Been, M.: A detailed analysis of rats' spatial memory in a probe trial of a Morris task. Behav. Brain Res. **154**(1), 71–75 (2004)
6. ELMostafi, H., et al.: Neuroprotective potential of Argan oil in neuropsychiatric disorders in rats: a review. J. Funct. Foods **75**, 104233 (2020)
7. Van Haaren, F.: Methods in Behavioral Pharmacology. Elsevier, Amsterdam (2013)
8. Wilcox, R.R., Keselman, H.J.: Modern robust data analysis methods: measures of central tendency. Psychol. Methods **8**(3), 254 (2003)
9. Li, H., et al.: Modulation of astrocyte activity and improvement of oxidative stress through blockage of NO/NMDAR pathway improve posttraumatic stress disorder (PTSD)-like behavior induced by social isolation stress. Brain Behav. **12**(7), e2620 (2022)
10. Boyko, M., et al.: Traumatic brain injury-induced submissive behavior in rats: link to depression and anxiety. Transl. Psychiatry **12**(1), 239 (2022)
11. Norma Oficial Mexicana NOM-062-ZOO-1999: Especificaciones Técnicas para la producción, Cuidado y Uso de los Animales de Laboratorio. Secretaría de Agricultura, Ganadería, Desarrollo Rural, Pesca y Alimentación. México, D.F. (1999)
12. Institute of Laboratory Animal Resources (US): Committee on Care & Use of Laboratory Animals: Guide for the care and use of laboratory animals (No. 86). US Department of Health and Human Services, Public Health Service, National Institutes of Health (1986)
13. Pellow, S., Chopin, P., File, S.E., Briley, M.: Validation of open: closed arm entries in an elevated plus-maze as a measure of anxiety in the rat. J. Neurosci. Methods **14**(3), 149–167 (1985)
14. Cohen, H., Matar, M.A., Joseph, Z.: Animal models of post- traumatic stress disorder. Curr. Protoc. Neurosci. **64**(1), 9–45 (2013)
15. Hall, C.S.: Emotional behavior in the rat. I. Defecation and urination as measures of individual differences in emotionality. J. Comparat. Psychol. **18**(3), 385 (1934)
16. Martínez, R.E.B., et al.: Árboles de decisión como herramienta en el diagnóstico médico. Rev. Médica Univ. Veracruzana **9**(2), 19–24 (2009)
17. Fernández-Demeneghi, R., et al.: Effect of blackberry juice (Rubus fruticosus L.) on anxiety-like behaviour in Wistar rats. Int. J. Food Sci. Nutr. **70**(7), 856–867 (2019)
18. Hassan, C.A.U., Khan, M.S., Shah, M.A.: Comparison of machine learning algorithms in data classification. In: 2018 24th International Conference on Automation and Computing (ICAC), pp. 1–6. IEEE (2018)
19. Arluk, S., et al.: MDMA treatment paired with a trauma-cue promotes adaptive stress responses in a translational model of PTSD in rats. Transl. Psychiatry **12**(1), 181 (2022)
20. Kraeuter, A.K., Guest, P.C., Sarnyai, Z.: The open field test for measuring locomotor activity and anxiety-like behavior. Pre-clin. Models: Techn. Protocols 99–103 (2019)
21. Bernal-Morales, B., Rodríguez-Landa, J.F., Ayala-Saavedra, D.R., Valenzuela-Limón, O.L., Guillén-Ruiz, G., Limón-Vázquez, A.K.: La conducta de acicalamiento en ratas estresadas y el efecto restaurador de la fluoxetina1, 2. Conductual **8**, 1–14 (2020)

Analysis of Proteins in Microscopic Skin Images Using Machine Vision Techniques as a Tool for Detecting Alzheimer's Disease

Sonia Lilia Mestizo-Gutiérrez[1]([✉]), Héctor Gabriel Acosta-Mesa[2], Francisco García-Ortega[3], and María Esther Jiménez-Cataño[4]

[1] Facultad de Ciencias Químicas, Universidad Veracruzana, Xalapa, Mexico
smestizo@uv.mx

[2] Instituto de Investigaciones en Inteligencia Artificial, Universidad Veracruzana, Xalapa, Mexico

[3] Laboratorio Nacional de Informática Avanzada (LANIA), Xalapa, Mexico

[4] Facultad de Medicina, Universidad Autónoma de San Luis Potosí, San Luis Potosí, Mexico

Abstract. Alzheimer's disease (AD) is a disease characterized by progressive loss of memory, orientation, judgement and language. Its progression is slow and there is no cure for this multifactorial disease, so it is of great importance the discovery of new methods of early diagnosis, as well as the development of more effective treatments. AD is the most common type of dementia. Currently, the incidence of dementia has increased and constitutes a priority health problem worldwide, so urgent measures focused on prevention and reduction of risk factors, as well as early diagnosis, are required. In this work, we proposed computer vision techniques as a tool for histological analysis to support experts in the search for a biochemical marker for early diagnosis of AD. We analized ten samples of skin tissue biopsies (5 controls and 5 AD patients) of microscopic images to find characteristic staining colors to differentiate the images of healthy subjects from Alzheimer's disease patients. In the immunohistochemistry process, the antibodies that revealed tau protein oligomers antigens were AT22. For lamin A, we used anti-lamin A. Results allowed us to find differences between healthy subjects and subjects with Alzheimer's disease.

Keywords: Alzheimer's disease · Machine vision techniques

1 Introduction

Alzheimer's disease (AD) is a progressive and irreversible neurodegenerative disorder characterized by progressive memory loss, cognitive impairment, and behavioral disturbances. It is the most common cause of dementia. Alzheimer's disease is multifactorial, and the main risk factor is advanced age. There is a neuronal loss in people with Alzheimer's disease. Two typical changes are present: the neuritic plaque or senile plaque, which is produced mainly by the beta-amyloid (Aβ) protein, and the neurofibrillary tangle, which forms in neurons and whose main component is the hyperphosphorylation of neuronal tau protein [1]. The World Health Organization (WHO) estimates that

H. Calvo et al. (Eds.): MICAI 2023 Workshops, LNAI 14502, pp. 432–438, 2024.
https://doi.org/10.1007/978-3-031-51940-6_33

more than 55 million people worldwide are living with dementia [2]. Unfortunately, there is still no cure, and early detection of the disease can be challenging due to the difficulty in accurately diagnosing such pathology [3]. The diagnostic certainty of Alzheimer's disease is around 85% and is only confirmed by post-mortem examination.

In medicine, a biochemical marker is a characteristic of cells or tissues that can be used to measure a biological change associated with a disease. Biomarkers (chemical, physiological, or morphological) are measurable and related to a disease. In Alzheimer's disease, neuritic plaques and neurofibrillary tangles can be detected decades before the onset of symptoms, providing an opportunity for early detection of the disease [4]. The main biomarkers for the diagnosis of Alzheimer's disease are biochemical (mainly blood, cerebrospinal fluid, retina, and urine), structural or functional neuroimaging (especially computed axial tomography, nuclear magnetic resonance, positron emission tomography -PET- and single photon emission tomography -SPECT-) and genetic markers [4–9].

The main characteristic of neurodegenerative diseases, including AD, is the presence of proteinopathies where structural alterations and protein folding errors occur, so it is possible to detect them in extracerebral tissues. Immunohistochemical techniques allow pathologists or experts in the field of neurodegenerative diseases to quantify the presence of the abnormal proteins characteristic of AD accurately [10]. Because the brain and skin share an ectodermal origin, it is plausible to find pathological alterations in the skin and the central nervous system (CNS). In AD, one study demonstrated the presence of tau protein in skin biopsies [11]. Additionally, lamin A protein pathology has been shown to play a causal role in tau protein-mediated neurotoxicity and thus may indicate the degree of tau protein involvement [12]. These findings opened the possibility of conducting new research that contributes to the search for early-stage diagnosis of AD. This work aims to propose computer vision techniques as a tool for histological analysis to support experts in the search for a biochemical marker as a new method of diagnosis of AD, applicable to living subjects and accessible to the population because the tests and techniques for the diagnosis of AD have a high cost.

2   Materials and Methods

Ten samples of skin tissue biopsies encapsulated in histologic slides (5 controls and 5 Alzheimer's disease patients) provided by experts from the School of Medicine of the Autonomous University of San Luis Potosi were used. In the immunohistochemistry process, the antibodies that revealed tau antigens (when they formed oligomers) and lamin A were AT22 and anti-lamin A (see Fig. 1).

The staining technique used was H-DAB (hematoxylin and diaminobenzidine) to stain the tissues gray/blue/purple and brown, respectively, according to the antigen-antibody reaction. An Olympus biological binocular microscope and a 12 MP AmScope digital camera, including its image acquisition and enhancement software, were used for the microscopic image acquisition process. MATLAB was used for coding machine vision algorithms. Figure 2 shows the proposed steps in the methodology used for the histological analysis tool.

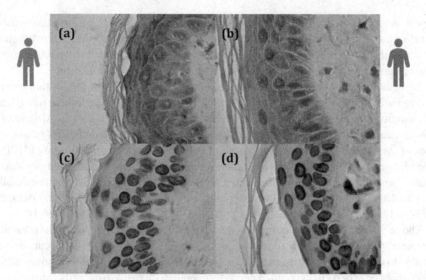

Fig. 1. (a) Expression of tau protein in healthy subject, (b) Expression of tau protein in subject with AD, (c) Expression of lamin A protein in healthy subject, (d) Expression of lamin A protein in subject with AD.

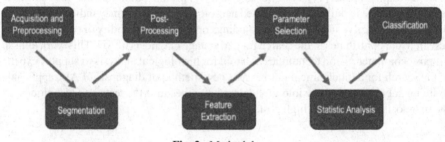

Fig. 2. Methodology

3 Experiments and Results

In the analysis of microscopic images of the skin to find characteristic staining colors to differentiate the images of healthy subjects from Alzheimer's disease patients, experiments were carried out with different machine vision techniques, which are described below and show three essential findings:

1. It was found that by using Otsu's thresholding method in the saturation channel of the HSI color space, segmentation of the diaminobenzidine (DAB) stain, which was used in the immunohistochemistry process, can be achieved (see Fig. 3).
2. Using the segmentation method by growing regions in the H channel of the HSI color space, it was possible to completely segment the epidermis and improve the segmentation result in the post-processing stage. Interestingly, it was possible to replicate the result, no matter where the growth seed was placed. For this algorithm,

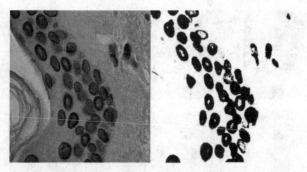

Fig. 3. Segmentation of lamin A protein in healthy subjects by Otsu method (x100).

an experimentally calculated maximum intensity distance equal to 0.2 was used (see Fig. 4).

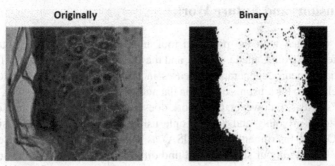

Fig. 4. Segmentation using the tau protein's regions growth algorithm in healthy subject (magnification x100).

3. The results were as follows: the research hypothesis was accepted, given that the means of both samples of healthy subjects and Alzheimer's disease patients for the selected colors were significantly different using the Wilcoxon statistical test, with a value of $p = 0.0079$, 95% confidence, and ±5% error. Visualizing the differences in cellular structures within the epidermis was also possible by reducing the thickness of the false-color dots used to mark the colors found exclusively in healthy subjects and Alzheimer's disease patients (see Fig. 5).

Confidence in the results of the work was based on expert validation, hypothesis testing, and evidence of differences found between healthy and Alzheimer's disease subjects.

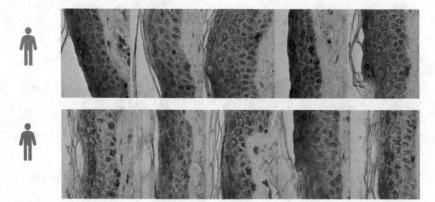

Fig. 5. Marking by applying exclusive color identification of healthy (green) vs. subjects with AD (red). (Color figure online)

4 Conclusions and Future Work

The results obtained with the proposed tool are comparable to those obtained by a group of scientists from Canada, France, and the United States in the search for proteins characteristic of neurodegenerative diseases in the skin, who were able to detect α-synuclein related to Parkinson's disease in the nucleus of skin cells, as well as tau protein (Alzheimer's disease), however, their work does not report the detection of differences of both proteins in young and adult people using IMARIS Image Analysis software (Bitplane (Oxford Instruments), MA, USA) as a tool to quantify the presence of the proteins [13], while in our study we did find differences between healthy subjects and Alzheimer's disease patients (see Fig. 6). In future work, we plan to increase the number of cases to validate our findings' robustness, improve the user interface's functionalities by applying software engineering and usability techniques, and explore machine learning techniques by adding additional clinical data such as clinical analysis and treatment.

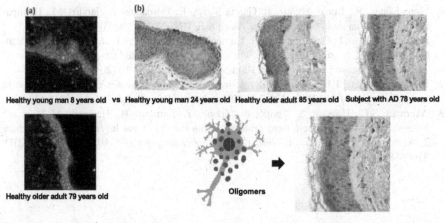

Fig. 6. (a) Research results in (Ackerman et al., 2019) vs. (b) Results obtained in our research.

Acknowledgments. To the Consejo Nacional de Ciencia y Tecnología (CONACyT) for the scholarship granted to Francisco García Ortega to pursue graduate studies under the Programa Nacional de Posgrados de Calidad (National Program for Quality Graduate Studies) with scholarship number 893200.

References

1. Khan, S., Barve, K.H., Kumar, M.S.: Recent advancements in pathogenesis, diagnostics and treatment of Alzheimer's disease. Curr. Neuropharmacol. **18**(11), 1106–1125 (2020)
2. World Health Organization Homepage. https://www.who.int/news-room/fact-sheets/detail/dementia. Accessed 23 Sept 2023
3. Ossenkoppele, R., van der Kant, R., Hansson, O.: Tau biomarkers in Alzheimer's disease: towards implementation in clinical practice and trials. Lancet Neurol. **21**(8), 726–734 (2022). https://doi.org/10.1016/S1474-4422(22)00168-5. Epub 25 May 2022. PMID: 35643092
4. Mantzavinos, V., Alexiou, A.: Biomarkers for Alzheimer's disease diagnosis. Curr. Alzheimer Res. **14**(11), 1149–1154 (2017)
5. Porsteinsson, A.P., Isaacson, R.S., Knox, S., Sabbagh, M.N., Rubino, I.: Diagnosis of early Alzheimer's disease: clinical practice in. J. Prevent. Alzheimer's Dis. **8**, 371–386 (2021)
6. Martínez-Rivera, M., Menéndez-González, M., Catalayud, M.T., Pérez-Piñeira, P.: Biomarcadores para la Enfermedad de Alzheimer y otras demencias degenerativas. Archivos Med. **4**(3), 3 (2008)
7. Barrera-López, F.J., López-Beltrán, E.A., Baldivieso-Hurtado, N., Maple-Alvarez, I.V., López-Moraila, M.A., Murillo-Bonilla, L.M.: Diagnóstico actual de la enfermedad de Alzheimer. Rev. Med. Clín. **2**(2), 57–73 (2018)
8. Kim, D.H., et al.: Genetic markers for diagnosis and pathogenesis of Alzheimer's disease. Gene **545**(2), 185–193 (2014). https://doi.org/10.1016/j.gene.2014.05.031. Epub 15 May 2014. PMID: 24838203
9. Hart, N.J., Koronyo, Y., Black, K.L., Koronyo-Hamaoui, M.: Ocular indicators of Alzheimer's: exploring disease in the retina. Acta Neuropathol. **132**(6), 767–787 (2016). https://doi.org/10.1007/s00401-016-1613-6. Epub 19 Sept 2016. PMID: 27645291; PMCID: PMC5106496

10. Mena-López, R., Luna- Muñoz, J., García Sierra, F., Hernández-Alejandro, M.: Histopatología molecular de la Enfermedad de Alzehimer. UNAM **4**(7) (2003)

11. Rodríguez-Leyva, I., Chi-Ahumada, E., Calderón–Garcidueñas, A.L., Medina-Mier, V., Santoyo Martha, E., Martel-Gallegos, G.: Presence of phosphorylated tau protein in the skin of Alzheimer's disease patients. J. Mol. Biomark. Diagn. S **6**, 005–10 (2015)

12. Frost, B., Bardai, F.H., Feany, M.B.: Lamin dysfunction mediates neurodegeneration in tauopathies. Curr. Biol. **26**(1), 129–136 (2016). https://doi.org/10.1016/j.cub.2015.11.039

13. Akerman, S.C., Hossain, S., Shobo, A., Zhong, Y., Jourdain, R., Hancock, M.A., et al.: Neurodegenerative disease-related proteins within the epidermal layer of the human skin. J. Alzheimers Dis. **69**(2), 463–478 (2019). https://doi.org/10.3233/JAD-181191. PMID: 31006686

Comparative Study of the Starting Stage of Adaptive Differential Evolution on the Induction of Oblique Decision Trees

Miguel Ángel Morales-Hernández[1], Rafael Rivera-López[2(✉)],
Efrén Mezura-Montes[3], Juana Canul-Reich[4], and Marco Antonio Cruz-Chávez[5]

[1] Laboratorio Nacional de Informática Avanzada, Xalapa, Mexico
[2] DSC, TecNM, Instituto Tecnológico de Veracruz, Veracruz, Mexico
rafael.rl@veracruz.tecnm.mx
[3] IIIA, Universidad Veracruzana, Xalapa, Mexico
emezura@uv.mx
[4] DACyTI, Universidad Juárez Autónoma de Tabasco, Cunduacán, Mexico
juana.canul@ujat.mx
[5] CIICAP, Universidad Autónoma del Estado de Morelos, Cuernavaca, Mexico
mcruz@uaem.mx

Abstract. This study describes the application of four adaptive differential evolution algorithms to generate oblique decision trees. A population of decision trees encoded as real-valued vectors evolves through a global search strategy. Three schemes to create the initial population of the algorithms are applied to reduce the number of redundant nodes (whose test condition does not divide the set of instances). The results obtained in the experimental study aim to establish that the four algorithms have similar statistical behavior. However, using the dipole-based start strategy, the JSO method creates trees with better accuracy. Furthermore, the Success-History based Adaptive Differential Evolution with linear population reduction (LSHADE) algorithm stands out for inducing more compact trees than those created by the other variants in the three initializations evaluated.

Keywords: Oblique Decision Trees · Differential Evolution · Initialization stage

1 Introduction

The growing interest in using machine learning (ML) techniques to solve prediction problems and support decision-making in almost any area of human activity

Mexico's National Council of Humanities, Science, and Technology (CONAHCYT) awarded a scholarship to the first author (CVU 1100085) for graduate studies at the Laboratorio Nacional de Informática Avanzada (LANIA).

H. Calvo et al. (Eds.): MICAI 2023 Workshops, LNAI 14502, pp. 439–452, 2024.
https://doi.org/10.1007/978-3-031-51940-6_34

is undeniable. Using artificial neural networks (ANNs) has allowed novel application development. However, ANNs are black-box models that require more means to explain how they make predictions. Therefore, their use has begun to be regulated to prevent wrong decisions supported using these methods in areas such as medicine and economics [29]. This has driven the use of post hoc methods to generate explanations of ANNs predictions [31]. Other ML strategies, such as decision trees (DTs), are white-box models with high interpretability and transparency levels. Among the DT types are those that use oblique hyperplanes in their internal nodes, producing more compact DTs than those that use a single attribute to make their partitions. However, traditional methods to induce DTs present some drawbacks, such as overfitting, selection bias toward multi-valued attributes, and instability to small changes in the training set [16]. For this reason, other inducing techniques have been proposed that, instead of applying recursive partitioning techniques, perform a search in the space of possible DTs. Evolutionary algorithms (EAs), such as genetic algorithms and genetic programming, have been widely applied to find near-optimal DTs that are more precise than those created by traditional techniques [16].

Differential Evolution (DE) is an EA that has successfully solved numerical optimization problems, and its standard versions have also been applied to induce DTs [6,9,18,24,25]. To the best of our knowledge, adaptive DE versions have not been applied to induce oblique DTs. On the other hand, it has been observed that an initial random population produces redundant internal nodes that affect the size and precision of the induced model. In this study, we analyzed the effect of using four adaptive DE versions to induce oblique DTs (JADE, SHADE, LSHADE, and JSO) and the effect on model precision and size when two additional strategies are used to create the initial DE population: dipoles and centroids. The experimental results establish that the four algorithms exhibit similar statistical behavior. However, using the dipole-based start strategy, the JSO method creates trees with better accuracy. Furthermore, the LSHADE algorithm induces more compact trees in the three initializations evaluated.

The rest of this paper is organized into four additional sections. Section 2 introduces the oblique DT characteristics and the adaptive DE approaches are described in Sect. 3. In Sect. 4, elements for the comparative study are detailed. Section 5 presents the experimental details and results. Finally, in Sect. 6, the conclusions of this study are presented, and some directions for future work are defined.

2   Oblique Decision Trees

DTs are classification models that split datasets according to diverse criteria, such as the distance between instances or the reduction of classification error. These models create a hierarchical structure using test conditions (internal nodes) and class labels (leaf nodes), allowing visualization of attribute importance in decisions and how they are used to classify an instance. DTs are the most popular interpretable algorithm for classification and regression [13]. Depending on the number of attributes evaluated in each internal node, two decision tree types are induced: univariate (axis parallel DTs) and multivariate (oblique and non-linear DTs). In particular, oblique DTs (ODTs) use test conditions representing hyperplanes having an oblique orientation relative to the axes of the instance space. ODTs are generally smaller and more accurate than univariate DTs, but they are generally more difficult to interpret [4]. ID3 [22], C4.5 [23] and CART [2] are the most popular methods for inducing univariate DTs, and CART and OC1 [19] are well-known methods for creating oblique DTs. Figure 1 shows an example of an ODT. On the right of this figure, the instance space is split using two oblique hyperplanes.

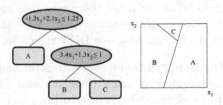

Fig. 1. Example of an oblique decision tree.

3   Adaptive Differential Evolution Approaches

DE is an EA evolving a population of real-valued vectors $\mathbf{x}_i = (x_{i,1}, x_{i,2}, \cdots, x_{i,n})^T$ of n variables, to find a near-optimal solution to an optimization problem [21]. Instead of implementing traditional crossover and mutation operators, DE applies a linear combination of several randomly selected individuals to produce a new individual. Three randomly selected candidate solutions ($\mathbf{x}_a$, $\mathbf{x}_b$, and $\mathbf{x}_c$) are linearly combined to yield a mutated solution $\mathbf{x}_{mut}$, as follows:

$$\mathbf{x}_{mut} = \mathbf{x}_a + F(\mathbf{x}_b - \mathbf{x}_c) \tag{1}$$

where F is a scale factor for controlling the differential variation.

The mutated solution is utilized to perturb another candidate solution $\mathbf{x}_{cur}$ using the binomial crossover operator defined as follows:

$$x_{new,j} = \begin{cases} x_{mut,j} & \text{if } r \leq Cr \vee j = k \\ x_{cur,j} & \text{otherwise} \end{cases} ; j \in \{1, \ldots, n\} \tag{2}$$

where $x_{new,j}$, $x_{mut,j}$ and $x_{cur,j}$ are the values in the j-th position of $\mathbf{x}_{new}$, $\mathbf{x}_{mut}$ and $\mathbf{x}_{cur}$, respectively, $r \in [0, 1)$ and $k \in \{1, \ldots, n\}$ are uniformly distributed random numbers, and Cr is the crossover rate.

Finally, $\mathbf{x}_{new}$ is selected as a member of the new population if it has a better fitness value than that of $\mathbf{x}_{cur}$.

DE starts with a population of randomly generated candidate solutions whose values are uniformly distributed in the range $[x_{min}, x_{max}]$ as follows:

$$x_{i,j} = x_{min} + r(x_{max} - x_{min}); i \in \{1, \ldots, NP\} \wedge j \in \{1, \ldots, n\} \tag{3}$$

where NP is the population size.

DE is characterized by using fewer parameters than other EAs and by its spontaneous self-adaptability, diversity control, and continuous improvement [7]. Although DE requires fewer parameters than other EAs, its performance is sensitive to the values selected for Cr, F, and NP [32]. In the literature, several approaches exist to improve DE performance using techniques to adjust the values of its parameters or combine the advantages of different algorithm variants. Methods that adjust the algorithm parameters can be considered global approaches when the parameters are updated at the end of each generation, and all population members use their values [5,17]. On the other hand, the most successful approaches are those in which each individual uses a different value of the control parameters [27,30]. Finally, there are other methods where different mutation or recombination strategies are combined within the evolutionary process [11,26]. In this study, four adaptive DE versions are used:

1) **JADE:** This DE variant introduces a successful mutation strategy and an adaptive parameter method using Gaussian and Cauchy distributions [30]. The *current-to-pbest* mutation, shown in Eq. (4), improves the balance between search space exploration and exploitation by allowing the selection of an individual ($\mathbf{x}_{pbest}$) from a subset of the p best individuals in the population to create a mutant vector ($\mathbf{x}_{mut}$). An optional external archive is also used to diversify the donor vectors. $\mathbf{x}_a$ is chosen randomly from the current population, and $\mathbf{x}_b$ are selected from this external archive that recorded solutions previously discarded during the evolutionary process.

$$\mathbf{x}_{mut} = \mathbf{x}_{cur} + F_i(\mathbf{x}_{pbest} - \mathbf{x}_{cur}) + F_i(\mathbf{x}_a - \mathbf{x}_b) \tag{4}$$

JADE uses F and Cr parameter values adjusted to each i-th individual in the population. F_i is selected from a Cauchy distribution, $F_i = randc_i(\mu_F, 0.1)$, and Cr_i is generated using a normal distribution, $Cr_i = randn_i(\mu_{Cr}, 0.1)$. μ_{Cr} is updated at the end of each generation using the arithmetic mean of the set of all successful crossover probabilities. μ_F is similarly computed, but the Lehmer mean of all successful scale factors is used.

2) **SHADE:** The Success-History based Adaptive DE (SHADE) is an enhanced JADE version employing a historical record of the pair of μ_{Cr} and μ_F values [27]. These values are randomly selected to create each new individual instead of using the same values in each generation.

3) **LSHADE:** A population size linear reduction strategy is applied in this SHADE variant [28]. The population decreases linearly in each generation until its size equals a minimum value.

4) **JSO:** This LSHADE improvement replaces the F_i parameter in the second term of Eq. (4) for a weighted F_i value updated as a function of the number of objective function evaluations [3].

Recursive-partitioning and global-search strategies to induce ODTs have been implemented using DE-based approaches. In the first case, OC1-DE [25], the Adapted JADE with Multivariate DT (AJADE-MDT) method [12], and the Parallel-Coordinates (PA-DE) algorithm [6] evolve a population of real-valued individuals to find near-optimal hyperplanes. In the other case, two methods implement a global search strategy to find a near-optimal ODT: (1) The Perceptron DT (PDT) method [9,18], where the hyperplane coefficients of one DT are encoded with a real-valued individual. Hyperplane-independent terms and the class label of leaf nodes are stored in two additional vectors. In each DE iteration, the mutation parameters are randomly altered. A group of new DTs randomly created replaces the worst individuals in the population. (2) The DE algorithm to build ODTs (DE-ODT) [24], where the size of the real-valued vector is computed as a factor of the number of internal nodes of an ODT estimated using the number of dataset attributes.

4 Comparative Study Details

In this study, we use the mapping scheme introduced by the DE-ODT method [24]. DE-ODT evolves a population of ODTs encoded in fixed-length real-valued vectors. Figure 2 shows the scheme for converting a DE individual into an ODT. The steps of this mapping scheme are described in the following paragraphs.

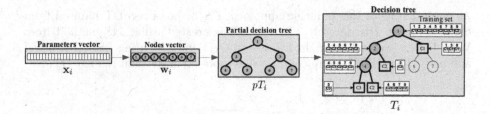

Fig. 2. Mapping scheme used on the DE-ODT method.

1) ODTs linear representation: Each candidate solution encodes only the internal nodes of a complete binary ODT stored in a fixed-length real-valued vector $(\mathbf{x}_i)$. This vector represents the set of hyperplanes used as the ODT test conditions. The vector size (n) is determined using both the number of features (d) and the number of class labels (s) of the training set, as follows:

$$n = n_e(d+1) \tag{5}$$

where $n_e = 2^{max(H_i, H_l)-1} - 1$, $H_i = \lceil log_2(d+1) \rceil$, and $H_l = \lceil log_2(s) \rceil$

2) Hyperplanes construction: Vector $\mathbf{x}_i$ is used to build the vector $\mathbf{w}_i$ encoding the sequence of candidate internal nodes of a partial ODT. Since the values of $\mathbf{x}_i$ represent the hyperplane coefficients contained in these nodes, the following criterion applies: Values $\{x_{i,1}, \ldots, x_{i,d+1}\}$ are assigned to the hyperplane h_1, the values $\{x_{i,d+2}, \ldots, x_{i,2d+2}\}$ are assigned to the hyperplane h_2, and so on. These hyperplanes are assigned to the elements of $\mathbf{w}_i$: h_1 is assigned to $w_{i,1}$, h_2 is assigned to $w_{i,2}$, an so on.

3) Partial ODT construction: A straightforward procedure is applied to construct the partial DT (pT_i) from $\mathbf{w}_i$: First, the element in the initial location of $\mathbf{w}_i$ is used as the root node of pT_i. Next, the remaining elements of $\mathbf{w}_i$ are inserted in pT_i as successor nodes of those previously added so that each new level of the tree is completed before placing new nodes at the next level, in a similar way to the breadth-first search strategy. Since a hyperplane divides the instances into two subsets, each internal node has assigned two successor nodes.

4) Decision tree completion: In the final stage, several leaf nodes are added in pT_i by evaluating the training set to build the final ODT T_i. One instance set is assigned to one internal node (starting with the root node), and by evaluating each instance with the hyperplane associated with the internal node, two instance subsets are created and assigned to the successor nodes. This assignment is repeated for each node in pT_i. Two cases should be considered: (1) If an instance set is located at the end of a branch of pT_i, two leaf nodes are created and designated as its successor nodes. (2) If the instance set contains elements for the same class, the internal node is labeled as a leaf node, and its successor nodes are removed if they exist. Furthermore, an internal node is labeled as a leaf when it contains an empty instance subset.

As one example of this inducing approach, Fig. 3 shows two DT induced from the well-known Iris dataset: The left DT is created using J48 method from Weka library [8], and the right DT is and ODT induced using LSHADE-based approach. The ODT is more compact and more accurate than the left DT.

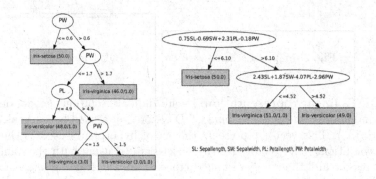

Fig. 3. DTs for Iris dataset using J48 (Left) and LSHADE (Right).

On the other hand, three initialization strategies are analyzed in this work:

1) Random initialization: This is the classic initialization strategy previously described in Eq. (3). This approach generates many hyperplanes that do not divide the instances into two subsets, producing redundant nodes impacting the induced model's size and precision.

2) Dipoles: A dipole is a pair of training instances. A mixed dipole occurs when these instances have different classes [1]. Dipoles were first used to induce ODTs through a recursive partitioning strategy [15]. A hyperplane h_i must split a mixed dipole to divide the instances into two nonempty subsets. Each individual of the initial population is built by creating hyperplanes splitting randomly selected mixed dipoles from the training set. h_1 uses a mixed dipole chosen from the training set, h_2 uses a mixed dipole from the first subset created by h_1, h_3 from the second subset, and so on until the number of internal nodes n_e is completed. In particular, a random hyperplane is created if an internal node contains an empty instance subset.

3) Centroids: This initialization strategy is proposed in this work. It is similar to the previous one, but the hyperplanes are created using the centroid of the instance set instead of a random mixed dipole. The centroid is a dummy instance where each value is the middle between the instance set's minimum and maximum values.

5   Experiments and Results

In this section, the experimental study conducted to analyze and compare the performance of the algorithms is presented. First, the datasets used in this study and the validation technique applied in the comparative analysis are described. Next, the experimental results and statistical tests are outlined. Finally, a discussion about the results is provided. Thirteen datasets (shown in Table 1), from the UCI machine learning repository are used in the experimental study [14]. These datasets have only numerical attributes since ODT internal nodes are a linear combination of their values.

Table 1. Datasets desçription.

Dataset	Instances	Features	Classes	Dataset	Instances	Features	Classes
Glass	214	9	7	Breast-tissue-6	106	9	6
Australian	690	14	2	Ionosphere	351	34	2
Iris	150	4	3	Balance-scale	625	4	3
Ecoli	336	7	8	Heart-statlog	270	13	2
Wine	178	13	3	Liver-disorders	345	6	2
Diabetes	768	8	2	Parkinsons	195	22	2
Seeds	210	6	3				

The methods used in this study are implemented in the Java languaje, using the JMetal [20] and Weka [8] libraries. The implemented algorithms run 30 times for each dataset and initialization scheme, using the ten-fold cross-validation sampling strategy to estimate the precision and size of induced trees. Subsequently, the Friedman test is applied to statistical analysis of the results obtained. Friedman test is selected since it has been demonstrated that the conditions to apply parametric tests are not satisfied to EA-based machine learning methods [10]. In the subsequent tables of this section, the best result for each dataset is highlighted with bold numbers, and the numbers in parentheses refer to the ranking reached by each method for each dataset. The last row in these tables indicates the average ranking of each method.

Tables 2, 3, 4, 5, 6, and 7 show the average accuracy and the average size (number of leaf nodes) of the trees obtained for each dataset and each DE variant, respectively.

Table 2. Average accuracy of DE variants using random initialization.

Datasets	DE	JADE	SHADE	LSHADE	JSO
Iris	96.31 (3)	96.04 (5)	96.53 (2)	**96.76** (1)	96.20 (4)
Glass	60.75 (3)	**61.12** (1)	60.34 (5)	60.37 (4)	60.78 (2)
Parkinsons	78.02 (2)	77.71 (4)	**78.12** (1)	77.95 (3)	77.28 (5)
Diabetes	67.82 (4)	**68.16** (1)	67.59 (5)	68.14 (2)	68.04 (3)
Australian	75.49 (3)	75.39 (4)	75.28 (5)	75.52 (2)	**75.60** (1)
Ionosphere	89.32 (3)	89.03 (5)	89.23 (4)	89.56 (2)	**89.65** (1)
Balance-scale	**90.59** (1)	90.35 (4.5)	90.44 (2)	90.35 (4.5)	90.39 (3)
Ecoli	**80.31** (1)	79.75 (5)	79.98 (4)	80.00 (2.5)	80.00 (2.5)
Heart-statlog	**74.04** (1)	73.40 (3)	73.35 (4)	73.91 (2)	73.27 (5)
Liver-disorders	**68.19** (1)	67.46 (5)	68.08 (4)	68.09 (3)	68.14 (2)
Wine	86.72 (2)	86.10 (5)	**86.91** (1)	86.39 (4)	86.70 (3)
Breast-tissue-6	55.47 (4)	**56.07** (1)	55.38 (5)	55.79 (2.5)	55.79 (2.5)
Seeds	90.75 (5)	90.90 (4)	91.10 (2)	**91.38** (1)	91.06 (3)
Average ranking	**2.538**	3.654	3.385	2.577	2.846

Table 3. Average accuracy of DE variants using dipole-based initialization.

Datasets	DE	JADE	SHADE	LSHADE	JSO
Iris	96.07 (5)	96.22 (4)	96.27 (3)	**96.40** (1)	96.38 (2)
Glass	63.80 (5)	64.38 (2.5)	64.38 (2.5)	63.96 (4)	**64.53** (1)
Parkinsons	**83.30** (1)	83.16 (2)	83.09 (3)	82.87 (5)	82.89 (4)
Diabetes	74.01 (3)	74.00 (4)	73.78 (5)	**74.21** (1)	74.02 (2)
Australian	84.52 (2)	84.26 (5)	84.43 (3)	84.37 (4)	**84.58** (1)
Ionosphere	88.74 (5)	88.76 (4)	**89.27** (1)	89.03 (2)	88.98 (3)
Balance-scale	90.22 (5)	90.34 (3)	90.41 (2)	90.31 (4)	**90.51** (1)
Ecoli	82.20 (4)	82.10 (5)	**82.38** (1)	82.25 (2.5)	82.25 (2.5)
Heart-statlog	78.75 (2)	77.95 (5)	77.96 (4)	**78.83** (1)	78.70 (3)
Liver-disorders	69.18 (2)	69.06 (3.5)	69.04 (5)	69.06 (3.5)	**69.33** (1)
Wine	81.16 (3)	80.39 (5)	81.24 (2)	**81.91** (1)	80.43 (4)
Breast-tissue-6	**54.43** (1)	52.74 (5)	53.30 (3)	53.14 (4)	53.71 (2)
Seeds	88.44 (4)	**89.22** (1)	88.38 (5)	88.51 (3)	88.56 (2)
Average ranking	3.231	3.769	3.038	2.769	**2.192**

Table 4. Average accuracy of DE variants using centroid-based initialization.

Datasets	DE	JADE	SHADE	LSHADE	JSO
Iris	**96.18** (1)	96.00 (4)	95.93 (5)	96.16 (2.5)	96.16 (2.5)
Glass	65.61 (2)	64.81 (5)	65.37 (4)	**65.62** (1)	65.42 (3)
Parkinsons	**82.44** (1)	82.19 (2)	82.15 (3)	91.97 (4)	81.88 (5)
Diabetes	**72.02** (1)	71.88 (3)	71.92 (2)	71.85 (4)	71.77 (5)
Australian	78.00 (5)	78.45 (3)	78.71 (2)	**78.81** (1)	78.38 (4)
Ionosphere	89.67 (5)	89.77 (4)	89.89 (3)	**90.12** (1)	90.11 (2)
Balance-scale	45.76 (3)	45.76 (3)	45.76 (3)	45.76 (3)	45.76 (3)
Ecoli	82.88 (2)	82.62 (4)	82.67 (3)	**82.92** (1)	82.45 (5)
Heart-statlog	77.35 (2)	76.46 (5)	**77.68** (1)	76.93 (3)	76.60 (4)
Liver-disorders	69.60 (2)	69.15 (4)	69.29 (3)	**69.81** (1)	68.62 (5)
Wine	81.24 (4)	**81.67** (1)	81.50 (3)	81.57 (2)	81.07 (5)
Breast-tissue-6	54.43 (3)	53.14 (4)	52.92 (5)	54.75 (2)	**54.94** (1)
Seeds	88.19 (4)	88.14 (5)	88.32 (2)	88.29 (3)	**88.71** (1)
Average ranking	2.692	3.615	3.000	**2.192**	3.500

Table 5. Average size of DE variants using random initialization.

Datasets	DE	JADE	SHADE	LSHADE	JSO
Iris	7.21 (4.5)	**7.16** (1)	7.19 (3)	7.21 (4.5)	7.17 (2)
Glass	11.93 (3)	11.91 (2)	**11.87** (1)	12.04 (4)	12.07 (5)
Parkinsons	12.64 (4)	12.48 (2)	**12.39** (1)	12.59 (3)	12.73 (5)
Diabetes	21.44 (2)	21.5 (4)	21.63 (5)	**20.99** (1)	21.47 (3)
Australian	20.08 (5)	**19.27** (1)	19.62 (4)	19.48 (2)	19.51 (3)
Ionosphere	22.18 (4)	21.94 (3)	22.37 (5)	**21.46** (1)	21.8 (2)
Balance-scale	12.53 (5)	12.32 (2)	12.49 (4)	**12.21** (1)	12.41 (3)
Ecoli	14.18 (4)	14.27 (5)	**14.03** (1)	14.16 (3)	14.12 (2)
Heart-statlog	**8.41** (1)	8.71 (5)	8.60 (3)	8.48 (2)	8.64 (4)
Liver-disorders	12.56 (5)	12.29 (3)	12.22 (2)	**12.19** (1)	12.31 (4)
Wine	6.84 (5)	6.82 (4)	6.78 (3)	**6.56** (1)	6.61 (2)
Breast-tissue-6	**17.47** (1)	17.53 (3)	17.5 (2)	17.74 (5)	17.67 (4)
Seeds	7.94 (3)	**7.73** (1)	8.06 (5)	7.90 (2)	7.95 (4)
Average ranking	3.577	2.769	3.000	**2.346**	3.308

Table 6. Average size of DE variants using dipole-based initialization..

Datasets	DE	JADE	SHADE	LSHADE	JSO
Iris	**6.56** (1)	6.68 (3)	6.70 (4)	6.62 (2)	6.77 (5)
Glass	23.93 (2)	23.96 (3)	23.98 (4)	**23.90** (1)	24.16 (5)
Parkinsons	30.61 (4)	30.90 (5)	30.44 (2)	**30.16** (1)	30.59 (3)
Diabetes	20.73 (5)	**20.22** (1)	21.43 (4)	20.48 (2)	20.53 (3)
Australian	**12.40** (1)	12.79 (5)	12.74 (4)	12.5 (2)	12.55 (3)
Ionosphere	24.74 (5)	**24.05** (1)	24.10 (2)	24.33 (4)	24.16 (3)
Balance-scale	12.76 (2.5)	12.78 (4)	12.92 (5)	**12.61** (1)	12.76 (2.5)
Ecoli	15.46 (2)	**15.45** (1)	15.48 (3)	15.62 (5)	15.57 (4)
Heart-statlog	5.41 (3)	5.56 (4)	5.58 (5)	**5.22** (1.5)	**5.22** (1.5)
Liver-disorders	11.44 (2)	11.49 (3)	**11.22** (1)	11.50 (4)	11.64 (5)
Wine	**8.93** (1)	9.85 (5)	9.73 (4)	9.14 (2)	9.61 (3)
Breast-tissue-6	22.06 (5)	**20.93** (1)	20.95 (2)	21.53 (4)	21.22 (3)
Seeds	11.52 (5)	11.32 (4)	11.19 (2)	**11.16** (1)	11.28 (3)
Average ranking	2.884	3.077	3.308	**2.346**	3.385

Table 7. Average size of DE variants using centroid-based initialization.

Datasets	DE	JADE	SHADE	LSHADE	JSO
Iris	7.00 (5)	6.97 (4)	6.89 (2)	**6.87** (1)	6.91 (3)
Glass	24.17 (3)	24.13 (2)	24.20 (4)	**24.05** (1)	24.21 (5)
Parkinsons	22.28 (4)	**21.87** (1)	22.07 (2)	22.34 (5)	22.26 (3)
Diabetes	15.76 (2)	15.81 (3)	15.99 (5)	15.84 (4)	**15.59** (1)
Australian	10.09 (4)	10.19 (5)	**9.95** (1)	10.04 (2)	10.08 (3)
Ionosphere	19.09 (5)	19.04 (4)	18.89 (2)	18.93 (3)	**18.79** (1)
Balance-scale	1.11 (3)	1.11 (3)	1.11 (3)	1.11 (3)	1.11 (3)
Ecoli	16.35 (4.5)	16.35 (4.5)	16.33 (3)	**16.19** (1)	16.31 (2)
Heart-statlog	**5.79** (1)	6.16 (5)	5.97 (2)	6.01 (3)	6.07 (4)
Liver-disorders	11.92 (2)	12.22 (5)	12.13 (3.5)	**11.91** (1)	12.13 (3.5)
Wine	9.93 (5)	9.54 (2)	9.65 (3)	**9.44** (1)	9.77 (4)
Breast-tissue-6	22.10 (3)	**21.77** (1)	21.84 (2)	22.3 (5)	22.16 (4)
Seeds	11.04 (4)	11.18 (5)	**10.84** (1)	11.02 (3)	10.96 (2)
Average ranking	3.500	3.423	2.577	**2.538**	2.962

From Tables 2, 3, and 4, it is observed that the methods getting better accuracy are: the standard DE using random initialization, JSO with dipole-based initialization, and LSHADE with centroid-based start strategy. On the other hand, Tables 5, 6, and 7 show that the LSHADE method produces more compact ODTs using the three start strategies.

Tables 8 and 9 show the best results obtained by the DE versions for accuracy and tree size, respectively.

Table 8. Accuracy for the best methods for each initialization strategy.

Datasets	Classes	Random DE	Dipoles JSO	Centroids LSHADE
Parkinsons	2	78.02 (3)	**82.89** (1)	81.97 (2)
Diabetes	2	67.82 (3)	**74.02** (1)	71.85 (2)
Australian	2	75.49 (3)	**84.58** (1)	78.81 (2)
Ionosphere	2	89.32 (2)	88.98 (3)	**90.12** (1)
Heart-statlog	2	74.04 (3)	**78.70** (1)	76.93 (2)
Liver-disorders	2	68.19 (3)	69.33 (2)	**69.81** (1)
Iris	3	96.31 (2)	**96.38** (1)	96.16 (3)
Balance-scale	3	**90.59** (1)	90.51 (2)	45.76 (3)
Wine	3	**86.72** (1)	80.43 (3)	81.57 (2)
Seeds	3	**90.75** (1)	88.56 (2)	88.29 (3)
Breast-tissue-6	6	**55.47** (1)	53.71 (3)	54.75 (2)
Glass	7	60.75 (3)	64.53 (2)	**65.62** (1)
Ecoli	8	80.31 (3)	82.25 (2)	**82.92** (1)
Average ranking		2.231	**1.846**	1.923

Table 9. Tree size for the best methods for each initialization strategy.

Datasets	Classes	Random LSHADE	Dipoles LSHADE	Centroids LSHADE
Parkinsons	2	**12.59** (1)	30.16 (3)	22.34 (2)
Diabetes	2	20.99 (3)	20.48 (2)	**15.84** (1)
Australian	2	19.48 (3)	12.50 (2)	**10.04** (1)
Ionosphere	2	21.46 (2)	24.33 (3)	**18.93** (1)
Heart-starlog	2	8.48 (3)	**5.22** (1)	6.01 (2)
Liver-disorders	2	12.19 (3)	**11.50** (1)	11.91 (2)
Iris	3	7.21 (3)	**6.62** (1)	6.87 (2)
Wine	3	**6.56** (1)	9.14 (2)	9.44 (3)
Seeds	3	**7.90** (1)	11.16 (3)	11.02 (2)
Balance-scale	3	12.21 (2)	12.61 (3)	**1.11** (1)
Breast-tissue-6	6	**17.74** (1)	21.53 (2)	22.30 (3)
Glass	7	**12.04** (1)	23.90 (2)	24.05 (3)
Ecoli	8	**14.16** (1)	15.62 (2)	16.19 (3)
Average ranking		**1.923**	2.077	2.000

From the results on in Table 8, the Friedman test shows a p-value of 0.5836, pointing out that the three methods behave similarly, but it is observed that, on average, the JSO algorithm gets ODTs with better accuracy than the other approaches. Also, Table 8 shows that JSO with dipoles finds more accurate ODTs from datasets with two or three classes (Parkinsons, diabetes, Australian, Heartstatlog, Iris), and LSHADE with centroids finds accurate ODTs from those with two (Ionosphere, Liver-disorders) or with more than three classes (Glass, Ecoli). On the other hand, Standard DE with random initialization induces better ODTs for imbalanced datasets: Balance-scale has three classes with 288, 288, and 49 instances, respectively, and the Wine dataset has three classes with 59, 71, and 48 instances, respectively. Breast-tissue-6 has six classes with 22, 21, 14, 15, 16, and 18 instances, and only Seeds is a balanced dataset with 70 instances per class.

The statistic value computed by the Friedman test from results in Table 9 shows a p-value of 0.926, indicating that the three initialization strategies built ODTs of similar sizes. The centroid initialization scheme produces the smallest ODTs for datasets with two classes, and the random initialization creates more tiny ODTs for multiclass datasets (Wine, Breast-tissue-t, Glass and Ecoli).

6   Conclusions and Future Work

The experimental results indicate that it is important to continue studying the application of adaptive variants of DE to solve non-numerical optimization problems because it is necessary to analyze in detail the search space's characteristics and the strategies for mapping between DE individuals and their tree-like representation. LSHADE and JSO obtained slightly better results than the other approaches, which is to be expected since they are improved versions of SHADE and JADE methods. In future work, we will integrate a tree-pruning strategy and a more effective method to remove redundant nodes into the decision tree induction process.

References

1. Bobrowski, L.: Piecewise-linear classifiers, formal neurons and separability of the learning sets. In: Proceedings of 13th International Conference on Pattern Recognition, vol. 4, pp. 224–228 (1996)
2. Breiman, L., Friedman, J., Olshen, R., Stone, C.: Classification and Regression Trees. Chapman and Hall (1984)
3. Brest, J., Maučec, M.S., Bošković, B.: Single objective real-parameter optimization: algorithm jSO. In: CEC 2017, pp. 1311–1318 (2017)
4. Cantú-Paz, E., Kamath, C.: Using evolutionary algorithms to induce oblique decision trees. In: GECCO 2000, pp. 1053–1060 (2000)
5. Draa, A., Bouzoubia, S., Boukhalfa, I.: A sinusoidal differential evolution algorithm for numerical optimisation. Appl. Soft Comput. **27**, 99–126 (2015)

6. Estivill-Castro, V., Gilmore, E., Hexel, R.: Constructing interpretable decision trees using parallel coordinates. In: Rutkowski, L., Scherer, R., Korytkowski, M., Pedrycz, W., Tadeusiewicz, R., Zurada, J.M. (eds.) ICAISC 2020. LNCS (LNAI), vol. 12416, pp. 152–164. Springer, Cham (2020). https://doi.org/10.1007/978-3-030-61534-5_14

7. Feoktistov, V.: Differential Evolution: In Search of Solutions. Springer, New York (2007). https://doi.org/10.1007/978-0-387-36896-2

8. Frank, E., Hall, M., Witten, I.: The WEKA Workbench. Online Appendix (2016). https://www.cs.waikato.ac.nz/ml/weka/Witten_et_al_2016_appendix.pdf

9. Freitas, A.R.R., Silva, R.C.P., Guimarães, F.G.: Differential evolution and perceptron decision trees for fault detection in power transformers. In: Snášel, V., Abraham, A., Corchado, E. (eds.) SOCO 2012. AISC, vol. 188, pp. 143–152. Springer, Heidelberg (2013). https://doi.org/10.1007/978-3-642-32922-7_15

10. García, S., Fernández, A., Luengo, J., Herrera, F.: A study of statistical techniques and performance measures for genetics-based machine learning: accuracy and interpretability. Soft. Comput. **13**, 959–977 (2009). https://doi.org/10.1007/s00500-008-0392-y

11. Ghosh, A., Das, S., Panigrahi, B.K., Das, A.K.: A noise resilient differential evolution with improved parameter and strategy control. In: CEC 2017, pp. 2590–2597 (2017)

12. Jariyavajee, C., Polvichai, J., Sirinaovakul, B.: Searching for splitting criteria in multivariate decision tree using adapted JADE optimization algorithm. In: SSCI 2019, pp. 2534–2540 (2019)

13. Kamath, U., Liu, J.: Explainable Artificial Intelligence: An Introduction to Interpretable Machine Learning. Springer, Cham (2021). https://doi.org/10.1007/978-3-030-83356-5

14. Kelly, M., Longjohn, R., Nottingham, K.: The UCI Machine Learning Repository (2023). https://archive.ics.uci.edu

15. Krętowski, M.: An evolutionary algorithm for oblique decision tree induction. In: Rutkowski, L., Siekmann, J.H., Tadeusiewicz, R., Zadeh, L.A. (eds.) ICAISC 2004. LNCS (LNAI), vol. 3070, pp. 432–437. Springer, Heidelberg (2004). https://doi.org/10.1007/978-3-540-24844-6_63

16. Kretowski, M.: Evolutionary Decision Trees in Large-Scale Data Mining. Springer, Cham (2019). https://doi.org/10.1007/978-3-030-21851-5

17. Liu, J., Lampinen, J.: A fuzzy adaptive differential evolution algorithm. Soft. Comput. **9**, 448–462 (2005). https://doi.org/10.1007/s00500-004-0363-x

18. Lopes, R.A., Freitas, A.R.R., Silva, R.C.P., Guimarães, F.G.: Differential evolution and perceptron decision trees for classification tasks. In: Yin, H., Costa, J.A.F., Barreto, G. (eds.) IDEAL 2012. LNCS, vol. 7435, pp. 550–557. Springer, Heidelberg (2012). https://doi.org/10.1007/978-3-642-32639-4_67

19. Murthy, S.K., Kasif, S., Salzberg, S., Beigel, R.: OC1: a randomized algorithm for building oblique decision trees. In: AAAI 1993, vol. 93, pp. 322–327 (1993)

20. Nebro, A.J., Durillo, J.J., Vergne, M.: Redesigning the jMetal multi-objective optimization framework. In: GECCO 2015, pp. 1093–1100 (2015)

21. Price, K., Storn, R.M., Lampinen, J.A.: Differential Evolution: A Practical Approach to Global Optimization. Springer, Heidelberg (2006). https://doi.org/10.1007/3-540-31306-0

22. Quinlan, J.R.: Induction of decision trees. Mach. Learn. **1**(1), 81–106 (1986). https://doi.org/10.1007/BF00116251

23. Quinlan, J.R.: C4.5: Programs for Machine Learning. Morgan Kaufmann (1993)

24. Rivera-Lopez, R., Canul-Reich, J.: A global search approach for inducing oblique decision trees using differential evolution. In: Mouhoub, M., Langlais, P. (eds.) AI 2017. LNCS (LNAI), vol. 10233, pp. 27–38. Springer, Cham (2017). https://doi.org/10.1007/978-3-319-57351-9_3

25. Rivera-Lopez, R., Canul-Reich, J., Gámez, J.A., Puerta, J.M.: OC1-DE: a differential evolution based approach for inducing oblique decision trees. In: Rutkowski, L., Korytkowski, M., Scherer, R., Tadeusiewicz, R., Zadeh, L.A., Zurada, J.M. (eds.) ICAISC 2017. LNCS (LNAI), vol. 10245, pp. 427–438. Springer, Cham (2017). https://doi.org/10.1007/978-3-319-59063-9_38

26. Sallam, K.M., Elsayed, S.M., Sarker, R.A., Essam, D.L.: Improved united multi-operator algorithm for solving optimization problems. In: CEC 2018, pp. 1–8 (2018)

27. Tanabe, R., Fukunaga, A.: Success-history based parameter adaptation for differential evolution. In: CEC 2013, pp. 71–78 (2013)

28. Tanabe, R., Fukunaga, A.S.: Improving the search performance of SHADE using linear population size reduction. In: CEC 2014, pp. 1658–1665 (2014)

29. Yeung, K., Lodge, M.: Algorithmic Regulation. Oxford University Press, Oxford (2019)

30. Zhang, J., Sanderson, A.C.: JADE: self-adaptive differential evolution with fast and reliable convergence performance. In: CEC 2007, pp. 2251–2258 (2007)

31. Zhang, Y., Tiňo, P., Leonardis, A., Tang, K.: A survey on neural network interpretability. IEEE Trans. Emerg. Top. Comput. Intell. 5(5), 726–742 (2021)

32. Zielinski, K., Laur, R.: Stopping criteria for differential evolution in constrained single-objective optimization. In: Chakraborty, U.K. (ed.) Advances in Differential Evolution. SCI, vol. 143, pp. 111–138. Springer, Heidelberg (2008). https://doi.org/10.1007/978-3-540-68830-3_4

Correction to: Use of IoT-Based Telemetry via Voice Commands to Improve the Gaudiability Rate of a Generation Z Pet Habitation Experience

Alberto Ochoa-Zezzatti, Jose De los Santos, Maylin Hernandez, Ángel Ortiz, Joshuar Reyes, Saúl González, and Luis Vidal

Correction to:
Chapter 9 in: H. Calvo et al. (Eds.): *Advances in Computational Intelligence. MICAI 2023 International Workshops*, **LNAI 14502, https://doi.org/10.1007/978-3-031-51940-6_9**

In the original version of this paper important information and one figure was missing. This was corrected.

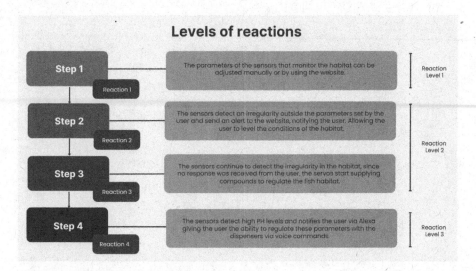

Fig. 13. Diverse levels of reaction in this project associated with habitat control of Aquarium

---

The updated version of this chapter can be found at
https://doi.org/10.1007/978-3-031-51940-6_9

Author Index

A

Acosta Roman, Jose Luis 152
Acosta-Mesa, Héctor Gabriel 397, 422, 432
Acosta-Mesa, Héctor-Gabriel 315, 337, 349, 360, 373, 409
Aguilar-Justo, Marving 170
Aguirre-Lam, Marco Antonio 326
Alcantar Alcantar, José Iraic 285
Aquino-Bolaños, Elia-Nora 373
Armenta, Ángel 191
Arreola Marín, María Esmeralda 285
Avendaño-Garido, Martha Lorena 422

B

Barradas-Palmeros, Jesús-Arnulfo 409
Barrón-Estrada, María-Lucia 3
Batiz-Beltran, Victor-Manuel 3
Bonilla Robles, Juan Carlos 29
Brambila-Hernández, José A. 326

C

Campaña, José Ismael Ojeda 259
Canul-Reich, Juana 439
Cardona, Ismael 242
Chávez Marcial, Mariela 285
Cornejo Monroy, Delfino 152
Cossio Franco, Edgar Gonzalo 242, 285
Cruz-Chávez, Marco Antonio 439
Cruz-Reyes, Laura 326
Cuevas-Chávez, P. Alejandra 12

D

De los Santos, Jose 102
Diaz, Alejandro Padilla 274

E

Escobar, Daniel Macias 132
Escobar, Roberto Macías 77, 132
Estrada, Blanca 61

F

Fraire-Huacuja, Héctor J. 326
Frausto-Solis, Juan 326
Fuentes-Tomás, José Antonio 397

G

Gallegos, Julio Cesar Ponce 274
García-Calvillo, Irma-Delia 385
García-Morales, Miguel A. 326
García-Ortega, Francisco 432
Ghasemlou, Shaban Mousavi 298
Gómez-Jiménez, Salvador 116
Gómez-Santillan, Claudia Guadalupe 326
Gonzáles-Franco, Nimrod 21
Gonzáles-Serna, Juan Gabriel 21
González, Saúl 102
González-Franco, Nimrod 41
González-Serna, Gabriel 41
Guerrero-Méndez, Carlos 116
Guzmán, Eder 159

H

Hernandez, Maylin 102
Hernández, Misael Perez 77
Hernández, Ricardo Perez 77
Hernández, Yasmín 12, 29
Hernández-Aguilar, José Alberto 29
Herrera-Meza, Socorro 422
Herrera-Sánchez, David 349

J

Jiménez, Rodolfo Hernandez 397
Jiménez-Cataño, María Esther 432

L

Lecona-Valdespino, Carlos Natanael 50
Lopez, Victor 298
Lopez-Betancur, Daniela 116
López-Herrera, Carlos-Alberto 337

© The Editor(s) (if applicable) and The Author(s), under exclusive license to Springer Nature Switzerland AG 2024
H. Calvo et al. (Eds.): MICAI 2023 Workshops, LNAI 14502, pp. 453–454, 2024.
https://doi.org/10.1007/978-3-031-51940-6

López-Lobato, Adriana-Laura   315
López-Sánchez, Máximo   21, 41

M
Macias-Huerta, Pablo Isaac   50
Magadan-Salazar, Andrea   41
Magdaleno, Diego   61
Marmolejo-Ramos, Fernando   50
Márquez-Grajales, Aldo   360, 373
Martinez, Erwin   298
Martinez, Luciano   159
Martínez, Marco Antonio   274
Mejia-de Dios, Jesus-Adolfo   385
Mestizo-Gutiérrez, Sonia Lilia   432
Mezura-Montes, Efrén   315, 337, 349, 360,
       397, 409, 439
Montes Rivera, Martín   77, 132, 152, 170
Montes, Martín   61, 159
Morales-Hernández, Miguel Ángel   439
Morales-Reyes, José-Luis   373

N
Narciso, Samuel   12
Navarrete Procopio, Miriam   201
Navarro-Solís, David   116

O
Ochoa Ortiz, Carlos Alberto   201
Ochoa-Zezzatti, Alberto   61, 102, 298
Ochoa-Zezzatti, Carlos Alberto   152
Ojeda Campaña, José Ismael   201
Ortega, Rosa Lizeth Estrada   218
Ortiz, Ángel   102
Ortiz-Hernandez, Javier   12

P
Perez Hernández, Misael   170
Pérez, Itzel Celerino   12
Ponce, Julio   159

R
Reyes, Joshuar   102
Reyes-Ortiz, José Alejandro   21
Rivera Cruz, Lizmary   29
Rivera-López, Rafael   409, 439
Rivera-Rivera, Leonor   21
Robles-Guerrero, Antonio   116
Rodríguez, Francisco Javier Álvarez   274
Rodríguez-Landa, Juan Francisco   422
Rodriguez-Martinez, Hector-Heriberto   385
Romero, Antonio   191

S
Salas-Martínez, Fernando   360
Sánchez, Luis Fernando Bernal   259
Sánchez-Jiménez, Eduardo   12
Sánchez-Zavala, Luis-Miguel   3
Santamaría-Bonfil, Guillermo   50
Saucedo-Anaya, Tonatiuh   116
Silva, Luis   116

T
Terrero Mariano, Rafael   201
Toribio-Candela, Maricarmen   41
Torres, Vianey   298

U
Uriarte-Portillo, Aldo   3

V
Valadez, Juan Martín Carpio   326
Valenzuela-Robles, Blanca Dina   21
Vargas-Moreno, Isidro   422
Velázquez-Cano, Jorge Enrique   21
Vidal, Luis   102

Z
Zatarain-Cabada, Ramon   3
Zezatti Flores, Víctor Manuel   201
Zezzatti, Alberto Ochoa   159, 259

Printed in the United States
by Baker & Taylor Publisher Services